GW00503649

FUNDAMENTALS OF MARINE BIOLOGY WITH PARTICULAR EMPHASIS ON THE TROPICAL REALM

K. V. Singarajah

MINERVA PRESS

LONDON
MONTREUX LOS ANGELES SYDNEY

FUNDAMENTALS OF MARINE BIOLOGY WITH
PARTICULAR EMPHASIS ON THE TROPICAL REALM
Copyright © K. V. Singarajah 1997

All Rights Reserved

No part of this book may be reproduced in any form,
by photocopying or by any electronic or mechanical means,
including information storage or retrieval systems,
without permission in writing from both the copyright owner
and the publisher of this book.

ISBN 1 86106 282 6

First Published 1997 by
MINERVA PRESS
195 Knightbridge
London SW7 1RE

Printed in Great Britain for Minerva Press

FUNDAMENTALS OF MARINE BIOLOGY WITH PARTICULAR EMPHASIS ON THE TROPICAL REALM

About the Author

K. V. Singarajah has had a distinguished career. He received his PhD from the University of Wales in September 1966. He investigated functional aspects of behaviour of marine planktonic animals which led him to develop interests on the nervous systems of a variety of animals, including whales and also humans. His continued interests in neural implications in behaviour enabled him to pursue further specialisation in Neurophysiology in a number of universities, including Edinburgh and Oxford. He has experience in both marine biological and neurophysiological approaches to comparative behavioural physiology. He has many awards and honours – a scholar, visiting fellow, visiting professor and visiting scientist at a number of universities both in the UK and abroad, and as a Professor in Marine Biology and Neurophysiology at the Federal University of Paraiba. The author has taught marine biology for more than two decades and neurophysiology for medical and postgraduate students for many years. He has published many original research papers in leading international scientific journals. He also provides consulting services on matters relating to marine biology and fisheries.

Preface

Teaching and researching in Marine Biology is a greater challenge, in modern times, particularly in tropical regions where there is a remarkably wide range of complex habitats and immensely varying number of species of inhabitants. We can notice the visible impact of the tremendous progress made during the last two decades. The marine science is growing rapidly and it is difficult to keep pace with its progressive developments, nevertheless efforts must be made to incorporate the new scientific advancements into teaching and research since they are fundamental for the students and the teachers alike. For many students marine biology is an important subject as it will serve as an introduction to their careers.

Nearly half of the globe's area is encompassed by the Tropics and subtropics, which, together with the scattered islands and archipelagos, constitute a land to sea ratio of almost 1:1. Although not many areas are fertile or habitable, and despite a greater diversity of complex habitats, more than one half of the world's population falls into these zones. Current human populations in these areas are increasing and will have doubled by the turn of the century, and their demand for food and other resources will also increase much faster. The world is faced with the serious problem of feeding the ever growing population. For a solution to food shortage, scientists have been looking with increasing interests to the sea. No part of the world is more richly endowed with such variety and proliferation of species that are endemic to the Tropics, subtropics, and the associated Archipelagos. The seas abound with an enormous supply of living as well as non-living resources, and it is the task of marine biologists and the other related scientists and administrators to determine the extent of these resources for the benefit of mankind. There has been an upsurge in the highly complex marine research during recent years, and a number of scientists have focused their attention on this relatively less explored horizon of tropical and subtropical realms.

Teaching tropical marine biology is no less a task because modern study needs to integrate many major scientific disciplines, particularly the tropical environment, the organisms, together with their structure, function and the interaction among the species themselves, and between the sea and the marine organisms. As far as possible, the book presents the concepts that are absolutely necessary for the understanding of marine biology, particularly the tropical and subtropical regions – it does not pretend to cover all fields of the realms in exhaustive detail.

There are three main reasons why the tropical realm was chosen: certainly, the tropical and subtropical regions are the least known; to provide an updated basic idea of marine biology with sufficient tropical marine organisms; that they are of considerable importance in many respects, such as the successive rising and falling of the seas during the glacial times; the presence of unusually large shallow seas and the extensive continental shelves between the continents and the neighbouring islands; the varied coastal characteristics; the wide diversity of species which are unmatched in any other marine environment; especially to realise the full potential of the seas' resources to feed the already impoverished population of this part of the world. It is to be expected that this text is intended primarily to provide an accurate and succinct account of the interdisciplines of marine biology which is still a less investigated scientific frontier.

Courses in marine biology are beginning to be introduced in advanced stages of undergraduate level, and there is a real need for a concise yet authoritative introductory textbook on the subject. Therefore, this book is written, though with limited scopes, as a balanced and fully integrated recent source of information into an introductory text for undergraduate students of the universities in both developing and developed countries, particularly identifying their needs within the entire domain of marine biology, but it can also serve as a specific reference to the fascinating tropical marine nature. Most examples discussed in the book are drawn from a wide range within the tropical and subtropical belts, but it can be used by students, teachers, researchers and scientists outside the tropical realm.

To understand how the subtle tropical marine ecosystems operate, it is necessary to bring together within a single volume the vast amounts of new and exciting data which are scattered spatially and temporally. Every effort has been made to explain the basic concepts in simple and concise terms avoiding, as far as possible, the often conflicting views and many of the mathematical hypotheses. Many original works have been referred to throughout the book: the bibliography does not contain all the pertinent literature, and only the selected references which are valuable to students are cited at the end of each chapter.

The book is divided into sixteen chapters and the subjects are arranged in some logical sequence, beginning with the most basic information and progressing to the most advanced and practical aspects. Earlier chapters detail the essential elements starting with a general introduction, followed by the origin and evolution of biosphere and marine environment, tropical and subtropical zones, physico-chemical properties of sea water, marine ecosystems and biotas, corals and coral reefs, tropical mangroves, economic resources, ocean currents, culture prospects of marine organisms, human impact and functional implications in behaviour of marine animals, which together are the evolution of ideas presenting the recent state of knowledge and technical advances in exploration, exploitation and management of the marine resources. At the end of the book are eight appendices that describe the comprehensive schedule for laboratory and field works which would be of immense value for students and instructors of the course. It also provides a comprehensive glossary for quick explanation of some of the terms and a good index for quick cross-reference.

Acknowledgements

The author wishes to thank many colleagues and friends who contributed freely in different ways to complete this relatively enormous task; it is a pleasure to take this opportunity to thank them all. In particular, I wish to express my appreciation to Professor G. A. Horridge, FRS, for his interest, encouragement, and helpful comments. I also wish to acknowledge the assistance of Mr S. C. Brown in printing some of the photographs. Much of the work was completed while I was a Visiting Fellow at the Australian National University during the academic years of November 1985 – November 1987, and at the Woods Hole Marine Laboratory as a Visiting Scientist during my sabbatical years 1991–1992, where I had access to most convenient and excellent libraries. As a long standing member, I also had the privilege of using frequently the excellent facilities of the Marine Biological Laboratory of the United Kingdom; and as a life member of the Indian Marine Biological Association I had access to many published works, especially based on the IIOE. This work was supported by a Senior Visiting Fellowship from the CNPq – National Research Council for Development of Science and Technology and partly by the CAPES – Ministry of Education, Brasilia.

Contents

9 Necton, and the Major Components of Tropical Fish Resources 223

xvi

List of Figures

List of Tables

List of Plates

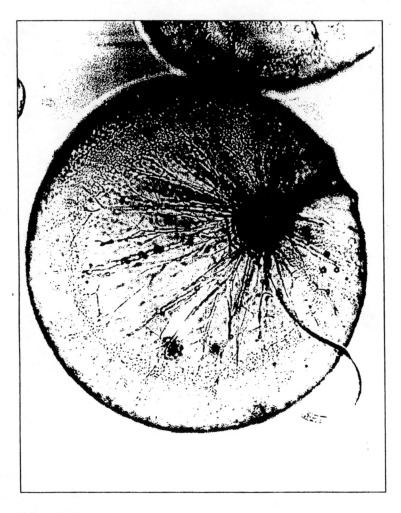

Living *Noctiluca miliaris* (by photomicrograph x 180). Planktonic flagellate (dionoflagellata) that often causes phosphorescence of the sea surface, recognisable especially during nights; it floats by ion exchange mechanism.

Chapter One
Introduction

1.1 Fundamentals and Concepts

Basic concept: marine biology is the science of life in the sea. It essentially deals with all vital life phenomena of an astonishing variety of plants and animals in the marine and the closely related environments. The exclusive dependence of life processes upon the physical and chemical factors of the sea inevitably tend to fuse the closely related disciplines of marine biology and biological oceanography. They grossly overlap each other and are often considered to be synonymous. Marine biology is a multi-disciplinary subject and it integrates many of the specialised areas of natural history, embryology, morphology, taxonomy and phylogeny, anatomy, physiology and physiological adaptations, behaviour, ecology and zoogeography, apart from chemistry, physics, and many other related subjects, including the dynamics of marine ecosystems and the flow of energy through the measure of the food chain. An understanding of tropical marine biology, however, entails the study of not only the constituent of living organisms but also the vast and constantly variable conditions of the tropical regions and habitats, together with the effects on the functional response, adaptations, subtle adjustments to abrupt changes in physico-chemical conditions, and distribution of the organisms. The unique characteristics of the tropical marine hydrosphere, delimitations of the tropical boundaries, diversity of habitats and species, specially the abundant growth of corals, mangroves, stromatolites; and the occurrence of unusually extensive bays, gulfs, estuaries, lagoons and mud-flats, mangroves and other coastal features will be considered in greater detail in the chapters that follow.

1.2 The Sea as the Potential Frontier of Resources and Support

From the time of human evolution, man always had an intrinsic curiosity about the sea. He looked at the sea for food, raw materials, easy transport, recreation and prosperity as the sea always offered its abundant resources and support in numerous ways. Man's continued dependence upon the food resources from the sea and his expanding scientific and technological abilities to alter his environment make it imperative to explore and exploit better the resources of the marine environment. Man's wise use of the sea will, no doubt, contribute profoundly to its capacity to meet the increasing demands for food and other material resources. His ability and responsibility to conserve the diverse forms of living organisms and to preserve the quality of their environment will greatly influence and shape marine science for many generations to come.

1.3 The Sea as the Origin of an Important Source of Food and Minerals

In addition to sustaining life, the marine environment has been an important source for food, minerals, transport, coastal commerce, aestheticism, recreation and tourism. One of the finest sources of a relatively cheaper form of excellent protein is that of marine origin and many nations will continue to rely increasingly on the food from the sea. Recent research and technological advancements in the sea have highlighted the national awareness of living and non-living resources of the sea. Offshore petroleum, natural gas, minerals, sands, gravels, precious corals, coral lime and a variety of other noble metallic resources have attracted a number of nations in developing their own modern technology within the exclusive economic zones. However, because of the 'greed' and short-term economic advantages, with limited goals and sophistication of gear, man often has overindulged in his ventures, so much so that the consequences are disastrous; there are already signs of depletion of resources and adverse effects on the marine environment. Apart from the highly industrialised Western nations, the Persian Gulf, Guinea, Nigeria, Brazil, Mexico, Venezuela, Argentina, India, Malaysia, China and

Australia are some of the fast developing countries that have encountered oil and natural gas with considerable success.

1.4   Regional Demographic Implications and National Product Per Capita

Demography is concerned with the geographical distribution of a variety of life forms in different regions of the earth. For our present purpose, we are concerned with human life. Demographic experts have predicted that the world population is growing rapidly. 'Every 30 seconds some 115 babies are born, and every thirty seconds only 45 die; and the population grows by 70', and so the present population of 5.54×10^9 would double by the year 2020. In the 'developing countries', mainly tropical and subtropical parts of the globe, the rate of population growth is even higher (Fig. 1.1).

The resources of the tropical regions are not inexhaustible and are bound to be squeezed out. This situation will seriously intensify the shortage of food problem, especially in the Third World countries (Fig. 1.2) of Asia, Africa, and Latin America, which are characterised by their paucity of resources and inadequate technology. The greatest challenge confronting many tropical and subtropical countries is the unpredictable rate of population increase. Population explosion and available resources are not balanced. At the current rate of about 1.8–2% population growth, many of the poor countries cannot achieve a reasonably high standard of living comparable to the rich countries.

Yet these countries cover almost 45% area of the global land surface, with the population of over 3.3×10^9. Consequently, the gross national product per capita (i.e. the total value of all products and services produced by the economy of a country per year divided by its population) is generally low. There is already a considerable disparity among nations in their ability and skills to exploit, utilise and conserve the marine resources. The wide gap between living standards and technological achievements of the developed and the 'developing countries' is ever more widening. Unless the impetus for economical growth and technological development in the rational use of the sea is accelerated, the poor nations will continue to be poorer and the rich nations will grow richer. Evidently, human misery also is becoming worldwide; and even the wealthier nations will be

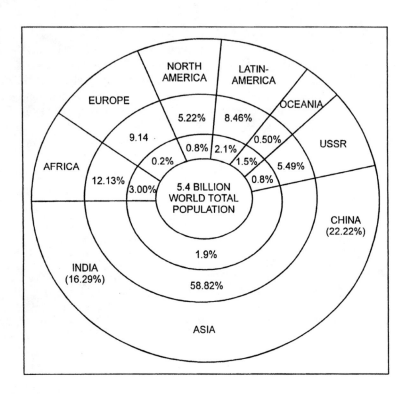

Figure 1.1   World total population, rate of annual increase, percentage of the major regional population distribution; the latter is expressed as a percentage of the total world population. China and India are the two most densely populated areas in Asia. The need for environmental resources will be enormous, especially in the developing countries. Data based on UN Demographic Year Book 1992.

affected by the consequence (as we have witnessed, for example, in Bosnia, Somalia, Rwanda, Iraq, and a few other countries, though these were man made). As a result, an instability in world power is inevitable. Military potential in the sea, particularly among maritime nations, will also vary. There are nearly forty-three nations at war today and just over one hundred and twelve countries are in conflict. In view of the exclusive economic zone of the sea (see Chapter Two), nearly 35% of the sea will now come under the jurisdiction of coastal states for exploration, exploitation, conservation, management, and to combat the ever threatening problem of pollution. The continued degradation of the natural environment by 'spills and dumps' and the rapid destruction of the natural resources of the sea in the Tropics and subtropics will have a devastating effect for the future generations, though they expect too much from the sea.

1.5 Early Developments and Recent Advances in Marine Biology

Marine biological research is often thought to have the beginning with the voyage of *HMS Challenger*. Among the many early expeditions, nothing is more global and inspiring than the *Challenger* expedition, particularly under those conditions which prevailed more than a century ago. This expedition was the outcome of an ambitious joint sponsorship by the British Admiralty and the Royal Society.

Challenger was the name of the ship. She was an elegant 2,000-ton Corvette, powered by both traditional sails and an auxiliary 1,234-hp steam engine. The conditions and scientific equipment aboard were the finest, even by modern standards. Sir Wyvile Thomption was the leader of her scientific team. *Challenger* left Portsmouth Harbour on 21 December 1872. She spent some three and a half years at sea. The scientists aboard made detailed investigation in all oceans, except the Arctic, under all different conditions. They also established 362 sampling stations in each ocean. At each station, they collected a variety of physico-chemical and biological data. The parameters mainly included: depth, salinity, temperature, pressure, weather conditions (and especially directions), and velocity of winds and oceanic currents. Their collection of biological samples was immense, and included plankton, benthic, demersal and pelagic organisms, together with bottom sediments and sea water samples.

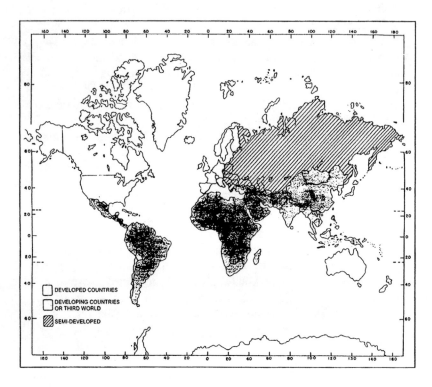

Figure 1.2   The geographical distribution of the 'third world' countries.

They covered a distance of 125,800 km and they returned to England in May 1876. The immense wealth of data collected during this expedition took nearly thirty-seven years to analyse. The written report, under the direction of Sir John Murrey, as '*Challenger* Report' stretches over fifty volumes. To their credit, 4,717 new species were discovered which are now fully classified.

The *Challenger* expedition was perhaps one of the few expeditions which contributed so much to the emancipation of this fascinating discipline of marine biology. Although many scientific expeditions under several flags soon followed during the succeeding decades, the other expedition to investigate and gain such profound knowledge of the sea in recent years was the 'International Indian Ocean Expedition' carried out during 1958–1963. The International Indian Ocean Expedition (IIOE), as it is called, was originally proposed by a special committee for Oceanographic Research (SCOR) at Woods Hole during 28 July to 30 August 1958, and later in September 1959 at the UN Headquarters in New York. The IIOE began in July 1962 with peak activities in 1963. At least twenty ships participated from more than a dozen countries. The operation lasted just over two years in the Indian Ocean for the purpose of studying different marine biological problems. The collective efforts with international scientists and different countries have contributed to increase the scientific knowledge which in the past was least known. The collection of samples and data was incredible. The results were spectacular. By such a concerted effort even this investigation had been confined to only a single ocean. The results were, and are still being, reported in a series of volumes under a unified system.

1.6   The Origin and Formation of the Marine Environment and Chemical Evolution

Although a few investigators deduce that the universe emerged from a single small dense hot region some 15 billion years ago (Halliwell, 1991), it is generally thought that our solar system was formed by condensation of a tenuous cloud of interstellar gas and dust, the 'Nebula' (= 'mist'), some 14×10^9 years ago. However, evidence from the study of ancient microfossils, rock samples of earth, moon and meteorites shows that they all have the same age of about 4.6×10^9 years. The earth would probably have been formed by

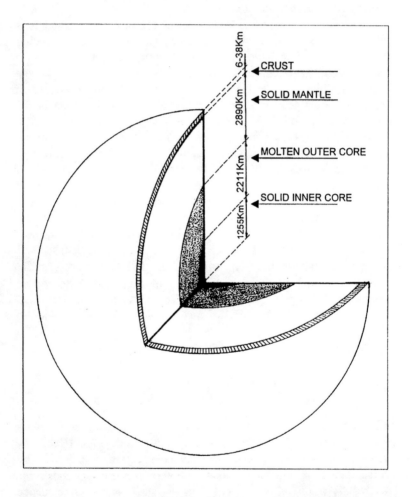

6-38Km
2890Km
2211Km
1255Km

CRUST
SOLID MANTLE
MOLTEN OUTER CORE
SOLID INNER CORE

Figure 1.3   The earth showing the major layers of crust, mantle, and the cores.

accretion of smaller solid particles rich in chemical components chiefly of hydrates, carbonates and nitrates. When the earth accreted, perhaps, most of these components had been trapped in the crustal zone of the earth (Fig. 1.3).

Thus, due to intense heat generated by radioactive substances such as uranium and thorium, some of these components escaped from the crustal layer by volcanic 'outgassing' or 'degassing' to form the primordial atmosphere of the earth. In this way, the atmosphere must have been derived from the changes of the earth's surface some time later than it was formed.

As the crust of the earth began to cool, the surface started to solidify. The primitive oceans might have been formed about 3.5×10^9 years ago. Sedentary detritus of basaltic rocks in an oxygen deficient environment began to leach with high pressure of carbon dioxide from internal as well as from the atmosphere; the pH was more on the acidic side then than the modern ocean. By a series of complex acid leach and chemical reactions the marine environment was formed. The primary minerals derived from the igneous rocks were more of alkaline than acidic nature. It is likely that they reacted with HCl and free CO_2 to form a fairly neutral environment. Chemical analysis of pre-Cambrian rocks shows that through influx and efflux of chemical substances, the ocean has achieved a steady state and, consequently, the relative proportions of the major individual elements have been fixed and remained constant. From the many gaseous components such as water vapour, carbon dioxide, nitrogen, and some rare gases released, water molecules were more abundant, which upon reaching the outer surface of the earth cooled and condensed as the primeval sea. Thus, both the primitive atmosphere and the hydrosphere were formed at the same time about a billion years later, after the formation of the earth (see Chapter Six on sea water).

Next in order of relative abundance were carbon dioxide, nitrogen and a few other rare gases. Of these, particularly carbon dioxide, being soluble in water, much of it dissolved; the nitrogen, being relatively inert, remained in the atmosphere; while others escaped or combined chemically with the crustal layer. Another inert gas of the atmosphere was argon which might have derived from the radioactive decay of heavy potassium (K^{40}). However, some geochemists strongly support the view that iron was also present in plenty on the surface of the crust. The chemical reactions between iron, carbon dioxide, and water molecules could have led to traces of free nitrogen and methane (CH_4), which, in turn, would have combined with nitrogen to form ammonia (NH_3). Consequently, the primordial atmosphere could have consisted of nitrogen, carbon dioxide, water vapour, argon, methane, ammonia and hydrogen.

The initial production of oxygen was probably by photodissociation of water of the primitive atmosphere by the photons of ultraviolet light:

$$2H_2O + hv \Rightarrow O_2 + 2H_2$$

The hydrogen escaped into the atmosphere and the oxygen would have reacted rapidly with other gases, especially hydrogen sulphide:

$$2H_2S + 3O_2 \Rightarrow 2SO_2 + 2H_2O$$

1.7 Early Life Forms in the Marine Environment

Despite its great importance, an appreciable amount of oxygen was absent until the decisive biological activities of photosynthesis begun by the prokaryotes, among which the most prominent were the cyanobacteria. These blue-green algae were in symbiotic colonies and had their origin mainly in the neritic part of the sea, some 3.5×10^9 years ago. Their prolific growth and concomitant widespread release of oxygen gradually saturated the sea and eventually started to escape into the atmosphere. Thus, the atmospheric free oxygen began to increase slowly and marked the dawn of the aerobic life, and remained crucial for the subsequent evolution of Eukaryotic life.

It was also possible, however, to a certain extent, that the free oxygen could have contributed photochemically during the slow decomposition of the water molecules in the upper layer of the atmosphere by ultraviolet radiation from the sun. When decomposed, the lighter hydrogen escaped, leaving the heavier oxygen to concentrate in the lower layer of the atmosphere.

The oxygen then began to increase gradually, sufficient to form ozone as a by-product, and to scan the lethal effect of the ultraviolet component of the sun reaching the lower layer of the atmosphere. However, all measure of evidence indicates that the concentration of oxygen during pre-Cambrian, about 600,000,000 years ago, was only about 1% compared with the present-day atmosphere. Since then, during the geologic time, free oxygen has gradually increased largely by the biotic activity to about 20.95% in the atmosphere of today (Table 1.1).

Table 1.1 The composition of (dry air) atmosphere

GASEOUS COMPONENTS		% BY VOLUME	* PARTIAL PRESSURE OF GASES (mm Hg)
Nitrogen	(N_2)	76.08	593.41
Oxygen	(O_2)	20.95	159.22
Carbon dioxide	(CO_2)	00.03	000.21
Argon	(Ar)	00.93	7.00
Neon	(Ne)	Trace	Trace
Helium	(H_4)		
Krypton	(Kr)		
Xenon	(Xe)		

* Based on NTP at sea level

1.8 Effects of Pollution on the Balance of Natural Marine Environment

Pollution problems have expanded out of all proportion. According to the UN report on pollution of the sea, 'the introduction by man, directly or indirectly, of substances or energy into the marine environment, including estuaries, resulting in such deleterious effects as harm to living resources, hazards to human health, hindrance to marine activities, including fishing, impairment of quality for use of sea water, and reduction of amenities'. Within the last two decades, there has been a steady annual increase of CO_2 in the atmosphere due to burning of fossil fuels, accidental and deliberate burning of vast areas of tropical rain forests and from the cement factories. Raw sewage is often discharged directly into the sea in many tropical and subtropical countries of the world. Although it contains some nutrients, largely harmful bacteria and pathogenic viruses often deplete the oxygen level for the marine organisms. Large-scale dumping of wastes into lakes, rivers, lagoons, estuaries, bays and coastal waters; and nuclear wastes in deeper waters; use of pesticides and insecticides and their inputs into the sea; fossil fuels, oil pollution, acid rains and gas pollutants such as nerve gas and mustard gas and massive amounts of cyanides have already modified the marine environment quite considerably. Much dumping of radioactive wastes

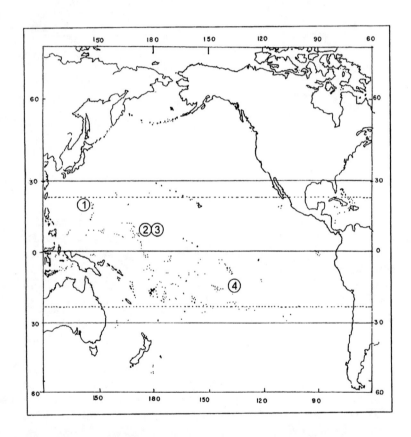

Figure 1.4   Radioactive pollution sites in the Pacific regions; (1). Marianas Trenches – site for dumping nuclear wastes, (2). Eniwetok & (3). Bikini Atoll – atomic tests were carried out and (4). Morrou Atoll and Fungi Atoll (French Polynasia) – site for French nuclear warhead tests.

and the nuclear tests occur around tropical islands in the south Pacific (Fig. 1.4). Despite the worldwide protest and condemnation, the French carried out their nuclear tests on the Polynesian Mururoa Atoll on 5 September 1995 and another one twice as powerful as the previous nuclear warhead at Fungi Atoll on 1 November 1995, a third and fourth followed, and a fifth rather more powerful one was exploded on 27 December 1995 at Mururoa Atoll and the sound shock wave could be registered in Canada. They exploded the sixth and largest ever 120 kilo-ton warhead at an atoll close to Mururoa on 27 January 1996 and they plan to carry out more tests in the year until they sign the international non-nuclear tests treaty. Nevertheless, China is still likely to continue nuclear proliferation. The combined effects of all these have profound impact on the marine environment. This unfortunate and lamentable situation calls for a united and conscientious effort, not only from maritime nations but from all countries of the world to control and preserve the seas and their coasts. Many shallow seas and their beautiful beaches can provide excellent resorts for recreation. The effective control of the land, the sea and the shore pollution is a common interest of all nations, and therefore a concerted effort to free the face of the earth from pollution will benefit future generations of all species on this planet.

Man's ability to alter his physical environment to suit his own needs may create long-term and catastrophic consequences that may be detrimental not only to the natural environment but also to life itself on this planet. As we have already seen earlier, due to burning of coal, oil and other fuels, the concentration of carbon dioxide has already increased from 270 parts per million (ppm) in 1958 to the present level of nearly 320 ppm. The continued consumption of fossil fuels, no doubt, will contribute substantially to the global concentration of carbon dioxide, which has the capacity to absorb the longer wavelengths of infrared light and, consequently, lead to the rise in the atmospheric temperature (see Chapter Fifteen for more details). This, in turn, will rise the sea level, resulting in serious changes of global climatic conditions.

References

Abell, G. O., *Exploration of the Universe*, New York, Holt Reinhart and Winston, 1975, pp.738

Bakus, C. J., 'Energetics and Feeding in Shallow Marine Waters', in *International Review of General Experimental Zoology*, (eds Felts, W. J. L. and R. J. Harison), vol. IV, New York and London, Academic Press, 1969, pp.275–669

Binham, R., 'Explorers of the Warth Within', in *Science*, vol. I (6), 1980, pp.44–55

Cherfas, J., 'The World Fertility Survey Conference: Population bomb revisited', *Science*, vol. I (7), 1980, pp.11–14

Dauvillier, A., *The Photochemical Origin of Life*, New York and London, Academic Press, 1965, pp.193

Friedrich, H., *Marine Biology: An Introduction to its Problems and Results*, London, Sidgwick and Jackson, 1973, pp.474

Gruin, J., 'In the begin', *Science*, vol. I (5), 1980, pp.44–51

Halliwell, J. J., 'Quantum cosmology and creation of universe', *Scientific American*, December, 1991, pp.76–85

National Council Board of Atmospheric Science and Climate, 'Changing Climate', Washington, DC, Academic Press, 1983

Suverdrup, H. U., Johnson, M. W., and R. H. Fleming, 'The Oceans', Indian edition, Bombay, Tokyo, Asia Publ. House, 1961 pp.1087

United Nations, *Demographic Year Book*, New York, UN, 1992, p.1020

Valiela, I., *Marine Ecological Processes*, New York, Springer-Verlag, 1984, p.547

Wagner, et al., 'The International Indian Ocean Expedition', *Science,* 3491, 1674–1676, 1978

Wald, G., 'The Origin of Life', in 'The Chemical Basis of Life', from *Scientific American*, San Francisco, W. H. Freeman and Company, 1973, pp.9–29

Chapter Two
The Nature and Range of Marine Environment

2.1 Spatial and Temporal Transition

Marine environment shows considerable spatial and temporal variations during its formation. Our knowledge of the present state of the sea is based on a very long and continuing process of physico-chemical evolution. The vast range of marine divisions, biological zones, habitats, and the fragile ecosystems of the Tropics and subtropics are immensely influenced by the variable biotic and abiotic factors. The interactions among all physical spheres of the earth, over several millions of years, contributed to form a fairly stable state of the sea. Of the total surface area of 510.066×10^6 km^2 of the earth (Fairbridge, 1966), the sea surface covers about 361×10^6 km^2 (71.00%) and contains a volume of some 1.381788×10^9 km^3 of sea water (97.351% of the hydrosphere). Experts on physio-chemical studies predict that the present 'steady state' of the sea, together with its chemical composition, can be sustained by the chemical equilibria through interactions of sea water and the marine sedimentary deposits. But there are still considerable gaps in our understanding. This chapter will be restricted to the spatial and temporal variations in relation to physico-chemical conditions of the marine environment.

2.2 The Life Realm and the Physical Spheres of Nature

The three primary components of the earth's surface are: the atmosphere, hydrosphere, and the lithosphere. There is a constant interaction between the atmosphere and hydrosphere. In the Tropics and subtropics, for example, the complex global wind pattern and the convergence of the trade winds near the equator – the inter-tropical zone – and rainfalls can have considerable effect on ocean currents and the lithosphere, which supports the continental and oceanic crusts.

2.2.1 Geosphere

Evidence from geophysical, bioecological, biogeochemical, and biogenesis shows that the earth was initially a 'sterile' environment with its ever evolving atmosphere, hydrosphere, and the lithosphere (the crustal part of the earth, including the deepest oceanic beds). Before the advent of any life on this planet, the primitive earth was formed of rocks, soil, shallow seas, and a thin atmospheric layer consisting, especially of gases such as methane, ammonia, hydrogen sulphide, and water vapour. This abiotic environment of inorganic nature of the primitive earth, devoid of any life, is called geosphere.

2.2.2 Biosphere

Biosphere then evolved from the geospheric stage. It is that part of the environment on earth where life forms, both plants and animals, can exist. Due to the continuous flow of energy from the sun and the constant interaction between the abiotic (non-living) and biotic (living) factors, habitats, ecosystems, or 'biocycles' were formed; and these 'bio-units' are able to sustain life in harmony with nature. Much simpler micro-organisms gradually evolved into more complex plants and animals. In this way, the major interacting system involving the living organisms and non-living components of the environment together constituted an ecosystem. In a broader sense, the sea and its inhabitants form the largest marine ecosystem. In the tropical and subtropical regions, especially, this largest aquatic ecosystem may be subdivided into: marine pelagic, marine benthic, coastal upwellings, corals and coral reefs, estuarine and marine mud-flats, mangrove swamps and marshes; and the littoral rocky, sandy and muddy ecosystems. These will be considered in more detail in chapters that follow.

2.2.3 The Earth and the Lithosphere

In order to understand the importance of the continental drift (see Chapter Three), the bottom topographic features and classification of the marine environment, a brief account of the earth, particularly the lithosphere, may be appreciated at this point.

The earth is nearly a spherical body, except for the slight flattening at the poles, and with an equatorial diameter of about 12,756 km and a circumference of about 40,075 km; its diameter along the polar axis is a little shorter, being 12,714 km. The earth consists of three concentric layers:

(1) The central core or centrosphere, which is subdivided into an inner solid metallic core about 1,255 km in radius and surrounded by an outer molten iron-nickel layer with a radius of 2,255 km (Fig. 2.1) – see also Fig. 1.3.

The density, pressure, and temperature are greatest at the central core, where the density is about ten times greater than that at the surface, and the pressure is about 3,640,000 atmospheres, while the temperature is about 2,475°C.

(2) The mantle which surrounds the core is about 2,890 km in radius and rocky (dunite) in nature, but the layer as a whole can yield to unequal stresses over a long period.

(3) The outermost crust is a relatively narrow zone which varied in thickness between 6 and 38 km, depending on the formation of continents, oceans and basins, troughs and sills. The mantle is separated from the crust by an extremely thin discontinuous lower boundary of the crust called Mohorovicic or simply 'Mohor' after his discovery of the interface between the crust and the mantle in 1940. This boundary is called discontinuous because of the major change in the characteristics of the rocks between the mantle and the crust. Consequently, an abrupt change in the primary seismic wave velocity (7.2–8.1 cms/sec) occurs at the boundary. The crust of the earth rests upon the outermost narrow zone of the mantle. These two rigid components – i.e. the crust and the outermost zone of the mantle – together constitute the lithosphere. Just beneath the lithosphere is Asthenosphere, perhaps, a 'hypothetical zone' within which the materials are believed to yield readily to persistent stress. The crust is further subdivided into a continuous lower basaltic rocky layer and an upper granitic rocky layer. Since granitic layer makes up the bulk of the land masses or continents, it may be extremely thin or

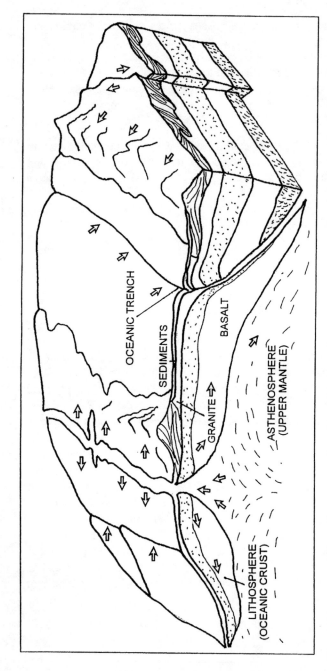

Figure 2.1   Schematic representation of the lithospheric layer and the tectonic plates. Note that when two such plates move toward each other, they collide and one sinks, beneath the other, into the mantle.

almost absent under deep ocean basins and trenches. The continental shelf part is mainly formed of sedimentary deposits just above the extremely thin granitic layer.

2.2.4 The Lithosphere and Tectonic Plates

The lithospheric layer is broken into several tectonic plates. Most of these plates are lined at the separation zone of the mid-oceanic edge. When two such plates move toward each other, they collide, and one sinks beneath the other into the mantle. The process of breaking, bending and sinking of the leading edges of the lithospheric plates into the mantle is called subduction (see Fig. 2.1).

2.2.5 The Hydrosphere

The term hydrosphere is generally applied to the total available global surface water, in different forms of liquid, solid and gas, and amounts to some 1.4×10^9 km^3. Of the global water, the marine hydrosphere- oceans and seas – constitutes about 97.351%; the rest being fastened in the polar ice, in crustal rocks, freshwater rivers and lakes, and as atmospheric vapour (Table 2.1).

Table 2.1   The distribution of the global hydrosphere

SOURCES	VOLUME $\times 10^9$ km^3	%
01 Marine Hydrosphere	1.381788	97.351
02. Polar Ice-caps and Glaciers	0.028196	2.014
03. Crustal rocks and Sediments	0.008750	0.625
04. Freshwater Rivers and Lakes	0.000126	0.009
05. Atmospheric Water Vapour	0.000014	<0.001
TOTAL	1.400000	100.00

2.2.6 The Hydrolic Cycle

The hydrosphere is maintained constant by the process of hydrolic cycle. Hydrologic cycle is the transference of the water from the surface of the oceans and continents as vapour into the atmosphere and the return of this atmospheric moisture to the earth as precipitation. Because of the difference between evaporation and precipitation at different regions of the oceans, the salinity also varies notably. In the Tropics, heavy rainfalls frequently occur and the salinity may be relatively less than in the subtropical oceanic waters. However, the hydrolic cycle can be greatly influenced by other factors such as radiation, altitudes, latitudes, temperature, tropical storms, saturation of humidity and bottom topography. Excluding the freshwater fraction of atmospheric, land and frozen water, the sea water in the two hemispheres is unequally distributed. Also, the proportion of land and water varies considerably in both hemispheres. The ratio of land to sea is 1:1.6 in the northern hemisphere and 1:4 in the southern hemisphere (Table 2.2).

Table 2.2   Distribution of global land and water

GLOBAL SURFACE	% LAND	% OCEANS AND SEAS	RATION
Northern hemisphere	39.44	60.56	1:1.6
Southern hemisphere	19.72	80.28	1:4.1
TOTAL	29.20	70.80	1:2.5

Yet, the overall potential for productivity, particularly fisheries, is far greater in the northern hemisphere than in the southern hemisphere.

2.3   The Marine Environment, its Range, Major Divisions and Marine Biological Zones

Of the global area, nearly 70.8%, or an area of 361.127×10^6 km^2, is covered with sea water, while only 29.2%, or 148.939×10^6 km^2, is occupied by the continents. The total volume of the marine environment consists of some 1381.778×10^9 km^3 with an average depth of 3,386 metres. Nearly 62.3% of the marine environment lies below 1,000 metres. In the tropical marine environment, except the fathomless trenches, the depth is relatively less and there are many shallow and tepid seas. The marine environment comprises two primary spatial media:

(1)   The enormous body of salt water mass above.

(2)   The underlying vast extent of solid bottom which supports the water.

However, despite its impressive size and greater capacity to support life, a major part of the sea, particularly the deeper areas, is less productive. This is largely because of the unequal distribution of nutrients, insufficient mechanism for mixing and variable conditions of illumination and temperature, and also due to changes during the geological time.

Physical, chemical and biological characteristics of the marine habitats change constantly with the dynamic complexity of the marine environment. There is no universally agreeable approach in classifying the marine environment; however, the most important factors involved in its classification is a matter of choice of criteria based on the amount of light penetration, proximity of the land to the sea, and the types of organisms that live. Although many criteria have been recognised, the classification is mainly based on the biological and the physical characteristics of qualitative and quantitative components of the light that penetrates into the water column, the depth, the pressure, distribution of temperature, available nutrients dispersed by the circulation, species diversity, nature of sedimentary deposits and the complex features of the bottom topography and boundaries.

As previously noted, the marine environment constitutes the largest of the global aquatic environment and is more favourable to life; and can be divided vertically and horizontally. The two main divisions are: the pelagic division, and the benthic division. These divisions (Fig. 2.2) are based mostly on the physical characteristics rather than on their endemic inhabitants.

2.3.1   The Pelagic Divisions

The pelagic division includes the entire mass of water which lies above the sea floor. It is subdivided into two main provinces: the Neritic province, and the Oceanic province.

2.3.1.1   The Neritic Province

The neritic province is the relatively shallow part of the pelagic division; i.e. the water proximal to the continent, which covers the littoral part of the sea floor, or from the mean high water mark of the sea shore to the edge of the continental shelf, roughly at about 200 metres depth. It consists of less than about 6% of the world's seas. The neritic province is characterised by penetration of sufficient light to carry out photosynthesis; the presence of abundant nutrients; especially phosphates, nitrates, silicates and suspended organic materials; strong wave actions and turbulence, coastal currents,

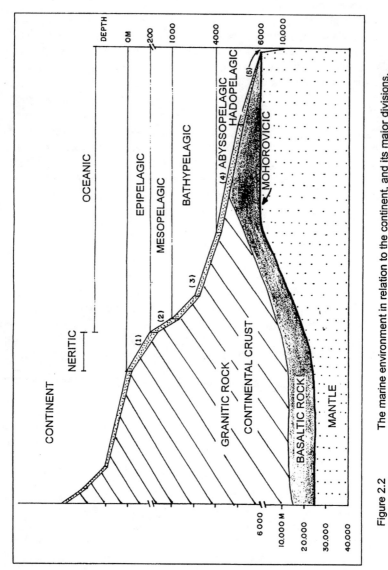

Figure 2.2   The marine environment in relation to the continent, and its major divisions.

dissolved oxygen; marked changes in temperature and salinity; and it is constantly subject to variable degrees of abiotic and biotic factors. Consequently, there is a greater potential for productivity, notably plankton, fish, shellfish, and free floating seaweeds.

2.3.1.2 Oceanic Province

The oceanic province is that portion of the pelagic division, which lies beyond the depth of 200 metres and above the deep sea systems. This surface-to-bottom depth limit, on a vertical plane, though conjectural, separates the neritic province from the oceanic province of the pelagic division on the one hand, and the littoral system from the deep sea systems of the benthic division on the other. Unlike the neritic province, except the surface layer, the oceanic province is relatively stable, fairly uniform, and less pronounced by regional and seasonal variables. There are broadly two ways in which the oceanic province can be classified:

(1) Essentially, based on quantity and quality of the light that penetrates the sea water.

(2) The light to sustain productivity, and the range of the depths of the sea floor.

In the following classification both these systems will be integrated. The oceanic province then can be subdivided into five horizontal, parallel to the surface, zones:

(1) *The Epipelagic Zone*, which extends from the shoreline seaward, where the light is quite sufficient for the photosynthesis. The epipelagic zone extends downwards to 200 metres depth and corresponds to the euphotic zone; although some experts prefer to limit the euphotic zone to a depth between 60 metres and 180 metres, but in the Tropics the depth will reach the conventional depth of 200 metres. The euphotic zone is the most prolific part of the whole marine environment. As already seen, the extent of this neritic province is also limited vertically from the sea surface downwards to the edge of continental shelf at about 200 metres. The high productivity of plankton supports a substantial proportion of marine organisms, including a variety of fish.

(2) *The Mesopelagic Zone* extends from this level down to about 1,000 metres, where light is poor or at great premium. It corresponds to the disphotic zone. Generally, the temperature and dissolved oxygen decrease, but the hydrostatic pressure increases with depth.

(3) *The Bathypelagic Zone* lies between 1,000 and 4,000 metres, where the light is virtually absent, though the bioluminescence is a common phenomenon of life by many benthic creatures. This zone is relatively colder with higher pressure and low biological activities.

(4) *The Abyssal pelagic Zone* descends from this depth to the sea basins and troughs of about 6,000 metres deep. The abyssal-pelagic zone is characterised by total darkness, increased pressure, temperature below 5°C and low concentration of dissolved oxygen. The creatures that dwell here have unusual adaptations – some with stalked eyes and some blind – and the fish species are usually with enlarged lateral lines.

(5) *The Hadopelagic Zone* extends from here to the deepest ocean trenches (see below). The spatial extent between bathypelagic and hadopelagic zones can also be regarded as aphotic.

2.3.2   The Benthic Divisions

The benthic division includes the whole of the sea floor from the high water mark of the spring tide to the unfathomable deepest bottom of the ocean trenches. On the basis of different criteria such as depth, sediment deposits, relief, gradient and faunal composition, the benthic division can be further subdivided into two major topographic components related to pelagic environment: the continental margin, and the ocean basin. Within the continental margin are the continental shelf, slope and continental rise. The ocean basin comprises abyssal plains, oceanic islands (e.g. Cape Verde Islands and Ascension Island in the equatorial Atlantic), island arcs (e.g. Marianas), abyssal hills, the most prominent submarine ridges and sea mounts and guyots formed of volcanic materials and the deep trenches. Like the pelagic division, the uppermost part of the benthic division is subdivided into an upper littoral system, and the lower deep sea system.

2.3.2.1   Littoral System

The littoral system is that part of the sea floor, which extends from the shoreline to a depth of 200 metres; i.e. to the seaward margin of the shelf, or continental edge, where it drops abruptly as the 'shelf break' below which the continental slope begins. The littoral system can be further subdivided into an eulittoral zone, and sublittoral zone.

(1) *The Eulittoral Zone* extends from the high tide shoreline (water mark) to a depth of about 60 metres, where active coral reef building can take place, and some attached algae can still flourish (Sverdrup et al., 1961).

Supralittoral is the uppermost part of the eulittoral zone, i.e. the habitat between the high tide and the spray zone, where extreme variations occur and the population of marine organisms is relatively sparse.

On the other hand, the upper fringe of the eulittoral zone is usually occupied by intertidal zone which varies spatially and temporally from a few centimetres to several metres in depth, depending on the wave break and surge, and the range of tidal ebb and flow. The intertidal zone is generally characterised by variable nature of substrates with rocky, gravely, sandy, silty, and muddy shores, and subject to seasonal variations and turbulence. In the Tropics, the littoral system supports the growth of coral and coral reefs, and along the coasts, where the large rivers and estuaries merge with the sea, often forming a variety of habitats, for example, swamps of brackish water areas, where the mangroves predominantly colonise and contribute to yet another rich ecosystem (see Chapter Four on mangroves).

(2) *Sublittoral Zone*: extends from this rather arbitrary depth limit of 60 metres to 200 metres, reaching the edge of the continental shelf. The range of the sublittoral zone is quite extensive and the fauna found here are of mixed type.

2.3.2.2   Continental Shelf General Characteristics

The relatively shallow and flat submerged plain of the sea of varying width and bordering the continent is the continental shelf. Based on the more recent geomorphological data and the orthodox ideas, the shelf extends from the shoreline to a depth of 200 metres, and it virtually corresponds to the littoral system. Overall, the continental shelf stretches from the shoreline seaward, sloping gently; and the gradient changes from 0.001 to 0.002 (=0.10°–0.20°), to an average depth of 200 metres. As already noted above, this criterion of depth coincides with the vertical limit of the neritic province too. Within this conventional depth limit, the range of width and depth of the continental shelf may vary considerably from different regions of the maritime continents. Besides the depth limit, the width of the shelf is largely determined by other factors such as submerged geological land mass from the continental coasts, bottom topographical features, glaciation, sedimentation and effluents from the rivers, tidal estuaries and bays; these topographical features are more diversified. In the lower latitudes, particularly in the Tropics and subtropics, much of the shelves are formed of massive deposits of sediments, several metres thick, which are mainly of continental origin.

The continental shelf of the entire continents worldwide comprises some 7.8% area of the marine hydrospheric environment. The continental shelf is narrow in some areas and considerably broader in other places. It is quite extensive off the north coast of Siberia, in the Arctic, New Foundland, Argentina, Falklands, and northern Australia with an average width of 75 km, while it is very narrow or almost absent in the south-eastern coast of Ceylon, Florida and Riviera (the coastal region between south-eastern France and north-western Italy bordering on the Mediterranean). But, within the tropical belt, the continental shelf consists of some 2.3% of the global continental shelf area, and with the exception of the north-eastern Australia, which is covered by the Great Barrier Reef, has an average width of 38 km. Although the extent, nature of bottom topography and depth of the continental shelf of a geographically different locality can determine productivity, other factors are also involved and will influence profoundly. For example, despite very extensive shelf area of the Arctic, which consists some 18.24% of the total area of the world's continental shelf (Table 2.3), the productivity is significantly affected by unfavourable conditions for life, particularly of factors such as cold and ice. On the other hand, the continental shelf areas are relatively narrower in the Indian ocean, the Atlantic Ocean, and the Pacific Ocean and yet the productivity is far greater.

Table 2.3   Global distribution of water, land and
continental shelves in major oceans,
including the tropical and subtropical zones

TOTAL GLOBAL AREAS: $\times 10^6$ km^2	SEA WATER $\times 10^6$ km^2	LAND $\times 10^6$ km^2
510.066	361.127	148.939
100%	70.80%	29.20%

The total continental shelf areas (27.161 $\times 10^6$ km^2; and their distribution in relation to the major oceans, expressed as a percentage of the total marine hydrospheric area.

	%	%
Atlantic Ocean	22.37	2.557
Indian Ocean	20.01	0.880
Pacific Ocean	45.03	2.603
Other Oceans	12.54	1.760
Total Continental Shelves		7.800%

After much debate and a recent UN resolution, it is currently recognised on an international principle that a coastal country may claim up to 200 nautical miles which corresponds to the Exclusive Economic Zone (EEZ) (Fig. 2.3).

Because of the proximity of the land, many coarse materials of terrestrial origin are drained off by rainfall, storms, streams, rivers, estuaries and shoreline erosion of beaches into the sea. The land derived materials generally contribute to the rich deposits of the bottom sediments of the continental shelf. Due to the past glaciation, the bottom topography and the relief of the continental shelves vary quite considerably from place to place. The surface is not always smooth – some areas have deep depressions formed by past glaciation, and other areas are irregular due to formations of coral reefs. The continental shelf is the most productive portion of the marine environment. Large quantities of oil deposits, natural gas, minerals such as soil nodules of manganese, iron, cobalt, nickel, gold, tin, diamonds, phosphorus, sand and silts are found here. The largest deposits of oil and gas reserves are found in Iraq, Saudi Arabia, Abudabi, the Persian Gulf, Mexico, Venezuela and in offshores of Caribbean countries. Above all, much of the commercial fish and the most important, but less exploited, natural resource of seaweeds are

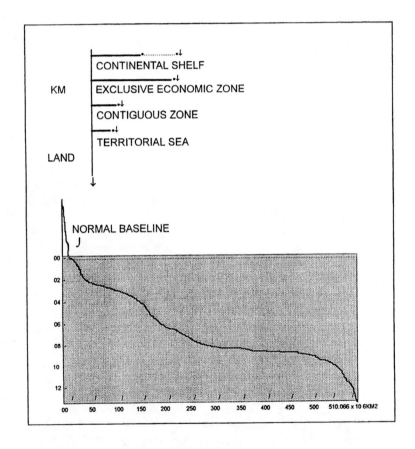

Figure 2.3   The extent of the continental shelf, and the newly proposed
UN exclusive economic zone (EEZ), all being measured from
the baseline.

confined to the continental shelf. Many parts of the tropical shelves are typically occupied by coral reefs (see Chapter Five on corals and coral reefs). Owing to the proximity to continents, relative shallowness, economic potentials and strategic importance, the continental shelf has become the most interesting and extensively explored part of the sea. The distribution of the sea floor is shown in Table 2.4. The UN convention, however, points out that regardless of the presence or absence of a geological continental shelf, the outer limit of the shelf zone extends at least 200 metres isobath irrespective of its distance from the coast (see below). The continental shelves end at a continental slope, where the gradient changes markedly in a steep slope that leads to the abyssal plains. The continental slope demarcates the true edge of the continents.

Table 2.4   The distribution of the sea floor, expressed as a percentage
of the world total marine area

GLOBAL SURFACE	ATLANTIC %	INDIAN %	PACIFIC %
Continental Margin	27.90	14.80	15.80
Deep sea ocean floor	38.120	49.30	42.90
Ridges, rises and fracture zones	33.30	35.60	38.40
Island arcs and Trenches	0.70	00.30	2.90
TOTAL	29.20	70.80	1:2.5

Sea floor related to depth range in metres		%
Sea level	0	–
Intertidal zone	0	2.00
Continental shelves	100–200	7.80
Continental slope	200–3,500	32.20
Continental rise	3,500–5,000	35.60
Hadal trenches	6,000–11.035	13.36

2.3.2.3   Regional Characteristics and Distribution of Tropical Continental Shelves

The effects of glaciation, the global tectonic activities, the physical extent, topographic features, the gradient of the sloping, the quantitative and qualitative nature of terrigenous effluents and sedimentary deposits, the periodical monsoons, which mix the coastal waters constantly and maintain the waters in turbid conditions, and the potential for productivity, all contribute to the formation and modification of the continental shelves in the Tropics. As if a natural

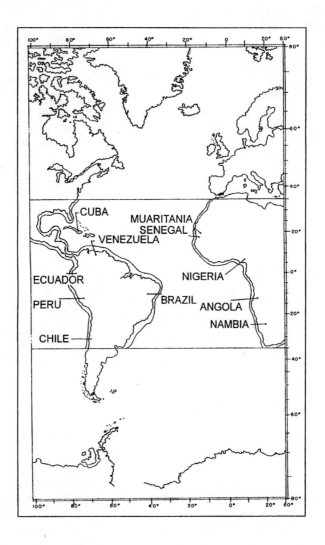

Figure 2.4   The continental shelves along the western side of Africa
                    facing the Atlantic Ocean.

rule, the continental shelves on the western side of the continents are usually more extensive than on the opposite side of the continents.

2.3.2.3.1 Tropical Shelves of Africa Facing the Atlantic

Within the tropical and subtropical realms, the continental shelves along the western side of Africa, facing the Atlantic Ocean, are bordered by Western Shara, Mauritania, Senegal, Gambia, Guinea Bassau, Sierra Leone, Liberia, Ivory Coast, Ghana, Nigeria, Cameroon, Equatorial Guinea, Gabon, Congo, Angola, Namibia and the south-western part of South Africa. These shelves, though relatively shallower, are generally quite broader (Fig.2.4).

The coastal ecosystems are also varied and more complex; because of the conflicting national interests and priorities of trading, navigation, aquaculture, industrialisation and other related activities, the coastal zones and their natural parameters are relatively less known. The continental shelves show large deposits of terrigenous origin. Typically, coral reefs are absent, except in a few islands in the Gulf of Guinea close to the Tropics. The major rivers, in the regions such as Nigeria, Camaroons and Congo, transport enormous amounts of inorganic and organic materials from the continent. The coastal zones of Congo, and the neighbouring Gabon and Zaire, covering the continental shelves to depths of 100–120 metres, where the sediment transports predominate with pelitic type, combined with 'green particles', calcareous deposits of carbonated faunal origin. The coastal zone is usually low and flat, fringed with indents of large lagoons and estuaries. The coastal regions are rich with a variety of fauna and flora, particularly the mangroves flourishing on marshy plains, though human population is relatively sparse. Except for the Guinea-Bissau and Niger Delta, mangroves are also scanty in most other places. Off the coast of Senegal is one of the rich fishing grounds, and the Spanish fishing fleet often goes there to take large quantities of fish illegally with no regard to international laws.

2.3.2.3.2 Tropical Shelves Fronting the Indian Ocean

The Indian Ocean is the third largest among the oceans of the world. It is bounded by Asia in the north, Africa in the east, and Australia and Antarctica in the south (Fig.2.5).

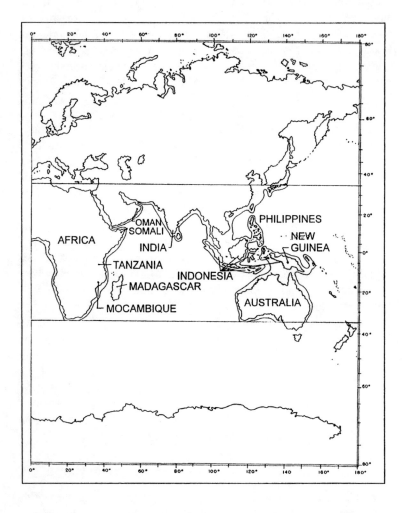

Figure 2.5   The continental shelf of Indian Ocean bounded by Asia, Africa, Australia and Antarctica.

There is no clear boundary between the three main oceans. However, the meridian through Cape Agulhas (20°E) between the Atlantic Ocean and Indian Ocean may imperceptibly separate the two; while the meridian line through South Cape, east of Indonesia, particularly Sumatra and Java, Timor and north of Australia, about 128° longitude and up to Cape Londonderrey, may separate the Indian Ocean and the Pacific Ocean. Along the coast of north-west of Australia, Timor Islands, Taiwan, Vietnam, Thailand, Malaysia, including Borneo, Indonesia, Burma and Bangladesh, the continental shelves widen to variable width and depth (see Chapter Six on sea water and ocean boundaries). The shelves bordering the coast of north-west Australia, fronting the Indian Ocean, are relatively narrow and fairly smooth with deposits of land-drained materials, but river drainage is scanty because of the lack of any large perennial rivers. On the other hand, off the south-eastern Indo-Burmese coast, the Irrawady and Ganges rivers fan out extensively and cover the shelf with terrigenous materials, including sand, silt and mud several kilometres seaward. India alone has a coastline of some 4,666 km and the extensive continental shelf slopes gradually down to the continental edge at an average depth of 183 metres. This triangular Indian continent consists of two wide arms of the Indian Ocean, notably the Bay of Bengal on the eastern side and the Arabian Sea on the western side (see Fig. 2.5). Many large and perennial rivers discharge their silt-laden waters into the sea. There are many other smaller gulfs, bays and rivers, extensive coral reefs and green belt of mangrove swamps, tidal estuaries and lagoons all along the coastline. Several groups of small and large oceanic islands are scattered in the south of the ocean. The southern limit of the Indian Ocean is close to the subtropical convergence at about latitude 40°S. Mixing of cold deep water with surface water takes place between Antarctic water and Red Sea water in mid-depth; and the deeper layers below 2,000 metres continue with Antartic cold waters (see Chapter Thirteen on ocean currents). The coast off the north of Ceylon (Sri Lanka) is fairly shallow, and large rivers are scanty, except in the north-eastern port of Trincomalee. The Palk Strait, off Mannar, between Talimannar and Mandapam Camp links Ceylon with the main land of India by a very shallow water sea; most parts of which are less than 10 metres deep and dotted with submerged fringing reefs and atolls of sandy bottom substrate. Many fishing grounds such as Pedru bank, Pead bank and Wedge bank are confined to this narrow sea between Cape Comorin and the Gulf of Mannar. Except the north-west coast, where the depth ranges from 10 metres to 100 metres, the rest of the continental shelf around Ceylon is relatively narrow; its width does not exceed 2-4 km at some points, and abruptly slopes down to the deep sea. Despite its coastline of nearly 1,800 km, the continental shelf resources are restricted to only a fraction of the coast. The coast is fringed with several inlets and bays. The coral reefs are usually absent in areas proximal to larger rivers' mouths. Much of the tropical part of the Indian Ocean has a continental shelf margined by Indonesia, Malay Peninsula, Thailand, Andaman Islands, Burma, Bangladesh, India, Saudi Arabia, Oman, South Yemen, Tanzania, Mozambique and south-eastern part of South Africa, Ceylon, Madagascar, the Seychelles and Scotra are some of the larger Islands proximal to the Indian continent (see Fig. 2.5).

Much of the coastlines are indented by the sea and covered with sedimentary alluvials. Bays, deltas, and lagoons are numerous. Coral atolls dominate the archipelagos of some of the smaller islands of Lacadive, the Maldives, Andaman and Nicobar, Amirante, Faraquhar and the Cocos Islands. Generally, the continental shelves are fairly narrow and extending seaward, at times up to 90 km, surrounding Andaman Island, the Malay Peninsula, the Gulf of Martaban and the Bay of Bengal, where the shelves exceed this width. The most extensive one is the Great Australian Bight, spreading up to New Guinea. Frequently, these shelves are cut longitudinally, especially off Burma, into an inner-shelf and an outer-shelf. Coral reefs are widespread as fringing, barriers and atolls (see Chapter Five on coral and coral reefs). Most shelves are predominated by silica of diatomaceace, calcarious radiolarian oozes, 'red clay' and terrigenous sediments. Deposits of manganous nodules and phosphorites are also abundant. The sedimental layer of the Red Sea beds have significant quantities of copper, zinc and silver and are currently exploited jointly by Sudan and Saudi Arabia as they have claims under the maritime jurisdiction of exclusive economic zone.

2.3.2.3.3   Tropical Shelves Facing the Pacific Ocean

The Pacific is the largest of the major oceans. The Pacific extends from the coast of Antarctica to the Bering Strait. The tropical coastlines include: Mexico, Guatemala, Salvador, Nicaragua, Costa Rica, Panama, and Columbia in the north of the equator; Quito and Equador, Peru and the upper end of Chile in the south of the equator. The separation between the Indian Ocean and Pacific Ocean is less than distinct but a line between Timor and Cape Londonderry in Australia (see Chapter Six on sea water and boundaries) divides the South Pacific from Indian Ocean. However, the tropical part of the Pacific coast facing the tip of California (North America) is fairly regular and relatively narrow. On the other hand, in the Malay Peninsula, Taiwan, Philippines, Papua New Guinea, a few islands of Indonesia and the eastern coast of Australia, and several scattered oceanic islands in the Pacific Ocean it is irregular. Many smaller tropical coral islands are of atoll, fringing and barrier types, and several hundred flat-topped guyots are also found (Fig. 2.6). Much of the bottom deposits are of land-drained red clay, together with diatom, radiolarian, Globigerina oozes. Thus the continental shelves are so varied in the different regions fronting the major oceans.

2.4   Deep Sea Systems

The deep sea system is divided into an upper archibenthic zone and a lower abyssal benthic zone.

The upper archibenthic zone extends from the edge of the continental shelf down to about 1,000 metres. The continental margin (shelf-break) and the upper part of the continental slope are confined to this zone. The lower abyssal benthic zone comprises the rest of the deep sea systems and extends roughly from this level to the deepest bottom, including the unfathomable trenches. The abyssal benthic zone can be further subdivided into the following different zones.

2.4.1   Bathybenthic Zone

The bathybenthic zone extends from 1,000 metres depth to about 4,000 metres and comprises largely the continental slope with a seaward gradient of about 4.25°, and at many intervals the slope is deeply cut into submarine canyons, which are primarily the consequence of the turbidity currents. Faults are also frequently encountered; the latter often exceed 45° in their declivity, as seen, for example, off the southern coast of Ceylon.

The continental rise is the transitional profile, which lies between

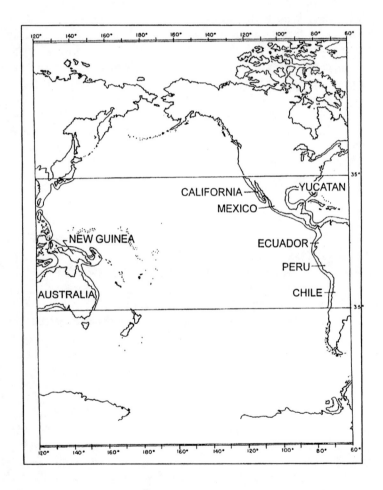

Figure 2.6   The continental shelf facing Pacific Ocean; and also note the
                  most extensive and relatively shallow continental shelf region
                  of Yucatan in the Atlantic side.

the continental slope and the ocean basins. Some of the distinct bottom topographical features are the relatively flat and smooth bottom due to accumulation of sedimental deposits of terrestrial origin, and hence the declivity hardly exceeds 0.5°.

2.4.2   Abyssal Benthic Zone

The abyssal benthic zone begins from 4,000 metres depth and descends down to about 6,000 metres, and comprises many of the great deep-ocean basins and troughs, and the mid-oceanic ridge system, the latter extends over 60,000 km and continues through all the oceans of the Arctic, Atlantic, Indian, Pacific and Antarctic (Fig. 2.7).

Besides sediment deposits, the sea bottom is characterised by many types of furrows and elevations. Numerous sea mounts and guyots dominate these ocean basin plains. Sea mountains are scattered submarine mountains reaching up to 900 metres or more from the bottom of the ocean basins and below 200 metres from the surface. More than 12,000 of them have been recorded throughout the oceans. On the other hand, the guyots, named after the Swiss-born American A. H. Guyot (1854), are flat-topped submerged elevations arising from the ocean bottom. They are thought to be ancient volcanic islands, but drowned about 300–1,000 metres below the surface during the geologic time. As the sea floor began to sink locally, the peaks flattened out as a result of subsidence. Data, based on dredging samples, suggest that their flat surface is less favourable to support coral growth to build atolls and to compensate the subsidence. Although the guyots are widespread in all oceans, they are quite common, particularly in the Pacific (Fig. 2.8).

2.4.3   Hadobenthic Zone, Including the Deep Sea Trenches

The Hadobenthic zone is the deepest part of the benthic division and descends from the abyssal plain into the deepest trenches. Mariana Trench, off the Philippines in the western Pacific Ocean, has been recorded as the deepest depression ever with a depth of 11,035 metres. It is now known that, despite increased pressure, lack of light and extremely low temperature (less than 5°C), many organisms dwell there. Recent underwater scanning TV photographs show a variety of

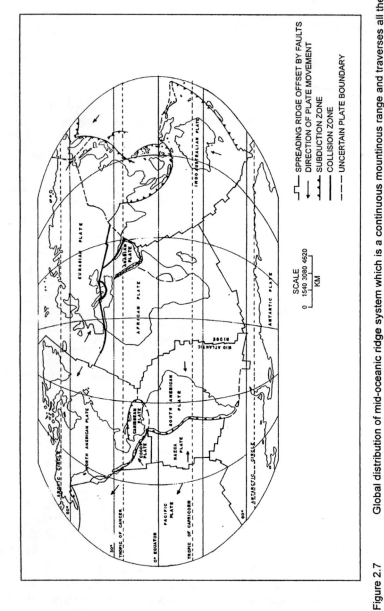

Figure 2.7   Global distribution of mid-oceanic ridge system which is a continuous mountinous range and traverses all the oceans and the crustal plates.

life forms in their abundance.

2.4.4 Deep Sea Trenches

Deep sea ocean trenches are elongated deep depressions in the ocean bottom. A good deal of great deep sea trenches are confined to the tropical and subtropical zones along island arcs, or far from the continents (Fig. 2.9).

Much of the knowledge gained about deep sea trenches is based on several expeditions. The major contribution came from the British *Challenger* (1872); the *Prince Alice of Monaco* (1901); the Russian *Vityaz* (1949); the Danish *Galathea* (1951); the American *Albatross* (1899), *Verna* (1958), and the *Eltanin* (1962), as well as of the bathyscape, *Triste* (1960); the French *Archimedes* (1962); and the more recent IIOE expedition (1962–1963). Many of the trenches appear to have formed during the Cenozoic period and have unique features. They are typically 'V' shaped and elongated. Their length varies from 100 to 300 km, while the width varies from 2 to 30 km. The narrow bottoms are filled with sediments derived largely from particles, settling from the surface, and transported from the neighbouring basins or sea floors, and by the turbidity currents due to intense seismic activity along the trench zones.

Hadal faunas seem to have developed a higher degree of endemism. The organisms that dwell in deep sea trenches are representatives of both major and minor phyla, principally of diverse forms of bacteria, several species of foraminefera, sponges of the class Demospongia and Hyalospongia, but calcareous sponges appear to be notably absent. Coelenterates of all major classes, corals, specially of the Madriporaria, are present; and the polychaete worms usually predominate. A few species of the minor phyla, such as nemertines, echiuroids of the family bonellidae and sipunculoids of phascolin, bryzoans, enteropeneusta, and pognophora are frequently found. Several species of gigantic crustaceans, including harpacticoid copepods, ostracods, and cirripeds of the family sgalpellidae; isopoda and amphipoda are commonly encountered, while mysidacea, cucumacea, and decapoda are usually absent at great depths. A variety of molluscs, especially gastropods, scapopod, bivalves and cephalapods are found; and of the echinoderms, holothurians, asteroids, ophiuroids are adequately represented. The protochordates

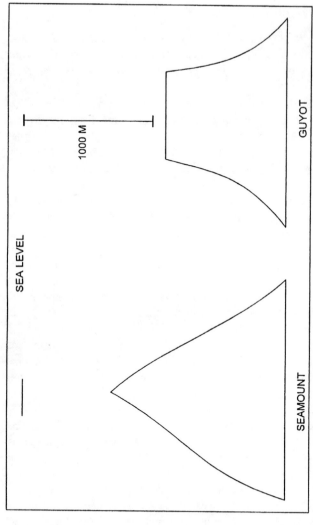

Figure 2.8   Seamount (A). and Guyot (B).   A semount is a submerged mountain elevation emerged from the sea floor by ancient volcanic activity.   Seamounts are of variable heights and more numerous in the Pacific Ocean.   The Guyot is a flat-topped seamount which lies about 1000m below the sea level.   Guyots are thought to be formed as a result of local subsidence of sea floor.   Though they are found throughout the oceans, many are scattered in the Pacific Ocean.

are typically represented by acidians, while chordates are poorly represented only by a few species of fish such as *Bassogigas profundissimus* and *Careproctus amblystomopisis* (Wollf, 1961). Most of the major trenches are located close to active volcanoes in the tropical and subtropical Pacific (Table 4.3).

Table 2.5   The major oceanic trenches of the world

TRENCHES	GEOGRAPHIC LOCATION	LENGTH km	WIDTH km	DEPTH m
Marianas	NW Pacific	2,550	70	11,035
Tonga	Eq. S Pacific	1,400	55	10,882
Kurile-Kamchatka	NW Pacific	2,200	120	10,542
Philippines	NW Pacific	1,400	60	10,497
Kermadec	SW Pacific	1,500	60	10,047
Peru-Chile				
Aleutian				
Romanche	Eq. Atlantif			7,865
Puerto Rico	N. Atlantic			
South Sandwich	Atlantic			
Java	Indian Ocean			

In summary, despite some conflicts in the terminology, the classification of the marine environment may be schematised for quick reference as follows:

Marine Environment

    Pelagic Division:
        Neritic Province
        and
        Oceanic Province

    Benthic Division:
        Continental Margin
            Littoral Systems:
                Eulittoral Zone
                Sublittoral Zone
        Ocean Basin
            Deep sea Systems:
                Archibenthic
                Abyssal benthic
                Bathybenthic
                Hadobenthic
                Deep sea Trenches

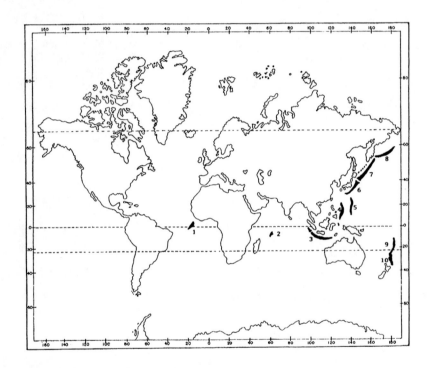

Figure 2.9   Some of the deep trenches; 1. Romanche Trench, 2. Vema Trench, 3. Java Trench, 4. Philippine Trench, 5. Marianas Trench, 6. Japan Trench, 7. Kurile-Kamchatka Trench, 8. Aluthean Trench, 9. Tonga Trench and 10. Kermadec Trench.

2.5 Exclusive Economic Zone and the Rights of Maritime States to Resources

LEGAL CONCEPT: Under the Geneva Convention of 1958 on the law of the sea, 'the outer limit of the continental shelf extends as the seabed adjacent to the coast but outside the territorial sea as a natural prolongation of its land territory to the outer edge of the continental margin, or to a distance of 200 nautical miles from the base line or shall not exceed 100 nautical miles from the 2,500 metres isobath, which is a line connecting the depth of 2,500 metres'. However, this entails some difficulties with the physical extent, particularly with island states. Inevitably, any conflicts or disputes in claiming a boundary of greater distance from the base line need to be resolved between the states involved, by negotiation based on 'equitable principles'. This idea is bound to be overlapped by the more recent convention of exclusive economic zone (see below). According to the UN conventions, most maritime nations have a certain degree of freedom of access to their continental shelves, territorial seas, contiguous seas, high seas, and sovereign rights to fishing and conservation of the natural resources of both living and non-living types in areas of the seabed, and ocean floor and subsoil. The UN conventional terms are defined as follows:

(1) *Normal baseline*: the low-water line along the coastal state (Article 5).

(2) *Breadth of the territorial sea*: up to a limit not exceeding 12 nautical miles, measured from base lines (Article 3).

(3) *The contiguous zone*: a zone contiguous to its territorial sea and the zone may not extend beyond 24 nautical miles from the baselines from which the breadth of the territorial sea is measured (Section 4, Article 33).

Since 1952, the maritime Latin American countries, in trying to keep the US tuna fishing fleet out, not only favoured the creation but sought sovereignty out to 200 nautical miles (370 km) from their coasts (I LM, 1970; 1977). However, by 1982, a legal consensus was reached by 153 nations on man's proper use of the sea and the sea floor beneath it. The exclusive economic zone is then defined as 'an area beyond and adjacent to territorial sea; shall not exceed 200 nautical miles from baseline (low water line) from which the breadth

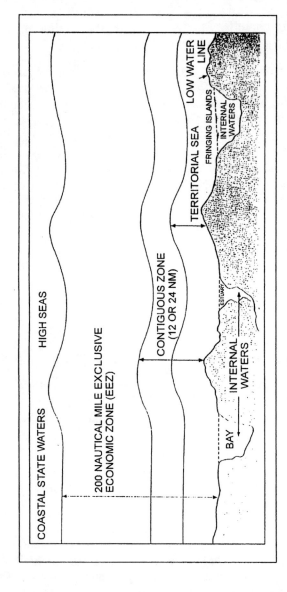

Figure 2.10 UN Conventional coastal sovereign rights: figure showing baseline, 2 n mile contiguous sea; 200 n mile exclusive economic zone (after A Couper 1983, Times world atlas of oceans, Copy right with kind permission of Times Ltd).

of the territorial sea is measured' (Art. 57, UN Conf. 63, 1982). The exclusive economic zone will extend, in most cases, far beyond the continental shelf, and the maritime nations will have jurisdiction and sovereign rights and greater advantage to control this zone, with respect to explore, exploit, conserve and manage all the natural resources, both the living and non-living, of the seabed and subsoil and the superjacent waters and with regard to 'other activities for economic exploitation and exploration of the zone such as production of energy from the water, currents and energy' (Fig. 2.10, also see Fig. 2.3).

2.6   Natural Resources

Natural resources include both non-living and living resources, for example, fish and seaweeds. In terms of economy, costs, benefits and potential, the exploitation of non-living resources entails much more advanced technology and funds, especially in the tropical and subtropical countries, than the exploitation of living-resources.

2.6.1   The Non-Living Types

Much of the potential deposits are petroleum and natural gas, which are confined to the Persian Gulf, Red Sea areas and Bastraits, Nigeria and Malaysia in the tropical region. Recently, large deposits of fossil fuel have been discovered along the east coast of India, and petroleum technology is being actively developed. Some of the rare constituents of other deposits are rutile, ziconium, monazite, diamonds, coraline lime stones; and sediments of iron, copper, manganese, and phosphorite nodules are also common.

2.6.2   Living Resources

Corals are quite common and widespread within the tropical shelves. They flourish well on shelves less than 60 metres deep and shallow areas have islands and atolls (see Chapter Five on corals). Much of the coasts are fringed with thick mangroves rich in both species and abundance (see Chapter Four on tropical mangroves). Others include in particular, crustaceans, molluscs, especially cephalapods; many species of pelagic and demersal fish: sharks, rays, a variety of highly coloured coral fish, flying fish, pomprets, anchovies, sail fish, mackerel, tuna, marine turtles, and marine mammals of all kinds.

2.7 Sea Floor and Sediments

The sea bottom is formed of sedimentary deposits of detrital materials of organic and inorganic nature. These are either of terrigenous or pelagic origin. Near the coastlines and over the continental shelves, the origin of much sediment is land-derived, particularly from the discharges of the major rivers and land erosion, wind blown or of biogenic and glacial past. Several pistion cores collected during the geological research cruises, including the sediments of the continental rise and Amazon Cone and abyssal plains off North Brazil, have revealed to contain hemipelagic clay rich in terrigenous silt, organic detritus, and pelagic plankton, especially foraminefera (Damuth, 1975), radiolarian oozes, globigerina, pteropods, pelicipods, gastropods, bryozons, corals, and phytoplankton. The pelagic sediments also include 'brown-clay', manganous nodule formed *in situ,* meteoric dust and volcanic ash. Sediments consist of detrital particles of different size and shape and, as already mentioned, are constantly subject to abrasive processes by wave action and tidal flow. When the particles eventually settle down, they form a variably thick or thin layer with several interstices. According to seismic surveys, most of the Atlantic and Pacific oceanic basins contain relatively thick layers of 300–500 metres. Silicate and carbonate sediments are commonly found over the continental shelves; the carbonate increases in the tropical belt. Sedimentation rates range from 5 to >50 cm per 103 years. It is of considerable economic importance since the petroleum is derived from the marine sedimentary deposits. Many tropical coastal nations are seriously concerned with investing large sums of money in exploring the oil reserves from the seabeds. The reserves generally contain massive fossilised skeletal materials of plant and animal origin. Many factors such as crustal movements, volcanic eruptions, tropical storms, river influx, exoskeletal materials of calcareous nature, and spicules of sponges are usually more in the Tropics and subtropics than in higher latitudes. Many marine organisms, especially diatoms and radiolarians, foraminifera, globigerina, pteropods, coarse coral rubbles, and many other organisms, and a host of marine bacteria, and sand, silt, mud, coastal erosion, submarine avalanches and turbidity currents all contribute to the formation of bottom sediments. In the coastal neritic waters, other factors such as storm wave and tidal actions also influence. On the surface of the intertidal sedimentary deposits, many autotropic diatoms and dianoflagellate and a variety of other organisms dominate. The globigerina ooze covers most of the ocean floor of the Atlantic and Indian Oceans, and greater parts of the Pacific. In some areas pteropod shells are also found mixed with calcareous oozes. The siliceous radiolarian shells are an important constituent of sediment at oceanic deeper layers in the tropical Pacific and Indian Oceans.

References

Belyaev, G. H., *Hadal Bottom Fauna of the World Oceans* (ed. L. A. Zenleevich), Israel Programme for Scientific Translation, Jerusalem, 1972, pp.199

Bischoff, J. L. and Manheim, F. T., 'Economical Potential of the Red Sea Heavy Metal Deposits', in *Hot Brine and Recent Heavy Metal Deposits,* (eds Degans, E. T., and D. A. Ross), 1969, pp.535–541

Bruun, A. P., 'Life and life conditions of the Deep sea', *Pacific Scientific Congress*, vol. I, 1955, pp.399–408

Couper, A., (ed.) *Times World Atlas of the Oceans*, London, Times Books, 1983, pp.227

Damuth, J. E., and R. W. Fairbridge, *The Equatorial Atlantic Deep Sea Arkosic and the Ice Age Aridity in the Tropical South America*, Geol. Soc. American Bull, 81, 1970, pp.189–206

Damuth, J. E., *Sedimentation on the North Brazilian Continental Margin*, An. Acad. Bras. Cienc., Suplemento, 1976, pp.43–50

Davis Jr. R. A., *Principles of Oceanography*, Reading, Massachusetts and London, Addision-Wesley Publishing Co., 1972, pp.434

Defant, A., *Physical Oceanography*, vol. I, Oxford, London, and New York, Pergamon Press, 1961, pp.729

Deshmuk, I., *Ecology and Tropical Biology*, Palo Alto, Blackwell Scientific Publishers, 1986, pp.387

Ewing, M. et al., 'Sediment Distribution in the Indian Ocean', in *Deep Sea Res.*, 16, 1969, pp.231–248.

74

Fairbridge, R. W., (ed.) *The Encylopedia of Oceanography*, vol. I, New York, Reinhold Publishing Corporation, 1966, pp.1021

Fisher, R. L., and Hess, H. H., 'Trenches' (ed. Hill, M.N.), *In the Sea*, vol. III, 1963, pp.411–436

Hedgpth, J. W., 'Classification of Marine Environment', pp.29–51, in *Treatise on Marine Ecology and Palaeocology*, vol. I, Me. Geol. Soc. Am., 67, 1957, p.1296

Hill, N. N, (ed.),. *The Seas*, vol. III, New York and London, John Wiley and Sons, 1963, p.963

Piccard, J., *Man's Deepest Dive*, Nat. Geogr. Mag., 118 (2), 1960, pp.224–239.

Russel, F. S., and Yonge, M., *The Seas*, London, Fredrick Warne and Co Ltd, 1975, pp.283

Sien, C. L., and Mac Andrews, C., *Southeast Asian Sea*, Singapore, McGraw Hill International Book Co., 1981, pp.375

Strahler, A. N., *Physical Geography*, John Wiley and Sons, Inc., 1960, pp.643

Sverdrup, H.V., Johnson, M. W. and Fleming, R. H., *The Oceans*, (Asiatic student edition), Tokyo, Prentice-Hall Inc., 1960, pp.1080

Tait, R. V., *Elements of Marine Ecology: An Introductory Course*, London, Butterworth, 1981, pp.356

U.N., 'Convention on the Law of the Seas', UN Conf. 63/WP10/Rev. 3, 1980, p.180

Weyl, P. K., *Oceanography: An Introduction to Marine Environment*, New York and London, John Wiley and Sons, Inc., 1970, pp.395

Wolfang, S., *Deep Sea Sediments of Indian Ocean*, (ed. Trask), 1939, pp.396–408

Wolff, T., 'Life in the oceans six miles deep' *New Scientist*, 24 (414), 1964, pp.241–244

Zeitzschel, B. (ed.), *The Biology of the Indian Ocean*, London, Chapman and Hall Ltd., 1973, p.549

Chapter Three
Origin and Evolution of Life in the Shallow Seas and the Impact of Continental Drifts by Global Plate Tectonics

3.1   Origin and Evolution of Life in the Shallow Sea

No one knows precisely how life began on this planet. For centuries, scientists have been trying to explain the intricate nature of living systems on the basis of physico-chemical laws. Most theories are speculative and none is still conclusive. Nonetheless, there is a remarkable degree of uniformity in the organisation of all living things. Life, *in toto,* including human beings, shares the fundamental properties common to all living organisms, despite their bewildering diversity. In spite of much speculation that life might have originated in an extraterrestrial environment, such as interstellar clouds, and colonised the earth, it is widely believed by many scientists that life on this planet might have originated through a chain of chemical evolution in the relatively warmer neritic part of the sea.

However, using radiocarbon dating techniques and extrapolating the data of the fossilised stromatolites of Isua rocks of Greenland and other subcontinents, particularly Swaziland in South Africa, Glinflint of Canada, Vempalle of India, Warrawoona of Australia, it can be inferred that the earliest sign of life, prokaryotes (Fig. 3.1), would have appeared some 3.5 billion years ago in the tropical seas. This is about a billion years after the primitive atmosphere had already formed. As the earth cooled, the atmospheric vapour began to condense, and the water accumulated to form the ocean. Prokaryotes, with no distinct nuclear membrane or chrosomal organisation,

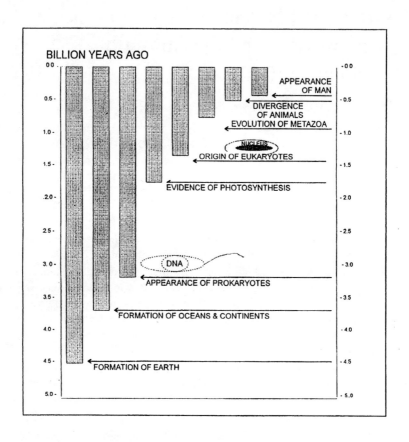

Figure 3.1   Evolution of life: the earliest sign of life as Prokaryotes appeared some 3.5×10^9 years ago.

certainly would have evolved in the absence of free oxygen and continued to live anaerobically for many millions of years, as seen, for example, in the anaerobic prokaryotic bacteria, *Kakabekia umellata*. Gradually, the anaerobic bacteria and blue-green algae (cyanobacteria) developed green pigments and produced oxygen as a photosynthetic by-product, using water and carbon dioxide. With the advent of oxygen in the atmosphere, the bacterial and the blue green-algal colonies of different kinds formed mutually beneficial symbiotic sub-colonies. The progressive symbiotic association between colonies of bacteria and the blue-green algae facilitated the incorporation of mitochondria and enriched the genetical resources, and probably led to the origin and development of more advanced cellular forms of life – the eurokaryotes. The latter, unlike prokaryotes, were with distinct nuclear membranes enclosing chromosomes containing gene pools and capable of division and with sexual potential. Some experts even claim that the cellular organelles of chloroplasts and mitochondria were once independent bacteria; chloroplasts to have descended from the anaerobic symbiotic photosynthetic cyanobacteria, while mitochondria originated from aerobiotic symbiotic non-sulphur purple bacteria. The first eukaryotes seem to have appeared about 1.5 billion years ago. As we can see today, the two most important eukaryotic cells, mitochondria and chloroplasts, share a number of unique characteristics with each other and with some of the bacteria from which they were probably derived. Among the eukaryotes only the *Pelomyx plaustris,* a primitive 'amoeba-like' protist, which lives in freshwater muddy bottom, lacks proper mitochondria. These seem to have two kinds of bacterial symbionts within them and they may play the same role as mitochondria as do in most of the other eukaryotes. On the other hand, based on distinct biochemical characteristics, the chloroplasts might have been derived from three distinct groups of symbiotic bacteria. The chloroplasts of red algae, which contain chlorophyll a, carotenoids and phycoblins are almost similar to those of symbiotic cyanobacteria. The chloroplasts of the brown algae, bacillariophyta and dianoflagellata, have chlorophyll a, carotenoids and fucoxanthin. These chloroplasts very closely resemble those of the symbiotic bacteria of the genus, *Heilobacterium*. The chloroplasts of all green algae, Euglenophyta, and higher plants, which contain chlorophyll a, b, and carotenoids, were almost certainly derived from the symbiotic bacteria of the genus, *Prochora,* which are predominantly found in the tropical and subtropical seas. This is one of the primitive prokaryotic groups of organisms and these blue-green algae have the ability to tolerate a wide range of environmental conditions. They are, however, distinct from the rest of the algae in their cellular organisations, characteristics, mode of reproduction, biochemistry and function; and spatial and temporal distribution in the ocean.

Some of these characteristics are shared by only a few other related groups of prokaryotic bacteria. Nevertheless, the bacteria and the blue-green algae have played an important role in enriching the primitive environment with oxygen and, consequently, the development of the eukaryotes by genetic recombination, reassortment and chromosomal crossing-over which gave rise to the evolution and diversity of many other more complex organisms. Because the sea provides the most suitable environment for diverse forms of life and there is a striking similarity in chemical composition between the sea water and the body fluids of the various life forms, it is generally inferred that life originated in the shallow warmer seas.

3.2   The Continental Drift Mechanisms by Global Plate Tectonics and the Sea Floor Spreading

By continental drift is meant the large-scale horizontal displacement of the continents relative to one another and to the ocean basins during geologic time. The overall theoretical concepts have developed over centuries into a true phenomenon now. A brief review of the highlights of the developments of the geological and geographical knowledge of the marine environment is not only pertinent at this point, but also may help to understand the profound changes that resulted from the moving continents.

Historically, around 1800, Alexander von Hamboldt, the German naturalist, thought that the continental shelf margins bordering the Atlantic Ocean on both sides had been joined together. Some fifty years later, the French palaeobiologist, Antonio Snider-Pellegris, suggested that the presence of identical fossil plants in both Europe and North American coal deposits could be possible if only the two continents were formerly connected. As early as 1908, the American geomorphologist, Frank B. Taylor, put forward the idea that the formation of the world's mountains could have been caused by continental collision; and that the matching of the west coast of Greenland to North America and the creep of Euro-Africa and the Americas formed the mid-Atlantic Ocean ridge.

Several lines of evidence emanating from modern research of geology, geophysics, geochemistry, geomagnetism and palaeontology certainly favours the original hypothesis of a continental drift. By comparing the striking similarities of the land masses of South America, Africa, the Indian peninsula, Australia and Antarctica in the southern hemisphere, an Austrian, Eduard Seuss (1909), speculated that these land masses were originally a single 'megacontinent' and he called it Gondwana or Gondwanaland, after a geologically significant province in the eastern part of India. However, at that time, Seuss could not explain the real mechanism involved in his geological correlation and the break-away concept.

3.3 The Concepts and Theories of Continental Drift

As early as 1912, the concept of continental drift, based on the apparent fitness of the coast lining the Atlantic Ocean came to Alfred Wegener, a German meteorologist; he first began to popularise the theory that the continents in both northern and southern hemispheres originated from a single parent 'supercontinent' called Pangaea[1]. Pangaea then split apart some time during the Triassic period, about 200 million years ago, into two great land masses. The continents were not always in the same positions in the past as they are today. In 1937, Alexandra L. Du Toit, a South African geologist, modified Wegener's theory and proposed the two primordial continents of Laurasia (northern and eastern Asia, Europe and North America) in the northern hemisphere, and Gondwanaland (Africa, South America, India, Australasia and Antarctica) in the southern hemisphere. These were partly separated by (1) the large marginal Tethys Sea, and (2) the rest of the area occupied by the vast ocean Panthalassa; the latter later became the Pacific Ocean. These two primordial continents or land masses began to drift apart slowly and became separated more and more by the then Tethys Sea, about 180 million years ago. As drifting continued, assisted by the various forces of the earth, the

---

[1] Greek, meaning 'all lands or all earth'.

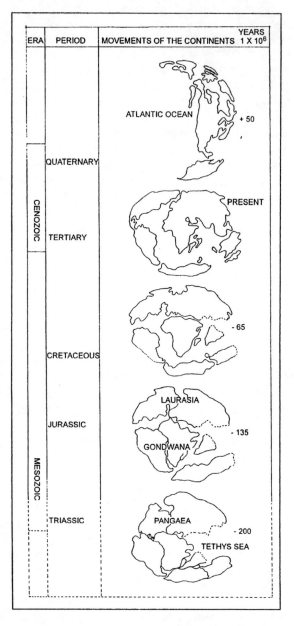

Figure 3.2   Distribution of the land masses during initial break-up and the subsequent global continental drift through geological time.

Laurasia began to split and carried with it the fragments of the continents of North America and Eurasia towards the north, while the Gondwanaland also began to split and took with it the fragments of South America, Africa, India, including Madagascar, Australia and Antarctica towards the south and the two primordial continents, together with their respective 'fragments', began to move away from one another – a westward drift of the plates of North and South Americas opened up the Atlantic Ocean, while the Tethys Sea began to be squeezed by subduction process, leaving almost a continuous equatorial seaway. The Indian plate moved towards the equatorial Tethys. As Australia and Antarctica also began to be detached from South America and Africa, the Indian Ocean began to be formed during the Jurassic period. Towards the end of Jurassic, the Atlantic Ocean began to widen and almost enclosed the Tethys Sea, forming the Mediterranean and leaving the remnant of the Tethys Sea, the present Bay of Bengal, to be integrated as a part of the Indian Ocean. Around fifty million years ago, Africa already drifted slowly northwards, and India joined with Eurasia; the collision resulted in formation of the Himalayas. Australia too split away from Antarctica about 40 million years ago, and North America and South America became connected and remained in their present positions (Fig. 3.2). These events are of special interest because of the continued process of drifts and the major changes in the evolution of the continents, the oceans and life.

3.4   Sea Floor Spreading

It was not until 1896 that the evidence for the earth's crustal motion became fully overwhelming. A new global phenomenon of plate tectonics was proposed. Essentially, the sea floor spreading mechanism by global plate tectonics – 'breaking and bending' – is said to work as follows:

As noted earlier, geologically, the outermost part of the mantle is fused with the whole of the crust to form a thick rigid layer, the lithosphere (see Chapter One). Underneath this lies the molten semisolid part of the mantle, the asthenosphere. Asthenosphere – a term coined by Joseph Barrel (1914) – for the underlying thick, hot, rather weak layer below the lithosphere. The asthenosphere is constantly subject to intense heat from the interior of the earth. As a result, the peripheral lithosphere is believed to be cracked into 6–12 major or primary oceanic plates, and each of these plates is about 108 km thick. Between them lie the minor or secondary plates. These lithospheric plates slide slowly or pull apart over the semisolid asthenosphere due to convection currents in the asthenosphere. Recent submarine investigations have produced convincing evidence that these plates border the mid-oceanic ridge system; the latter forms a zone of separation between the lithospheric plates, traversing through the Arctic, Atlantic, Indian, Pacific and Antarctic Oceans.

The mid-oceanic ridge system is a continuous chain of submarine mountains of varying heights of 1–3 km from the sea floor and separated by a cleft of 3 km deep and about 38 km wide and extending over 60,000 km (see Fig. 2.6). This zone of oceanic ridge is associated with many major earthquakes and volcanic activities of the past and the present.

Due to various forces of the earth, including the heat generated by the radioactive materials from interior, the lithospheric plates, interact at boundaries, diverge, converge, or even slide away from each other along the mid-oceanic zone. New materials well up and, on reaching the surface, outflow, thus spreading the continents on either side. As more and more materials are brought out, the older materials become deposited in the ocean basins far from the mid-oceanic ridge axis. The spreading rate of the continents is inconceivably finite– the major plates spread faster than others – and it has been estimated to be 1 centimetre a year for the Afro-Arabian plate, 2 centimetres a year for the Eurasian-American, and 12 centimetres a year for the Pacific-Nazca plate.

On the other hand, where two plates move towards each other, they collide and one plate bends and is thrust beneath the other to sink down into the earth, particularly at deep trench-sites (see Fig. 2.9). The plate tectonics theory has received much support since 1960 and has now become almost universally acceptable.

3.5 The Continental Drift and the Subsequent Change

As already noted, North America, a part of Laurasia, began moving away from Europe and Asia about 165 million years ago. This was followed by the separation of South America from Africa during the

Cretaceous period about 130 million years ago. Africa, India, Australia, and Antarctica began separating as well. Around fifty million years ago, Africa already drifted slowly northwards and the Indian plate moved towards the equatorial Tethys. As Australia and Antarctica also began to be detached from South America, Africa and the Indian Ocean began to be formed during the Jurassic period. As Australia and Antarctica also began to be detached from South America, and Africa and India moved several degrees towards the north, colliding with Asia (see Fig. 3.2). The northward movement and the collision of India with Asia, at the end of the Oligocene period about 40-53 million years ago, had been the major factors responsible for the closure of the Tethys and the upthrust of Himalayas and the Tibetian Plateau, which became interlocked with Eurasia. Australia too split away from Antarctica about 40 million years ago and North America and South America drifted and became connected, and remained in their present position (see Fig. 3.2).

3.6 Evidence in Support of the Continental Drift

Evidence in support of the original concept of a supercontinent and the subsequent continental drift comes from a variety of sources. Based on the recent geophysical investigations and increased knowledge, proponents who advocate the theory of continental drift receive much support along the following lines:

(1) The earliest marine deposits along the Atlantic coasts of South America and Africa are of the Jurassic age and the present-day oceans did not exist before.

(2) Data from studies of rocks taken from the margins of South America and Africa closely agree in both age and structure.

(3) The widespread glaciation during the late Palaeozoic, about 225 million years ago, was evident in all continents of Anarctica, Australia, India, South America and Africa.

(4) The glacial 'tillites' – lithified rocks – of the same age and characteristics caused by glaciation in the southern hemisphere are found in South America, Africa, Madagascar, India, Australia, Tasmania, New Zealand and Antarctica suggest the origin of these continents in Gondwanaland.

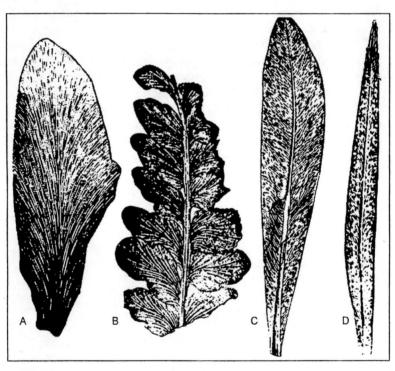

Figure 3.3   The fern-like seed bearing plants with long leaves of the
genera: (A) *Gangamopteris*, (B) *Glossopteris*, and (C & D)
young leaves of Glossopteris with mid-rib.   They are of
Precarboniferous age, well preserved in sedentary rocks of
Gondwana, an ancient Province of India. (Modified from E W
Berry, 1920).

(5) The presence of identical plants and animals as fossils on widely separated continents across great oceans would not have been possible by land bridges. For example, the fossils of well preserved 'fern-like' seed bearing plants with big leaves of the genera, *Glossopteris* and *Gangamopteris*, found in coal beds of the pre-carboniferous age were discovered in sedentary rocks of the Gondwanaland province of India (Fig. 3.3). Similar strata of rocks were later discovered in South America, Africa, Australia and Antarctica.

(6) Fossils of smaller reptiles of the genus, *Mesosaurus,* smaller sea-going Permian reptiles with a long slender head and delicate sharp teeth were found beautifully preserved in mudstones on both sides of Atlantic; in Brazil and recently in South Africa (Fig. 3.4). It is unlikely that these small reptiles could have crossed the broad stretches of open ocean; and their spatial distribution indicates that the continents in the southern hemisphere were once joined together.

(7) The freshwater fish of the two closely related genera, *Cleithrolepidina* and *Cleithroplepsis,* are found only in South Africa and Australia, respectively.

(8) The closely related tongueless frogs of the genus, *Agiossa*, are found both in Africa and in the Amazon regions.

(9) Studies of the magnetic properties of rocks of different ages from different continents indicate that the magnetic poles of the earth were in different places during different geological times. Since the core of the earth consists of iron, it is believed to act as a magnetic field; the magnetic effect is maximal at poles and minimal at the magnetic equator. These suggest that the continents, which are now separated were once joined together. The magnetic curves from Europe and North America suggest that North America has moved about $30°$ westwards relative to Europe since the Triassic period.

(10) Sediment dating and analysis also provide strong evidence in support of plate tectonics. As late as 1960, the theory of plate tectonics which combines the concept of sea floor spreading, as explained above, with the continental drift supports the view that as the lithospheric plates move away

CLEITHROLOPIDINA

MESOSAURUS

ARMADE

OPOSSUM

LHAMA

TAPIR

Figure 3.4   Fossils of the past and the living fossils of the present, especially the marsupials in the South American Continent.

from the mid-oceanic ridges, they carry the continents with them.

As we have seen above, from these and other reasons, it is possible to infer that the original single 'megacontinent' once became divided into two supercontinents. Due to very slow movements, they broke up into subcontinents and have been brought to their present positions. The slow process of movements and displacements of the continents and the oceans continues into the future geologic time.

3.7 The Impact of Continental Drift on Distribution of Continents, Oceans and Living Organisms

The impact of tectonics activities on evolution of life had been profound. The changes in the continental positions would have had marked effect on the global temperature, ocean size, ocean currents and distribution of nutrients. These and the fluctuation of temperature and periodic glaciation seemed to have brought about significant evolutionary changes in the oceans and land. A vast variety and highly specialised fauna and flora are now confined to the Tropics and subtropics. The prolificity, diversity of species and their numerical abundance are now found significantly more in the tropical marine realm than elsewhere. The diversity of species increased as the result of breaking, bending and the movement of the continents. The emergence of the Isthmus of Panama during Pliocene, about 5 million years ago, made it possible for the highly evolved forms, especially marsupial in the South American subcontinent. Many herbivorous land animals from the North American subcontinent migrated to South America and became well adapted and more successful than their endemic and contemporary species. However, some of the South American mammals became extinct due to predation by the immigrants from the North American subcontinent. As few animals, such as armadillos and opossums, began to move northwards, some of the rare mammals, such as llamas and tapirs, which came from the northern subcontinent, found the pristine southern subcontinent their favourable living environment. The closer of the isthmuses also caused natural barriers, which restricted the movement of many marine animals, including cetaceans such as grey and sei whale species.

3.8 Sea Level Changes and the Effect on Speciation

Recent evidence confirms the original view that the global mean temperatures have risen from 0.3°C to 0.6°C during the last 130 years (Houghton et al., 1992); and a rise of 0.8°C during the last 42 years in the upper layer of 100 metres of the sea. The increase was accompanied by an annual rise of sea level of between 1 and 3 millimetres (Roemmich, 1992). Global temperatures have fluctuated by 2°C since the end of the last ice age. During an ice age, sea level can fall and rise by about 100 metres (see Chapter Fifteen). Thus, time-scale has involved both geological and evolutionary changes; the continental drift has been affected profoundly over geologic time-scale, but played an important role in speciation of both marine and terrestrial organisms by creating the natural barriers for the free flow of genetical materials and had a significant influence on the distribution of many plants and animals across the major continents and many subcontinents of the world. This is in spite of their separation by vast and veritable oceans.

References

Berry, E. W., *A Sketch of the Origin Evolution of Floras*, Annual Rep. of Smithsonian Institute (1918), 1920, pp.289–409

Binham, R., 'Explorers of the Earth Within', *Science*, 1 (6), 1980, pp.44–55

Cloud, P., and M. F. Glassener., 'The Edicaran Period System: Metazoa Inherit the Earth', *Science*, 218, 1982, pp.783–792

Dauviller, A., *The Photochemical Origin of Life*, New York and London, Academic Press, 1965, pp.119

Dietrich, G., et al., *General Oceanography: An Introduction*, John Wiley, 1980 pp.620

Gould, S. J., *Ontogeny and Phylogeny*, Cambridge, Massachusetts and London, The Belknap Press of Harvard University, 1977, pp.380

Gray, M. W., 'The Bacterial Ancestry of Plastids and Mitochondria', *Bioscience*, 33, 1983, pp.693–698.

Grosbestein, G., *The Strategy of Life*, San Francisco, W. H. Freeman and Co., 1974, pp.174

Gruin, J., 'In the Begin', *Science*, 1 (5), 1980, pp.44–51

Hough, J. T., Callander, B. A. and Varney, S. K. (eds), Intergovernmental Panel on Climate, Cambridge University Press, UK, 1992, p.17

Margulis, L., *Symbiosis in Cell Evolution*, San Francisco, W. H. Freeman and Co., 1992, p.448

Menard, H. W., *The Oceans of Truth: A Personal History of Global Tectonics*, Princeton, Princeton University Press, 1986, p.353

Miller, S. L and Original, L. E., *The Origins of Life on Earth*, Inglewood, N. Jersey, Prentice-Hall, Inc., 1974, p.229

Romemmich, D., 'Ocean Warming and Sea Level Rise along the South-west Coast', *US Science*, vol. CCLVII, no.5069, 1992, pp.373–375

Schopf, J. W., 'The evolution of earliest cells', *Scientific American* (September), 1978, pp.45–103

Ursula, B. M., 1975. *Continental Drift, the Evolution of Concept*, Washington, Smithsonian Institution Press, p.239

Valentine, J. W., and Moores, E. M., 'Plate Tectonics and the History of Life in the Oceans', *Scientific American*, 1982, pp.19–28

Vidal, G., 'The Eukaryotic Cells', *Scientific American* (February), 1984, pp.48–57

Wilson, J. T., 'Geophysics and the Continental Growth', *American Scientist*, 47, pp.1–24

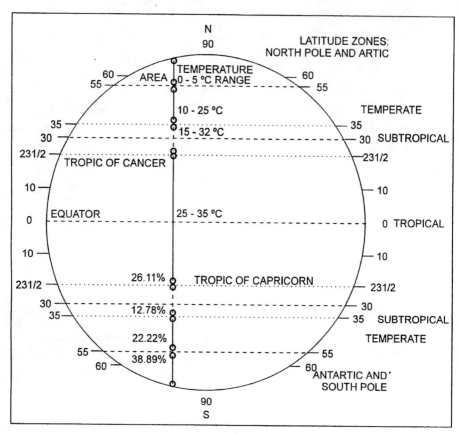

Figure 4.1   The major latitudinal zones of the earth associated with a range of sea surface temperature. The frigid zone includes the North pole and arctic and the South pole and antarctic. The distribution of animals and plants is largely determined by the variation in temperature, The areas are expressed as a percentage of the total area of the earth surface.

Chapter Four
The Tropical Marine Realms

4.1   Special Tropical Features

Although the tropical region is an imposing and formidable part of the globe, economically, it is not booming significantly. Yet, marine scientists and ecologists can find unrivalled variety of beauties and splendour, curiosities and surprises. Some of the most important features of the tropical region are: the shallow and tepid seas stretching between the twisted chain of archipelagos of islands and the continents; frequent volcanoes; and great continental shelves which are known to have submerged relatively recently during the end of Pleistocene; the marked importance of the tropical equable climate and its effects on the ecosystems and distribution of fauna and flora. Until recently, it was often proposed that the high tropical diversity resulted from a long period of constant and benign climate that allowed time for niche diversification and for the existences of different species. So, what constitutes the tropical environment is largely determined by the prevailing conditions. These conditions range quite considerably. There is always an enormous influence of the environment upon living organisms, i.e. there is a constant interaction between living organisms and the non-living environment. In the Tropics, however, this influence is exaggerated significantly since the interactions are directly related to control the temporal and spatial variation. This chapter considers briefly some of these conceptual bases of the tropical and subtropical marine aquatic ecosystems and their characteristics.

4.2   The Major Latitudinal Zones

These zones are generally associated with temperature ranges and the extent of the solar energy conditions. Depending on the path of the sun, and consequently the solar energy that reaches the earth, thelithospheric surface can be divided into three major latitudinal zones (Fig 4.1):

(1) The tropical zone (and subtropics).

(2) The temperate zone.

(3) The frigid zones (North and South Poles and the Arctic and Antarctic).

4.3 Tropical Zone

The Tropics had relatively an explosive geological past and are contrastingly different from the other temperate and frigid zones. The term Tropics is broadly used to denote the climatically fairly uniform belt of enormous stretches of continents, oceans, seas, and the widely scattered islands and archipelagos of amazing sizes and shapes of Asia, Africa, Americas, and Australia. Although these lie between the Tropic of Cancer and Tropic of Capricorn about 23.5° on either side of and parallel to the equator, where the sun reaches its greatest declination, a clear limit on the boundary cannot be fixed because of the tropical regional features that may extend a few degrees, at least up to 30°, further on each side of the equator into subtropical transitional zones where the climatic characteristics change very little from the Tropics as the latitude increases. For example:

(1) The conditions that favour the spread of the reef-building corals extend from the Gulf of Papua along the coast of Queensland (Australia) and just beyond the limit of the Tropic of Capricorn.

(2) The isothermal delimitation with an average annual temperature of between 18–20°C on both sides of the equator around 30°N and 30°S.

(3) The limit of distribution of tropical marine plankton in the seas.

(4) Zoogeographically, too, as Ekman (1953) pointed out, there is a greater similarity between the tropical and subtropical fauna than the latter and the temperate fauna.

(5) Also the distribution of some of the tropical species, for example, an inhabitant of tidal mud-flats and sand, the fiddler crab, *Uca*, extends from Florida through Brazil into the

temperate zone of Argentina. Surprisingly, the larvae are more resistant to thermal limits than the adults, although the reverse is true with regards to salinity.

(6) Plant ecologists also tend to generalise the geographical zones of the Tropics, temperate etc. on the basis of distribution of climatic and latitudinal 'biomes' i.e. vegetation recognisable by structural attributes and species composition of a geographical region.

(7) The distribution of mangroves and corals also characterise the tropical zone distinctly. Based on some of these characteristics, the global tropical belt is shown in Fig 4.2.

Obviously, besides the relatively tiny variations in temperature and salinity, the exogenous disturbances, such as the most destructive and severe tropical cyclone, hurricanes, typhoons, tsunami, tidal waves, and several other factors throughout the tropical belt, will contribute substantially to the typically tropical conditions. For example, the periodic strong hurricanes in the China Sea, Florida, the Caribbeans and Bangladesh; the heavy seasonal rainfall of India, Ceylon, Burma, Malaysia, Borneo, Singapore, Indonesia, the Philippines, Papua New Guinea; east and west coasts of Africa, Brazil, tropical Mexico, Colombia, Peru, Ecuador, and tropical east and west coasts of the USA; and the relatively little precipitation of the tropical part of Australia; and the enormous drainage of terrestrial materials and the sedimental effluents, exceeding 12×10^6 ton$^{s \cdot y \cdot 1}$, by major rivers into the seas (see Table 4.1).

However, the tropical marine environment is of considerable interest and of economic importance, particularly: in the use of the seas and their resources; because of the glacial past; the great variety of complex habitats; the uniformly hot and humid climate. Examples are the tropical shallow waters and shores are often dominated by the vast stretches of complex reef-building corals and extensive mangroves, gulfs, enclosed bays, deltas, lagoons and hypersaline lagoons with infinitely diverse forms of endemic life; the rich and prolific source of faunal and floral species and the variations in tropical coastal characteristics, many of which remain still unknown (see below). In the Tropics and subtropics, the oceanic waters between 30°N and 30°S are warmer with high evaporation; salinity often varies as precipitation varies with westerly tradewinds. The

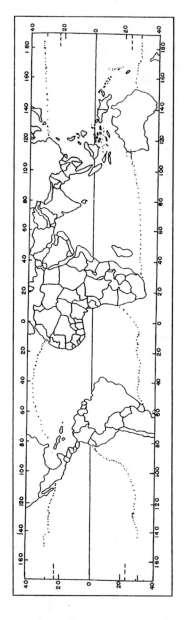

Figure 4.2   Global distribution of typically tropical and subtropical marine isotherm with an average of 20°C at sea level. The tropical belt may be strongly influenced by regional climatic and latitudinal range, but these are relatively less within the tropical seas.

Tropics, together with subtropics, from 30°N to 30°S, will cover approximately 40–50% of the global surface area, of which the marine environment constitutes more than one half and forms the centre of the major oceans and seas. From a viewpoint of atmospheric circulation too, the Tropics is an important part of the globe as the major energy input is derived from there.

4.4 The Tropical and Subtropical Regions, Populations and the Need to Share the Marine Resources

The impact of population on the environment is a fundamental concern of many nations, both the developed and developing countries of the world. The global resources are already degrading. In the Tropics and subtropics poverty retards any progressive transition from the rapid growth to slow down the population growth. People living in the tropical and subtropical countries will need high levels of consumption and will claim the largest share on food, fish, fuel and the other natural resources, both renewable and non-renewable. As the population increases, the requirement for resources may far exceed the sustainable natural sources that can be normally obtained.

The first man: the appearance of the first man, the *Homo sapiens* (derived from *Homo erectus),*with a brain weighing about 1,450 g, was about 100,000 years ago in the tropical part of Africa. Though confined to tropical Africa, they spread out to many geographical ranges, including Asia, Oceania, and Europe. Tropical Asia is one of the oldest and richest grounds for man's ancestors. *Pitheocanthropus erectus* (Java man), nearly 500,000 years old, was found in Djetis beds of Java. Fossils of early and modern man, *Homo sapiens,* were also discovered in Sarawak and the Philippines. These date back between 40,000 and 20,000 years. However, as the ice-caps of the Arctic Circle finally melted away, the ice retreated as did the land bridges that existed in the Bering Straits region at certain periods during the ice age. The relatively small group of hunting and fishing population penetrated into North America and into South America, and others to Indonesia and Australia along the Indonesia archipelagos through the shallow Bering Straits and several settled in many of the tropical and subtropical islands during the last ice age, about 10,000 BC. The global population was then around 4×10^6. By around 5,000

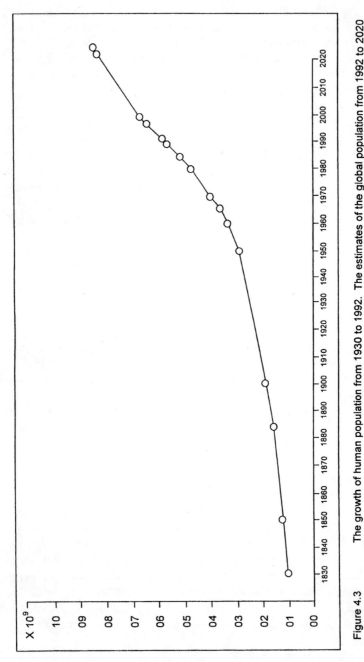

Figure 4.3   The growth of human population from 1930 to 1992.   The estimates of the global population from 1992 to 2020 (broken line) are predicted.   Data based on a number of sources, the principal being UN.

BC, the rate of growth began accelerating partly because of the good weather and partly due to the colonisation of Oceanias and the Americas through the shallow Bering Straits (McEvedy et al., 1978). Since the last ice age, the human population has been growing steadily, and the 4 million mark soon moved upwards to 100×10^6 and, with some occasional sluggish growth, the population reached some 200×10^6 by AD 200. This increase is reflective of improvement in agriculture, fishing and good weather. However, in 1575, the world population reached around 500×10^6; in 1826 it had nearly doubled to 1×10^9; in 1925 it was nearly 2×10^9 and, perhaps, due to modern innovation in technology, improvement in health conditions, lower death rate, and the higher living standards, during the last decade (Fig. 4.3) the global population grew rapidly, particularly in the tropical and subtropical areas. For example, the Philippines relative to its land size is one of the most densely populated tropical parts in the world (see also Table 4.3).

The relentless population growth is reflected more within the Tropics and subtropics. More than half of the world's population is living within this belt today, and nearly 22% of this live along the coastal countries (see Fig. 4.3). Many population scientists predict that by 1920, the world population will grow to 8×10^9 (see also Fig. 1.2).

4.5   Glacial Past

The facts of the rise and fall of the sea level during the Quatenary Pleistocene epoch of the ice age (20,000–10,000 years ago) and that the globe once endured an ancient glaciation has been well-known. Its existence is now taken for granted. Although much of the arguments are of historical and theoretical interest, practical contributions came from such famous names as Louiz Agassiz, Croll, Minkovick and others over the past century. In essence, about 20,000 years ago the earth was covered with glacial ice from the North Pole to the South Pole. The formation of ice, by the piling up of snow, layer after layer, and the repeated glaciation was experienced more by higher latitudes, but the tropical regions were also affected by enormous climatic variations: much water was withdrawn, resulting in the fall of about 100 metres from the original sea level, large areas of the

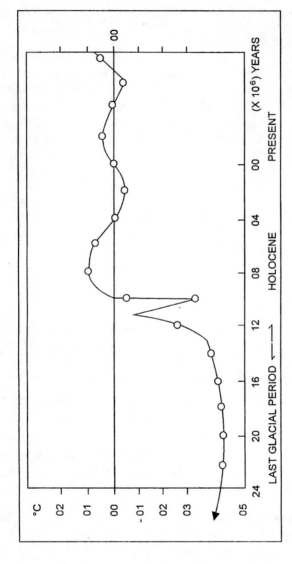

Figure 4.4   The glacial past; there is overwhelming evidence of a major shift from a warm to a cold climate about 3000 years ago in the Atlantic, Indian and Pacific Oceans, not once but several times in the past.

continental shelf became dry land and the width and depth of the most tropical continental shelves were affected considerably. Deep sea sediments were collected by making several hundreds of cores from all the oceans of the world during more than fifty expeditions, and the greater majority of them show post-glacial stratification of about 60–200 centimetres; all sediments consist of foraminefera lulite, a mixture of fine mineral particles from the continents and particles of calcium carbonate secreted by planktonic organisms. About twenty short climatic fluctuations, with two oscillating periods of about 550 years, were superimposed upon the long term climatic trend of post-glacial time. There was evidence of a major shift from a warm to cold climate, about 3,000 years ago, in the Atlantic, Indian and Pacific Oceans. It seems useful to briefly mention that about 14,000 years ago, the huge sheets of ice began to retreat and the 'melt-water' slid back during the first phase of glaciofluvial time, about 7,000 years ago. Then, the ice age came to an end and the sea level has since risen to the present-day level. There is overwhelming evidence that glaciation occurred not once but several times in the past (Fig. 4.4). The glaciation in the equatorial region contributed much to a colder climate.

However, it has been estimated that the present-day earth supports some 26×10^6 km^3 glacier ice (Flint, 1969) which occupies some 10% of the available land area. These ice masses are mainly locked up in the Antarctic, Greenland, Arctic and smaller glaciers in the Alphines. The major proportion of the ice, about 25×10^6 km^3, which accounts for nearly 90% of the total volume of the glacier ice of the earth, is confined to the Antarctic, which greatly influences our marine environment. The average annual temperature of the earth has fluctuated only very slightly (0.4–0.5°C), yet these subtle variations have had a major impact on the global climate. It seems that for the past half a billion years, the average annual temperature of the earth has fluctuated only slowly and to a very small range of degrees (Holum, 1977). From 1950 to 1980 worldwide emission of CO_2 rose at an average of 4% per year. However, future estimates based on energy models predict sizeable climatic changes: a mean rise in surface temperatures of 3°C ± 1.5°C by the first quarter of the next century. The implications of such shifts in temperature are uncertain but substantial, since 5°C represents the entire temperature variation

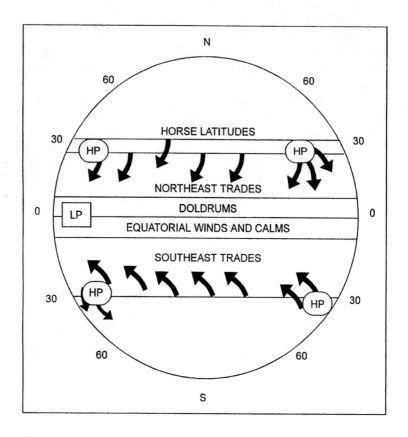

Figure 4.5   The general distribution of high pressure (HP) and low
pressure (LP), and the trade winds for the tropical and
subtropical belt.

of the last 125,000 years (Repetto, 1985). The consequences could include major shifts in the precipitation and run-off that could imperil agricultural production in important temperate farming regions, more frequent occurrences of damaging temperature extremes, and possible rises in sea level of 0.60–3.66 metres, enough to cause major flooding and storms in coastal areas of the Tropics.

4.6 Tropical Marine Environment

As seen above, the marine environment is influenced by the constant interactions between principally the geological, climatical, physical, chemical and biological factors. The marine environment in the Tropics and subtropics, between 30°N and 30°S from the equator, has a sea surface uniformly warmer due to the nearly constant quantity of insulation received. Because of temperature difference and the nature of chemical composition, the sea waters tend to form layers. Generally, the temperature of the surface waters within the equatorial belt is highest, about 25°C–32°C down to a depth of 100–200 metres; below which the temperature falls rapidly, characterising the beginning of the discontinuous or thermocline layer; the range of variation of temperature in some regions is about 5–8°C. Below the thermocline is a third layer extending to deepest sea floor where the waters are very cold with temperatures 0°C–5°C (see Fig. 12.6). Although the precipitation is unevenly distributed between continents and oceans, evaporation always exceeds precipitation, and salinity often fluctuates, with prevailing monsoon rains strongly influenced by trade winds. Over the equatorial troughs of low pressure, between about 5°N and 5°S latitude is the equatorial belt of doldrums, with variable winds and calms. North and south of the doldrums, covering roughly between 5° and 30° north and south, are the trade winds belt. The trade winds are the product of a pressure gradient – i.e. from the subtropical cold zone of high pressure to the equatorial trough of low pressure due to hot air. Generally, the trade winds blow continually toward the equator from the north-east in the belt between the northern horse latitudes and the doldrums and from north and south. Because of the earth's rotation, the air moving towards the equator in the northern hemisphere is deflected to turn westwards and the prevailing wind is from the north-east, hence the winds are termed north-east trade winds. In the southern hemisphere, deflection of moving wind to the left causes south-east trade winds (Fig.4.5). Trade winds are more well developed in the Pacific and Atlantic Oceans than in the Indian Ocean because of the enormous and continuous Asiatic landmass fringing the Indian Ocean. On the other hand, the monsoons have very strong influence on the north-eastern parts of the Indian Ocean and the extreme north-eastern regions of the Pacific.

4.6.1   Tropical Marine Ecosystems

The tropical marine ecosystems are complex and dynamic habitats. The process of interaction between land discharges, sea water, sea bottom, and the diverse forms of living organisms is in a continuous state. The product of the interactions at different stages between abiotic (non-living) factors and the organic (living) organisms pivoted by the flow of solar energy within a given unit of 'ambient' and 'time' may be considered as an ecosystem. On a geological time-scale of the significant global and evolutionary changes, for example, the continental drift and the changes in sea level due to past glacial impact which have profoundly affected the species, genera, and families that have evolved at different regions and times; and the transformations that had occurred in the tropical marine environment; the great range of marine interactive habitats and the endemic organisms can change; for example, the continental and oceanic restrictions; the formation of volcanic islands and the subsequent barriers created by the islands; exogenous disturbances such as the monsoons; high coastal turbidity during tropical cyclones, which may cause tremendous destruction in coastal areas; and hurricanes, tsunami, tidal waves which often disturb the coral growth; the vulnerability of the regional seas to pollution by sea-based effluents through human interventions or activities and, as a result, a new type of ecosystem may come into existence in those natural areas with the need for organisms to respond to the new environmental factors imposed, particularly in terms of their energy transfer, interdependence, distribution, abundance and adaptation. Thus, the tropical marine ecosystem, though more complex, may be divided into bioecologic units of smaller ecosystems (i.e. systems within a system) of the spatially largest of the marine aquatic ecosystem; as we shall see in coral, mangroves, mud-flats, intertidal pools, etc. First, let us consider at the outset some of the abiotic and biotic parameters and their interaction with the units and subunits of the marine ecosystems of the Tropics and subtropics.

4.6.2   Tropical Marine Coastal Ecosystems

In the Tropics, however, this influence of the environmental effect is exaggerated since the interactions are directly related to temporal and spatial variation. The rise and fall of sea levels during the ice ages; the coasts that are sheltered as inlets or bays or open directly to the tidal variations and rough wave actions; the extensive coral growths; and accumulations of sediments from land drainage and water spray into the atmosphere all combine in altering the structural characteristics and configuration of the coastline.

4.7   Tropical Marine Environmental Parameters

The prime parameters and characteristics to the Tropics and subtropics are unique. The subtlety and complexity vary in different localities. However, they relate to natural marine habitats and a variety of interacting marine life; hence an overview of the major and important environmental factors will be briefly discussed now.

4.7.1   The Climatic Regimes

Oceans play an important role in controlling the climate. Climatic changes are largely due to latitudinal variability of the sun, changes in the atmosphere and ocean circulation. The variation in the solar radiation that is received by the surface of the earth produces the seasons. The temperature diminishes from the equatorial Tropics polewards. The more localised tropical zone is greatly influenced by the environmental parameters. These mainly include: heavy rain fall, monsoons, floods, wind patterns of cyclones, hurricanes, tsunami, and super tsunami; turbidity currents; intense fluvial activities and the influx of major turbulent rivers such as the Amazon, La-plata, Orinoco, Niger, Nile, Brahmaputra, Ganges, Indus, Mekong, Columbia, Mississippi and Yangtze (Table 4.1) and their tributaries and many other unpredictable natural calamities. The Tropics is the well-lit part of the globe varying with seasons and latitudes; it is a buffer zone where a greater source of heat is absorbed and transferred to higher latitudes and there is always a higher rate of evaporation above precipitation.

Table 4.1   The Major Rivers of the Tropics

RIVER	LOCALITIES	LENGTH in km	FLOWS INTO	SEDIMENT discharged (m3/S)
Amazon	Brazil	6,275	Atlantic Ocean	238,667
Paraiba	Brazil	24	Atlantic Ocean	4,205
Paraná	Brazil	3,943	Atlantic Ocean	25,416
São Pablo	Panama			373
Brahmaputra	Bangladesh	2,896	Indian Ocean	66,857
Congo	Zaire	4,666	Atlantic Ocean	41,300
Galougo	Senegal	–	Atlantic Ocean	4,083
Noun	Cameroon	–	Atlantic Ocean	267
Ogowe	Gabon	–	Atlantic Ocean	9,295
Man	Sierra Leone	–	Atlantic Ocean	–
Daly	Australia		Indian Ocean	2,887
Fitzroy	Australia	2,816	Indian Ocean	181
Ganges	India	2,494	Indian Ocean	58,633
Godavari	India	1,022	Indian Ocean	22,666
Damodar	India	–	Indian Ocean	11,442
Brahmaputra	India	1,625	Indian Ocean	44,614
Mahavali [Ganga]	Ceylon	138	Indian Ocean	1,608
Kelani [Ganga]	Ceylon	90	Indian Ocean	2,310
Indus	Pakistan	1,166	Indian Ocean	20,414
Minab	Iran	–	Indian Ocean	552
Icopa	Madagascar		Indian Ocean	2,464
Orinoco	Venezuela	2,896	Atlantic Ocean	60,689
Perak	Malaysia		Indian Ocean	1,134
Kapua	Borneo	108	Indian Ocean	–
Gagayan	Philippines		Indian Ocean	4,771
Yangtze	China	6,300	Yellow Sea	34,000
Mekong	Thailand		China Sea	33
Zambezi	Zambia	3,540	Indian Ocean	7,000*

Data UNESCO and other sources. The calculation of average discharge is based on 3-4 years monthly recordings.
*Calculated.

4.7.2   Topography

As already noted previously, many natural processes are involved in altering the topography of the sea bottom.   In coastal shallow waters wave actions, tidal flows, volcanic eruptions and coastal currents can shift the sediments.   The nature of the ocean floor and variations in the relief shape the smoothness or irregularity of the bottom. Turbidity current can transport sediments from the continental shelves into the deep ocean and fan out. Bottom topography of the continental shelves is an important physical factor that might influence the productivity of a region. Some are narrower in extent and relatively shallower than others. There is an upwelling and vertical mixing of the water on the shelf and slope; the greater the slope of the continental shelves the greater is the tendency for more stable upwellings and sinking of the waste (Moisey, 1971). Despite the fact that continental shelves are narrower in the Tropics, the productivity is greater in some areas. The most highly productive regions with relatively narrow continental shelves are those of Benguela off the west coast of south-west Africa, Saudi Arabia, Thailand, the Gulf of Kutch, Peru on the Pacific coast and California.

4.8   Tropical Abiotic and Biotic Factors

The principal factors which affect the interactions are: physical, chemical, and biological.

4.8.1   The Physical Factors

Physical, together with chemical and biological, factors contribute to controlling the marine environment.

4.8.1.1   Solar Radiation

Solar radiation is the most basic factor in the tropical marine ecosystems. The sun radiates some 10^{26} calories of energy per second. This energy from the sun traverses the atmosphere and is finally intercepted by the surface of the land and the sea. The energy emitted by the sun consists of all forms of light such as ultraviolet, visible, and infra-red, including the cosmic rays. The electromagnetic waves of the spectral components of solar radiation and their relative importance associated with the distribution of energy are shown in Fig. 4.6.

The interaction between the air and the sea surface takes place at the interface. The penetration of light in sea water is determined by a number of factors. In the Tropics, where the water is clear, the amount of solar radiation incident on the sea surface varies considerably depending on the geographical latitude, elevation of the sun and time of the day. Within the equatorial belt of 10°N and 10°S,

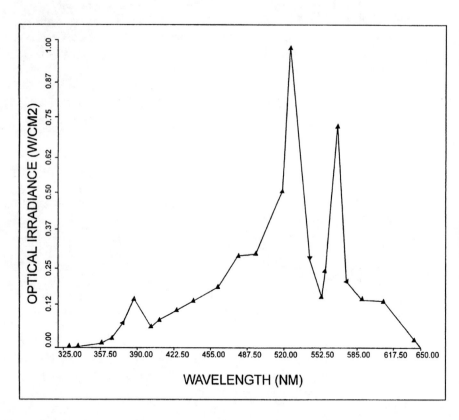

Figure 4.6   Solar energy and the distribution of energy at different wavelengths.

the sun provides sufficiently intense insulation all the year round and the day and night are of nearly equal duration; while in the other regions the insulation varies considerably as the latitude increases towards the poles. On an average, about 295 cals/cm^2/day of solar energy is transmitted to the sea surface. Of this, 42% is absorbed by the atmosphere, about 51% reflected back from the sea surface and only 7% is absorbed by the sea. In the Tropics there is a relatively high degree of transparency of water which enables the penetration of light to greater depths than in the higher latitudes. Because day and night are nearly equal in the Tropics, the amount of light that penetrates the water depends on the position of the sun and the angle of incidence. If the angle of incidence exceeds 60° a greater percentage of light is reflected. However, it should be pointed out that the light which enters the sea is never uniform, but the intensity attenuates logarithmically with an increase in depth at about 1% to its surface intensity. Below this critical level, the transparency of the water depends on the suspended organic and inorganic materials and mixing process of the ambient. Because of these inherent variables, the spectral components of the light is sequestered and much of the visible part is absorbed leaving only the blue-green part of the spectrum to the deeper waters. Based on the energy distribution, Jerlov (1969) showed that wavelength of maximum light transmission in the sea water, under different conditions, varies from 473–650 nautical miles, depending on the selective absorption by water itself. Molecular scattering plays an important role and the 'yellow substance' is largely responsible for the change towards longer wavelengths in turbid waters. In very clear tropical waters, light diminishes from the surface to greater depths quite rapidly. Nearly 62% of the radiation is absorbed within the top 1 metre depth; and the light decreases to about 1% within the next 30–150 metres; and only very insignificant light reaches to a depth of about 1000 metres, below which virtually no light is available, except for bioluminescences. The small percentage of unreflected light is the source of energy which contributes to the evaporation of the ocean's surface, the photosynthesis in the sea, for temperature gradients and the thermal energy of the oceans (see also Appendix III).

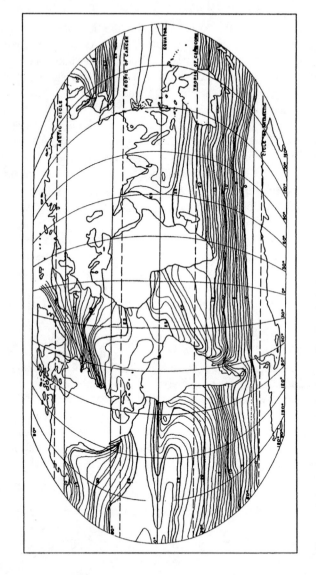

Figure 4.7   Characteristic temperature distribution profiles, particularly within the tropical and subtropical surface waters, in relation to temperate and polar regions.

4.8.1.2 Temperature

Thermal structure in the oceans of the Tropics and subtropics shows stratification. There is an upper isothermal layer, where temperature varies between 26°C and 36°C. The temperature falls rapidly between 100 to 200 metres below that, where the temperature may be less than 9°C and becomes a colder layer with a temperature lower than 4-5°C. The distribution of sea surface temperature for tropical and subtropical regions is shown in Fig.4.7. Between the surface and deeper layers is an intermediary transitional or thermocline layer, where the temperature gradients vary with depth (Defant, 1961). The physical mechanism which contributes to form and sustains the transitional zone is thought to be vertical motion, the upwelling and sinking. Data based on recent studies in Cape Comorin areas indicate that the vertical variation of temperature is clearly associated with vertical movements of water, but depending on seasonal and spatial factors, especially depth (Sharma, 1968). In the Tropics, the surface waters are warm and the temperatures are relatively uniform throughout the year, and the temperature plays an important role in the distribution of tropical marine organisms. Organisms living in high and fairly stable temperature are usually diverse in species, but each species is represented by a small number of individuals within a species in warm tropical waters. Temperature influences the metabolic rate; sexual maturity also occurs much earlier, although lifespan may be shorter. The rate of egg production is increased and may stimulate the migration of many species. Some species can tolerate a broad range of temperature, eurythermals while others can only tolerate a very narrow range of variation, stenothermals; the invertebrates and certain lower vertebrates are piokilothermic and are able to change their body temperature with that of the environment. Unlike the high latitudes, the seasons in the Tropics are often less predictable and relatively of different characteristics. Water temperatures are highly stable and uniform at depths greater than 2,000 metres. Temperature changes between 1.15°C to 4.0°C at depths of 6,000 to 11,034 metres, especially in different deep sea trenches. For example, the Banda Trench in the Indian Ocean has a relatively high temperature while the Java Trench and Tonga Trenches are colder due perhaps to the Antarctic cold currents entering these trenches. The point of compensation in the equatorial region depends on the seasons, regions, illumination, turbidity and many other factors and it varies from 80-120 metres.

4.8.1.3   Density and Specific Gravity

The density of a substance is defined as its mass per unit volume; and usually expressed as 'sigma t' and symbolised as σ_t. Since temperature affects the density of a substance, the temperature is usually specified. The density increases with a concentration of salts but decreases with temperature. The temperature of sea surface water in the Tropics is usually higher than in temperate latitudes, and is dependent much on the variations of salinity, temperature and pressure. An increase of 1‰ in salinity increases the density by about 0.80 ppt. For example, sea water with 35‰, will have a density of 1.0280 g/cm^3, measurement being taken at 0°C and at a pressure of 1 bar (atm); likewise, the tropical surface waters with an average salinity of 36‰, will give a density of about 1.0288 g/cm^3 at 0°C at 1 bar. In tropical waters, where the salinity and temperature vary with regions and seasons, the temperature has an important influence on the density. These characteristic properties of density and the specific gravity are related; i.e. specific gravity of a substance is defined as the ratio of the density of a given substance to the density of water at 4°C. The value of specific gravity is expressed as a pure number. Therefore the specific gravity of any substance will be numerically equal to its density. For example, since maximum density of pure water at 4°C is 1 g/cm^3, the specific gravity of water is simply equal to 1. When density is specified at 1 bar as sigma t (σ_t), the density of 1.028 is written as $\sigma_t = 28$, i.e. the difference between the density and 1 times 1,000.

4.8.1.4   Currents

Currents, assisted by the wind, submerged volcanic explosion and other factors, are the main transporting agents in mixing water masses of different regions. Direct measurement by the French bathyscape *Archimede* showed the current velocity mainly at the bottom of the Kurile, Kamchatka and Iszu-Bouin Trenches to be 0.2 knots. The geostropic current studies in the eastern part of the Arabian Sea cover a large area 11°-17°N within the meridian of 70° and 75°E at depths of 100-150 metres close to the coastal waters; the relatively smaller currents are similar in pattern and low in velocity (Ramirtham, 1966). However, at the depth below 2,000–3,000 metres the pattern is changed and with good velocity (see Chapter Thirteen on currents).

4.8.1.5 Hydrostatic Pressure

In the sea, hydrostatic pressure increases steadily with depth; i.e. for every 10 metres depth of sea water the pressure increases equivalent to 1 atmospheric pressure. At great depths, pressure increases in magnitude and only some animals have adapted to such enormous pressure. At abyssal benthic and hadal benthic zones, even the bacterial decomposition may be inhibited.

4.8.1.6 Tides

Tides are a natural phenomenon of the sea. The rhythmic rise and fall of the water level is related to the combined effect of the gravitational force of the sun, the moon and the rotational effect of the earth. In most of the tropical coastal countries, the diurnal range of tidal ebb and flow is very low, between 1–2 metres. Although the tidal cycle involves twenty-four hours and fifty minutes in most places, where as in Tahiti and Vietnam the cycle completes in twenty-four hours. The cycles have extremely variable impacts on living organisms. Tidal currents often contribute to form the configuration of tropical coasts. Because of the alternative exposure and submergence of the intertidal zone, the organisms have developed special adaptations. Unlike the higher latitudes, the tidal variations are relatively small. Tidal predictions are made based on mean tidal levels and published as Tide Tables and are available in most countries today. Due to the narrow latitudinal range within the Tropics, the spring tide and neap tide are less noticeable.

4.8.1.7 Waves

Like tides, waves are important to marine habitats. Most of us are familiar with tidal waves which sweep the entire coastal habitats and life and inundate vast coastal areas, resulting in loss of much of the coastal configuration. When the wind blows over the sea surface, there is a transfer of heat between the air and sea water, and air movement creates waves which drive the surface water in the direction of the wind. For example, tidal waves are occasionally encountered along the coast of Bangladesh with devastating consequences. Huge waves breaking are easily recognisable along the rocky shores of the Tropics. Most waves consist of speed, length and height. The height has two components – the crest and the trough. Although the waves are caused mainly by volcanic explosion, violent storms, winds, depth of water, and other factors, in the Tropics the waves have relatively smaller heights. Depending on their nature and origin, the waves are classified as microwaves, internal waves, standing waves, etc., but these will not be considered further here.

4.8.1.8   Acoustics

Acoustics are related to the sound and animals in the sea. Whales, fish, and crustaceans often produce sounds. The modern fishing industry uses fishing vessels equipped with electronic devices. Some of these are echo sounders which can help find the depth of the water at which fish are found and hence 'fishfinders', and radio-telephones. The sound travels at a velocity of 1,500 metres/sec, depending on salinity, temperature and pressure. The range of frequencies at which the echo sounders operate is 20–50 kHz and they are capable of scanning down to a depth of about 1,000 metres.

4.8.2   Chemical Factors

The marine environments are modified by a number of chemical factors over geological time. The salinity of the oceans has not changed significantly over geological time and is now considered to be fairly 'stable'. Evolution of the first marine cyanobacteria led to much of the oxygen on this planet. Some of the chemical factors concerning life process in sea water are briefly discussed.

4.8.2.1   Salinity

Sea water is a complex solution containing different salts in water. One of the important findings of the *Challenger* expedition led to the establishment of relative proportion of the major chemical components of sea water. Throughout the Tropics, discrete layers of sea water with special physico-chemical characteristics can be recognised. The surface salinity is very much determined by latitudes and weather conditions. As noted above, both salinity and temperature can influence the density of sea water. In much of the areas in the Tropics, due to transient difference in temperature, the layer below 200 metres, usually becomes discontinuous and is called thermocline (see Chapter Twelve). The surface salinity in most tropical waters is relatively higher; it varies from 34 to 36.5‰ at the surface layers and in some deep sea trenches, between 6,000 metres to 11,034 metres, the salinity varies very little from 34.58 to 34.82‰ (see Chapter Six on sea water).

4.8.2.2   Oxygen

Oxygen is of particular interest for living beings; many organisms in the sea use dissolved oxygen in sea water. In the sea oxygen concentration in dissolved form is of great significance in relation to metabolic process and primary production. If the productivity is greater, the surface water may become supersaturated, especially in upwelling areas. The value of oxygen concentration in the Tropics is generally low. It varies from 2 to 12 ml/l in the surface waters and decreases with depth, in some tropical trenches such as the Banda Trench in the Indian Ocean where concentration of O_2 is as low as 2.0-2.12 ml/l (see also Appendix II).

4.8.2.3   Carbon Dioxide and the Buffering Effect of pH

In most tropical areas, the surface water is relatively warmer than in higher latitudes. Quantitatively, the concentration of CO_2 depends on: rate of exchange between the surface water and the atmosphere; organic assimilation; organic dissimulation. CO_2 is also found in dissolved form, undissociated carbonic acid (H_2CO_3), and as bicarbonates and carbonates. Because of the well-lit euphotic layer in the Tropics, the absorption of CO_2 is greater, corresponding to the greater production of organic matter. Accordingly, the quantitative differences in concentration of CO_2, by removal or addition of carbonic acid, also influence in determining the amount of hydrogen ions which help the tropical water to maintain a pH within the range of 7–8.2, from the surface to up to about 300 metre depths, favourable to most organisms. Significant variations in pH values can also give clues as to the extent of the photosynthetic zone, oxygen content, thermocline, vertical mixing process and stratification. The regions where oxygen minimum are found show a low pH, a peculiar correlation attributable to photosynthetic activity.

4.8.2.4 Phosphates

Nutrients such as phosphates, nitrates and silicates are of considerable importance for productivity, particularly of phytoplankton. Phosphate is derived from the organic materials such as the metabolic wastes of living organisms, and the dead tissues of plants and animals. Unlike the temperate or polar regions, the concentration of phosphate in the tropical waters varies due to monsoons and is just sufficient to support phytoplankton growth. Notable exceptions are those upwelling areas where the nutrient supply is usually plentiful and phytoplankton productivity is highly enhanced. The concentration of PO_4 in upwelling areas reaching the continental shelf varies from $2.15\mu g$–$3.25\mu g$ at. P/l. While in the off shower waters, it is relatively less. Based on one year's observations in the bottom mud off Calicut, the interstitial phosphate content varied, it was lowest during the SW monsoon, but increased soon after the monsoon (Seshappa and Jayaraman, 1956), and thus the phytoplankton and phosphate became abundant. Following the bloom, both phosphate level and phytoplankton have been diminished due to grazing by zooplankton until a regenerative process sets in.

4.8.3 Biotic Factors

Basically, biotic factors are concerned with the complex life processes such as feeding, reproduction, survival and interactions, which, in turn, are intricately influenced by the abiotic environmental factors. All living organisms have to derive their energy to sustain life by biological process. Many are able to obtain their basic nutritional requirements, i.e. carbohydrates, proteins, fats, vitamins, etc. by synthesization from the surrounding media. The organic production contains the most important elements of water and CO_2. Biotic factors often tend to limit the distribution of organisms. In a range of tropical coastal fringe, where the sea meets the rivers, deltas, estuaries and lagoons, many and varied biological factors are evolved. The depth, illumination, temperature, salinity, and pressure, together with bottom topography, are the principal determinants for the basic biological process of organic production, coral growth and the enormously rich mangrove species. Competition among different species for limited sources of food and space in a given locality, and the interactions between species are also some important biotic factors recognisable to establish the hierarchy of marine organisms. Predator-prey relationships, species diversity and adaptations are more pronounced in the tropical areas than in higher latitudes. These warrant separate treatments and will be discussed in sections that will follow shortly (see also Chapter Twelve on primary production).

4.9   The Tropical Beach Profile

The narrow physical and dynamic interfaces between the land and sea are the beaches. A number of factors significantly alter the nature of the beaches. The beaches vary in character and are constantly subject to breaking wave action, rise and fall of the sea level, seasonal rainfall, and fluxes of larger rivers, which together play an important role in the ecology of the seas and their shores. Frequently, the physical characteristics of the same beach can change from one locality to another. The beach usually extends from the highest level of spray zone on the coast to the low tide water level. Tropical beaches are characterised by materials which comprise the sea coasts. The bulk of the beach is usually made up of sands. The part of the beach proximal to the coast is the backshore or spray zone because of the spray by wave action, coupled with rise and fall of the twice daily tide. Consequently, the nutrients and oxygen are constantly replenished. The part which extends up to low water level is the foreshore. The important processes of constant energy exchange and the wave actions with greater impact, even to the extent of grinding down the pebbly stones into coarse and, eventually, to silt and finer sand over several years time take place here. Frequently, some of the tropical beaches, especially of fine sandy nature, are formed into long continuous ripples parallel to the beaches when under low tides. The offshore extends seawards from the low line of the low tide level (Fig. 4.8). The beaches where the effluents are discharged from larger rivers such as the Amazon, Ganges, Brahmaputra, Mahavali Ganga, Congo, Niger, Paraná, Indus, Orinoco and Dalington, are usually sedimentary. These coupled with wave action generally support a very large population of microfauna. In the tropical waters, up to

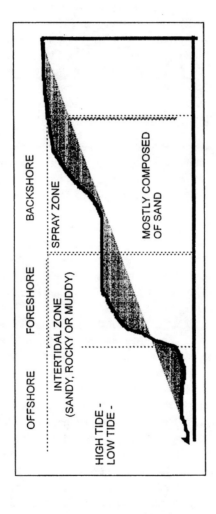

Figure 4.8   Tropical beach profile.

about 60 metres depth, corals frequently grow in abundance. High intensity of light has a profound effect on coral reef growth (see Chapter Five on corals). The tropical shores are typically very wide and are of either sandy, muddy or rocky nature, each type is rich in its own endemic and specific floral and faunal dominance. The rocky shores are usually scattered with tidal pools which harbour a variety of colourful algae and a number of sedentary and many mobile animals; occasionally, many oceanic animals get stranded in these tidal pools during low tide (see below). Where the great and turbulent rivers meet the shallow sea areas, the riverine effluents, including granitic sediments, settle down to form the mud-flats and some of the larger deltas.

Thus the beach plays an important part in determining areas vulnerable to erosion. Wide sandy but stable beaches may contribute to some extent to the protection of low-lying coasts. In many tropical countries, the coasts are excessively eroded, and steps are now being taken to protect them by laying rocks to lessen the destructive effects of storm waves. Incursions of the sea along many tropical beaches have the consequence of loss of coastal lands. The seas within the tropical region are important as they have changed quite considerably in sea levels and bottom topography during the successive glacial epochs, since the end of the Pleistocene, resulting in the beach which extends up to low water level. This is the foreshore, where there is a constant exchange of energy as well as greater impact of the wave actions.

4.9.1   The Beaches

Beaches are dynamic zones where many types of natural forces operate together. For example, due to constant wave actions, tidal movements, coastal erosion and hurricanes, clay, sand, silt particles, cobbles and debris are transported from the shores far into deeper waters. The sea level changes also affect the beaches; they are minimal around some equatorial islands but may increase up to more than 1 metre along the coast of the Bay of Bengal. One of the theories advanced by many scientists around the world over is the possibility that the significant rise in the sea level and change in the patterns of precipitation are related to the production of excess of CO_2 by the

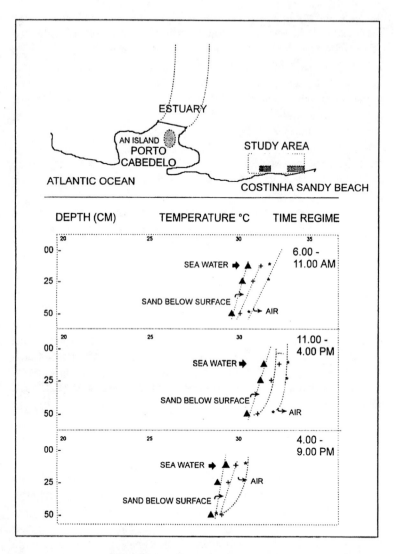

Figure 4.9   Temperature variations of three time regimes (each average of 5 hourly recordings of two stations) of a tropical sandy beach, Costinha, Brazil, 12m from the edge of the water level (✚ sandy beach; ▲ Sea water; and ☆ air temperature) during January – March 1988.

fossil fuel (coal, oil and gas) combustion, the so-called green house effect (see Chapter Fifteen).

4.9.2   Tropical Sandy Beach

Much of the coastal fringe of the Tropics and subtropics is formed of sandy beaches. Conditions of the sandy beach vary considerably with seasons. Many coastal regions of the Tropics have broad sandy beaches subject to physical forces. The effect of tides on the beach is the single most important attribute which demarks the boundaries of the intertidal zones, while the constant wave action shapes and controls the physical nature of the tropical beach such as its extent, slope and the size of the sand grains. When wave action is severe, the slope of the beach becomes high and the intertidal zone becomes narrow with particles of sand of medium and coarse type. Where the wave action is relatively less violent, the slope becomes low and the intertidal belt also widens with components of finer sand, and the size of sand particles usually increases progressively from the ebb of the high tide mark to the flow of the low tide level. The retention of water within the sandy layer depends on the distance from the sea water, the slope of the beach, and grade and interspaces between the sand particles. Accordingly, micro-organisms that occupy the interspaces of the sandy beaches vary. As in any other environment, temperature has a greater influence on the sand dwelling organisms. The surface temperature will fluctuate depending on the seasonal monsoon rains, overhanging clouds and atmospheric conditions, and with the progress of the day. Temperature usually increases from the existing low water level towards the relatively dry high tide water mark. As mentioned above, the slope and thickness of the sandy beach together help to retain relatively constant temperature in subsurface layers while the surface is subject to vicissitudes of external factors, especially surface temperature which, usually though high, are always a little less than that of the air, except at dusk. In most of the tropical countries, the day and night are of equal duration and the sun rises (dawn) at 6 a.m. and the sun sets (dusk) at 6 p.m. Therefore, it is not surprising that the surface sand gets cooled by changes in the atmospheric condition while the deeper layers can still retain relatively higher temperature. However maximum surface temperature is reached during midday; the subtle variations are shown in Fig. 4.9.

Salinity too varies, but to a negligible extent in relation to the proximal sea water. If one sinks a well on the sandy side of the beach on a very sunny day, due to excessive evaporation of interstitial capillary water, the salinity will show an increase relative to the adjacent sea water but on rainy days the salinity may fall considerably. There is again an average difference between the salinity of the sea water and the interstitial water (i.e. 34.8‰ ± 2‰) of a sandy beach, the interstitial being on the higher side. Oxygen content, pH, and organic debris are other important factors that contribute to sandy habitat. Except on rainy days, the oxygen concentration will remain lower in the sandy layers than in the sea waters near the margin of the beach. Given the normal conditions, the depletion of oxygen in the sandy layers is most likely to be due to evaporation and also due to consumption by micro-organisms during their respiratory process. The reduction of oxygen content will lead to an increase in the CO_2 and consequently it reflects on pH, being generally less in the interstitial waters than the sea water. The organic debris is also usually less on the sandy beach compared with the flooded edge of the sea.

Habitat and Organisms: sandy beaches are some of the most difficult habitats for the organisms to live in. In an ecosystem of this nature, the organisms have to surpass the most rigorous conditions such as tropical heat, unpredictable weather, seasonal changes, wave actions, tidal effects, texture of the substrate, etc. On a vertical plane, the range of life often varies and can extend from the surface to a depth down to 25–50 centimetres. Accordingly, many animals have developed special adaptations and gregarious habits to thrive on sandy beaches. Most organisms show morphological peculiarities with their bodies reduced in size, often elongated, adhesive pads, devoid of visual organs, but well developed special senses. Among the inhabitants, the pennale type of diatoms can be found on the surface of the sand with sufficient moisture. The faunal representatives include a number of species of invertebrates, especially flagellates, ciliates, nematodes, tubularians, polychaete worms, crustaceans, particularly the most characteristic and fast running ocypodiae shore crabs of the species, *Ocypoda*, commonly found in their thousands, the mole crabs, *Emerita*, and *Natuata*, and most of the tropical species of the fiddler crabs, *Uca*, found in Florida also inhabit the coastal areas of

Brazil, and their distribution extends to many other coasts, including the temperate southern coast of Argentina; isopods, harpacticoid copepods, ·ostracods; the bivalve molluscs of the genera, *Donax, Siligua, Tellina*, and the gastropods of the genera *Oliva, Harpa, Tonna*, and occasionally acraniates; the greatest numbers usually seek to occupy the subsurface layers about 15–30 centimetres below the surface near the sea water margin of the beach where conditions are optimal to avoid the danger of desiccation. The beaches with fine sand grains are occupied by numerous polychaetes predominantly eunicids, glycedds, and capittellids and, especially the euryhaline species such as *Arenicola brasiliensis, Nergine agilis, Laeonereis culvert*. However, pollution seems to affect the distribution of polychaete worms along the tropical sandy beaches.

4.9.3 Tropical Rocky Beaches

The rocky beaches are the most rugged and variably difficult habitats for most organisms to inhabit. Much of the marine rocky coastal environment in the Tropics characteristically show much richer fauna and flora in number as well as variety, than that of sandy coasts. Some are sheltered while others are totally exposed, except for the tidal spray which keep the organisms wet. The spray zone fringing the coast, the shaded part of the rocks, the rocky pools, and the varied size and shapes of crevices provide shelter and protection for common sessile organisms; many sluggish molluscs, engebenthic crustaceans, including isopoda, amphides and pycnogonids; the sessile sponges, hydroids, ascidens and agile fish. The rocky surface also forms an excellent substrate for many unicellular algae, multicellular seaweeds and some sciaphilic or shade loving benthic animals depending on the surface texture of the rocks. Some of the faunal communities characteristically able to resist desiccation due to the typically hot tropical rocky beaches are of the genera: *Mytilus, Ostrea; Littorina, Patella* – the latter between tide marks – *Balanus, Chthamalus, Sertularia, Penneria, Obelia, Bimeria, Plumeria; Herdmania, Polycarpa, Epizoanthus*, etc. Because many of them are either sessile or with limited mobility, they have evolved various kinds of structural adaptations such as thick shells, long and short spines, for example, the sea urchins under the shady part of the rock, and other mechanisms to obtain their food often in suspension.

4.9.4   Tropical Tidal Pools

Tidal pools are some of the most complex ecosystems and are subject to extreme daily changes of physico-chemical conditions. They are often distinguishable as special from the natural marine environment and found along most of the rocky shores of the Tropics and subtropics. They are formed by wave action, cyclones or tidal waves and become sheltered. Because of the diurnal exposure of the shallow pools to the ebb and flow of the tides, depending on the size, shape and depth, the tidal pools are subject to high evaporation, they often vary in salinity between 30‰ and 40‰; sometimes they become hypersaline with a salinity range between 40‰ and 80‰. Other hydrological characteristics are the slight increase in silicates and decrease in phosphates. The content of oxygen also varies from 2 ml/l to 7.8 ml/l, being maximum during high tide and minimum in low tide. Usually the maximum is less than the surrounding sea water and obviously temperature variation depends on the time of the day, being 25–26°C early in the morning at 6 a.m. and at a maximum, by 15–17 hours, of about 32–35°C, then the temperature again falls gradually to about 25-26°C by 6 p.m. The pH and dissolved CO_2 also vary considerably; during clear days the pH increases but the concentration of CO_2 decreases due to photosynthetic activities. However, these variations are slightly reversed after the sunset. Planktonic organisms include: some of the common tropical coastal water phytoplankton, especially the diatoms of the genera, *Nitzia, Cymbella, Pleurosigma,* and *Navicula* and occasionally the blue-green algae *Oscillatoria*; and the representatives of dianophyceae, especially of the genera *Pyrocystis, Peridinium, Ceratium* and *Gymnodinium,* can be seen. These exhibit bioluminescence during rough weather. With increasing salinity some flagellates of the genera *Carteria,* together with the saprophytic *Chilomonos* too can be noticeable. Among the zooplankton, many crustaceans, especially the littoral copepods, and the prawn larvae dominate. Some of the economically important prawn species often encountered are the genera of *Penaeus* and *Metapaneus,* also crab species of *Neptunus pelagicus* and others are amphipods and isopods. Squid larvae of the genus *Sepioleuthis* may be found occasionally. Fish species, perhaps because of their greater agility, can vary quite considerably. Several species of fish, including the larvae or the ichthyoplanktonic forms and adults predominantly of the coastal habitats are well represented; some of the common species are *Nematalosa, Liza, Mugil, Gerres, Leiognathus, Anchoviella,* etc.

4.9.5 Tropical Bays

With some subtle differences in bottom topography, bays and gulfs share many common features. The bays are coastal indentations or inlets usually of semi-circular configuration. One of the most popular bays is the Bay of Bengal (Fig. 4.10, A). It is an inlet of the Indian Ocean delineated by the coasts of Burma, India, Ceylon, and the group of the Andaman Nicobar Islands. Recently, some submarine canyons, which stretch out seawards from the Ganges-Brahmaputra Delta, have been reported. Other interesting tropical bays are the Campache Bay along Mexico and Yucatan, as have already been noted in Chapter Two.

4.9.6 Tropical Gulfs

Like the bays, the gulfs are connected with the sea. Some typical examples are that of the Persian Gulf, between Iran, Kuwait, Bahrain, Quator, and United Arab Emirates; the Gulf of Oman, which lies between Makaran of Iran in the north and west and east Hajar of Oman; the Gulf of Aden bordered by Southern Yemen and Oman in the north-east and Somalia Republic and Socota Islands in the south. All three of these gulfs are somewhat elongated and open into the Arabian Sea, a part of the Indian Ocean. From the point of view of an ecosystem, the Persian Gulf and the Gulf of Oman, despite their physical connection and exchange of waters through the Straits of Hormuz, show considerable differences, particularly in factors such as depth, salinity and temperature. Other well-known gulfs are the Gulf of Mexico, the Gulf of California, the Gulf of Thailand (Fig. 4.10, B), the Gulf of Martaban below Burma facing the Andaman Sea and the Gulf of Mannar in the relatively shallow sea between Tutucorin of South India and Puttalum of Ceylon (see Fig. 4.10, E). The Gulf of Khampart and the Gulf of Cutch along the north-western coast of India are also significant.

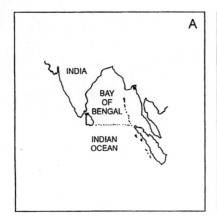

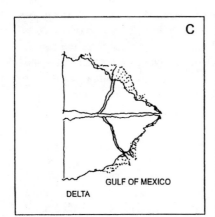
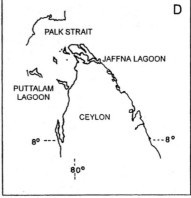

Figure 4.10   Some of the tropical coastal habitats of (A) Bay of Bengal; (B) Gulf of Thailand; (C) Delta and (D) Lagoon.

4.9.7 Deltas

Deltas are frequently encountered at large river mouths. They are of different shapes and are usually formed by the alluvial deposits of sediments, silts, sand, mud, and gravel. The continual accumulation of these often extends beyond the coastline. While the size of a delta is defined by the size of the river mouth from which it originates, the shape varies, depending on the amount of sediment deposits and the interacting forces between the river discharge and the tidal and wave actions. The Ganges-Brahmaputra Delta at the Indo-Bangladesh border, Cauvery Delta (Bay of Bengal), Sao Franscisco Delta in Brazil, the Orid Delta in northern Australia, the Mississippi in North America, the Nile Delta in Egypt, the deltas of Tjimanuk and Solo in Java, Mekong, and the Mexican Gulf Delta are some of the notable examples (see Fig. 4.10, D).

4.9.8 Tropical Estuaries

Estuaries are worldwide in distribution. Within the tropical band, aridity influences greatly the physical, chemical and biological features of an estuary and a transitional specialised ecosystem is developed. Estuaries are usually embayments, where the sea invades a river system. Consequently, estuaries become complex but an important and dynamic ecological habitat. Most rivers and streams flow into the estuaries and mix with sea water by back-and-forth tidal exchange through the entrances of the estuaries. The tropical estuaries normally support a rich variety of living organisms, which are constantly subject to tidal effects due to the rhythmic rise and fall of the sea level and thus making the hydrological conditions more dynamic and turbulent. Because of the variable abiotic and biotic factors, organisms living there are well adapted for life in such a rigorous environment. Some estuaries have large mouths and are deep enough to form natural harbours. Some may have smaller islands in the centre.

4.9.9 Tropical Lagoons

Although lagoons are widely distributed throughout the world, in some of the tropical and subtropical coastal areas they are dominant and well represented by elongated, often undulating, shallow water channels of tidal inlets which communicate with large bodies of sea water. No two lagoons are alike either physiographically or in environmental parameters and conditions, but each lagoon is specific in itself. However, direct access to the sea is somewhat obstructed and they become separated from the open ocean by the sedimental deposits, sands and silts which are built into smaller dotted-barrier islands. Yet free exchange of water with adjacent open sea can take place through the narrow passages. Coral lagoons are usually scattered within the tropical belt of the Indian and Pacific Oceans, and less frequently in the equatorial Caribbean sea. They are very diverse in sediment composition – for example, the coral lagoons on the banks of the Bahamas and Florida Bay. Coastal lagoons occur in great numbers along the shallow waters of South-east Asia (see Fig. 4.10), Africa, South and North America, and these are restricted to some extent in the western Pacific and northern Australia. There are many lagoons and bays along the West African coast. The Lagoon Ebde (Abidjan, Ivory Coast) is one of the most productive ecosystems in the world. In this lagoon, despite considerable seasonal variation, the salinity fluctuates between 14 and 43‰, the shrimp, *Penaeus duararum*, grows well. Since the mangrove muddy bottom soil is capable of acting as a reservoir and restoring the salt content during the rainy season, there seems to exist a correlation between the catch of the shrimp and the pluviometric condition; the productivity of the shrimps can be predicted and, on some occasions, over 1,500 tons per year can be harvested. The shrimps are reported to prefer those areas penetrated by freshwater; the fine sediments, deposited by the rivers and streams as they enter the lagoon, perhaps, are used by the shrimps for burrowing; detritus derived from the organic plant materials, as the favoured diet of the young penaeidae. Lagoons are mainly two types: the open and the closed. In all lagoons, the rainfall, evaporation and tidal movements have a combined effect on the variations of the temperature and salinity; the concentration of oxygen is greater in open lagoons than in the closed lagoons; the organic productivity is also higher in the open than in the closed lagoons; the fish biomass is usually greater than in closed lagoons. Fish resources include the marine fish species, crabs, shrimps – the latter are always abundant – and bivalves, besides many others in the lagoons. Many species of anadromous fish – for example, striped and white bass – and catadromous fish – for example, eels – enter the tropical lagoons for spawning. There is a tremendous energy storage in the form of fine mud, detritus, and associated microfauna which are constantly enriched by the drainage of incoming sediments with rainfall in lagoonic ecosystems. Most of the sediments are transported by water currents which may assist to build sand banks and marshy places. Very large lagoons often show some characteristics of 'inner and lower' waters masses. The inner being rich in indigenous fauna of foraminefera and the lower lagoon water mass contains almost cosmopolitan neritic fauna. The productivity has also been reported to be high. Gessener and Hammer (1962), using C^{14} method, found the primary production to be 2.6 $g/cm^{-2}Y^{-1}$ in the Unare Lagoon, Venezuela.

4.9.10   Hypersaline Conditions

Hypersaline water bodies are rare. In some extreme conditions, depending on the salinity and temperature, when evaporation exceeds the natural influx of run-off water, over a long period of time, hypersaline lagoons result. Hypersaline lagoons occur in places where sea water movements are restricted and salinity increases several times the normal average limit. The bottom sediments are usually formed of sand, silt, clay and organic matter. In spite of hypersaline conditions, many faunal species, foraminifers, polychaetes, copepods, amphipods, isopods, and some molluscs are found. A few well-recognised examples are: Lagua Madre of Texas and Mexico, Sivash at the edge of the Black Sea (or Sea of Azov), the Bitter Lakes along the Suez canal in Egypt, and a few lagoons on the south-east coast of India. The productivity of fish, when salinities are moderate, is often high. Invertebrates have stunted growth while fish may be large in size.

4.9.11   Tropical Muddy Beaches

Muddy shores are a discrete, highly complex and dynamic ecosystem of the marine coastal environment of the Tropics and subtropics. Despite the importance of the nature of substratum which determine the distribution of organisms, the muddy habitat at the edge of the sea in the Tropics have not been fully investigated. The most distinct major habitats of the tropical and subtropical coastal regions are: the mangrove dominated coastal low-lying mud-flats and swamps, and the

Plate 4.1   Mangroves growing on either side of a small estuary in the Northeastern part of Brazil.

tidally exposed subcoastal coral reefs. Accordingly, the fauna and flora which inhabit these habitats are very varied and of highly specialised kinds. Perhaps the heavy rainfall, the diurnal ebb and flow of the water levels, the pounding wave action, turbulent rivers and their effluents into the sea, and the bacterial action on the leaf litter produced by the mangroves themselves all contribute substantially to the mangrove and coral ecology.

4.10   Mangrove Environmental Characteristics

Mangroves and coral are the two most distinct characteristics of the coastal zone of the tropical and subtropical regions. Formation of many volcanic islands and the vast stretches of shallow and tepid seas between the main continents and the islands, provide an extensive and relatively shallow continental shelf. The geological evolution has contributed quite significantly to the climatic conditions which, in turn, have dramatically altered the pattern of floral and faunal distribution and their process of evolution. Regular seasons of torrential rains and intense heat have undoubtedly resulted in the formation of the most characteristic mangrove swamps, marshy peats, and mud-flats along much of the tropical coasts. Mangroves are an important and most productive ecosystem. Mangroves grow in the intertidal zone in the tropical and subtropical belts and they occupy some 15×10^6 hectares worldwide (Lacerda and Diop, 1993), and of which 46% are in the Indo-Pacific region, 27.33% in South and Central America (including the Caribbean region), about 3.33% in North American coasts, and 23.33% along African coasts. Mangroves inhabit the coastal low-lying muddy fringe which is constantly inundated by water through tidal dynamics (Plate 4.1). The global distribution of mangroves is shown in Fig.4.11. In recent years, extensive research on mangrove vegetation has been conducted in many countries such as South-east Asia, Africa, Australia, North and South Americas and the Caribbean.

4.10.1   Mangroves and the Muddy Tidal Flats

Besides coral, mangroves are an important and fragile ecosystem of the Tropics. The mangrove swamps are an extension of the intertidal zone between high and low tide marks. Mangroves are the most conspicuous and characteristic plants that occupy large coastal areas

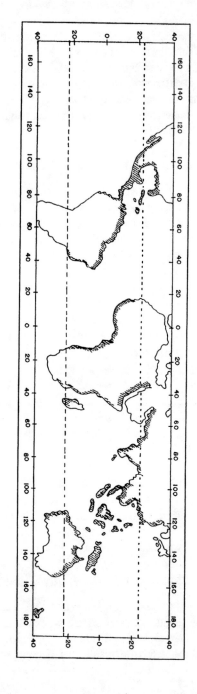

Figure 4.11   Global distribution of mangroves along the tropical and subtropical coastal regions.

with optimum climatic conditions in the Tropics and subtropics. Nearly 68% of tropical and subtropical coasts are fringed with a variety of species of mangroves; some grow as short shrubs and others as tall trees (see below). Mangroves are halophytic (living in salty water), but they are also xerophytic and their leaves are usually thick to retain water. For example, *Acanthus ilicifolius* has thick leathery leaves. On the other hand, some mangroves have fleshy leaves, and yet a few others, *Salicornia* and *Arthrocenemum*, are leafless but their cylindrical stem is jointed, fleshy and mostly green. Under flooded conditions, the complexly arched stilt roots have links with long spongy tissue to provide not only a sufficient supply of oxygen for their submerged parts but also to balance the salt intake. The succulent leaves usually contain large quantities of water sufficient to dilute the salt that enters. Though diverse in adaptive structural mechanisms for respiration and salt balance, mangroves are ubiquitous and perennial plants with an imposing series of stilt roots; they thrive best in muddy flats of swamps, marshy lands, embayments, the margins of large estuaries, lagoons and deltas throughout the tropical coasts of Africa, Asia, the Americas, and Australia. In some areas, where river effluents are of granitic origin and soon disintegrate, the mangroves flourish better than on sandy beaches. A few species of mangroves also grow naturally along the arid coasts of the northeastern coast of Brazil, north Africa, the Arabian Peninsula, Pakistan, and the west coast of Australia. Most mangroves are viviparous; they have buoyant fruits and seeds. The seeds begin to germinate while still on the parent plants and the embryos are ready for dispersal when washed away to another suitable silty shore. Their geographical limit of distribution is regulated mostly by temperature; they prefer a temperature minimum of 20°C–24°C and have greater tolerance for variations in salinity. Vigorous movements of water due to wave actions, tidal flushes and currents often assist in forming fine sediments of mud and help in anchoring the mangroves. The mangroves are of immense importance in preventing coastal erosion. Their complex root systems help to build up land along the coastlines by trapping silts, mud, and debris. Some of the mangrove trees provide timber for building huts or firewood as fuel, and the bark is often used as a good water soluble tanning organic substance. Most areas are used as nurseries for fish culture. Mangroves provide excellent and vital nursery grounds for a variety of larvae; juvenile stages and adult fish, shrimps and prawns and filter-feeders with rich food and protection from predators. The tiger prawn, *Penaeus monodon*, inhabits marine coastal waters but is often cultured in mangrove environments with great success (see Chapter Fourteen on aquaculture potential) as are bivalves, oysters, mussels, clams and cockles – especially in Latin America, Indonesia, the Philippines, India, and West Africa. The wide distribution of mangroves, from Florida to Mexico facing the Caribbean Sea i.e. Malaysia, including Borneo, Taiwan, Burma, the Indian subcontinent and many other Indo-Pacific coasts in the Tropics and subtropics (see Fig. 4.11) is influenced by global coastal climate, temperature, tropical storms and sea level variations.

Mangroves support relatively high rates of primary production. Overall, the mean annual carbon production is variable in different geographical localities. Over 80% of the global carbon production from mangroves are within the equator between $10°N$ and $10°S$ (Twilley et al., 1992). The productivity of mangroves is reported to be about 300–500 $gcm^{-2}yr^{-1}$ in Florida compared with the productivity of phytoplankton 50–75 $gcm^{-2}yr^{-1}$ in the surrounding salt water. Mangroves are generally viviparous sphanorogamous plants which characteristically grow as shrubs or trees, particularly all along tidal estuaries, in salt mashes, and on the muddy coasts subject to tidal variations in the Tropics. Most of the mangroves belong to principally the families Rhiphoraceae, Verbanaceae, Sonneratiaceae, and Areraceae. When fully grown, some trees, especially *Rhizophora* and *Bruguiera* can attain heights of up to 30–40 metres; they are characterised by the shiny green leaves and tiny flowers, and partially submerged tangled prop or stilt roots which help the silt to settle into muddy flats and build up extensive lands along the coast. The prop roots provide rhizophores or pneumatophores which are special adaptations for respiratory purposes as the level of mud-flats rise at times and with seasons. Respiration is a relatively difficult problem to solve in swampy habitats, but mangroves have evolved different devices to surpass this problem, and especially to tide over in their special habitats. In some other species of mangroves, for instance *Avicenia indica*, as noted above, the underground roots give rise to aerial respiratory roots which stand above the water level. On Indo-Pacific and in particular, on in West African coasts, for example, large areas of the Senegal, Congo, Gabon and others, especially during dry seasons, the mangroves grow well even if the salinity increases nearly two to three times more than that of the neighbouring sea water (UNESCO, 1981). The pH also varies between 7 and 8.5. Mangrove sediments, though usually water logged by the tidal cycle, are relatively nutrient-rich. There is no consistency in estimates of productivity made in different localities of mangrove areas; however, an average figure may be about 450 $gcm^{-2}yr^{-1}$, which is higher than that of offshore or oceanic waters. This significantly higher productivity supports a variety of micro, meio and macro fauna, especially the detritivores: bacteria, protozoons and polychaetes; predatory coelenterates, molluscs, crustaceans, fish, birds, marine reptiles, and mammals. The meio fauna with size range between 5μ–1,000μ occur in great density within the surface sediment layer of 0.5 centimetres (UEP, 1994) (see below, Section 4.12).

4.10.2   Epiphytic Diatoms on the Mangrove Prop Roots

The mangroves provide diverse habitats for many organisms; their prop roots are an excellent substrate for a variety of diatoms. Still other micro and filamentous algae grow on eelgrass. The periodic tidal fluctuations associated with changes in salinity and temperature stimulate the luxuriant growth of the diatoms. Living mangrove roots offer a better substrate, and frequently large concentrations of many species of diatoms thrive better than on the non-living substrates (Navaro, 1982). Some of the widely reported epiphytic diatoms in the tropical mangroves are of the genera *Mastigonia, Cysclostella, Thalasiosira, Melosira, Cosinodiscus, Actinocyclus, Biddulphia* and *Tricocentrum* of the order centrales, and the genera *Navicula, Fragilaria, Nitzia, Crammatophysia, Lincomophora* and *Delphines* of the order of pennales.

4.11   Diversity and Adaptations of Life in the Mangroves

The mangrove environmental impacts associated with coastal dynamic factors such as climatic conditions, temperature fluctuations, salinity regimes, rate of precipitation and evaporation, coastal currents, tidal ebb and flow, and rainstorms all influence the life of endemic organisms and migratory or non-migratory animals. As will be seen shortly, there has been a general trend of a greater diversity of marine species in the Tropics and subtropics, but the density of population of the individual species is reflected relatively less compared with temperate waters. The composition of marine organisms is also more diverse in the warmer Indo-Pacific regions than in the tropical Atlantic. The most common mangroves found along the muddy coast of South-east Asia, Africa, South and North America are a number of species differently adapted and specialised for the same type of ecosystem, for example, *Avencinia indica*, *A. marina*, *A. officials*, *A. asiatica*, *A. mucronata*, *Sonneratia alba*, *Rhizophora mucronata*, *Rhizophora mangle*, *Bruguiera parviflora*, *B. sexangluia*, *B. gymnorhiza*, *Xylocarpus spp*, *Lumnitzera recemosa*, *Nypa fruticans*, *Acanthus ilicifolius*, particularly in Ceylon, and occasionally *Pandanus*; the latter occurs particularly along sand mixed mud-flats with low salinity, while the others have the remarkable ability to tolerate a wide range of salinity. The prop or stilt roots (Plate 4.1) are often covered with rich flora and fauna; and also provide shelter or support for a wide variety of marine life. Often, a number of species of green and blue-green algae firmly settle and colonise in the roots. Among animals, the barnacles (*Balanus* and *Amphitrite*), shrimps and crabs, oysters, mudskippers, ascidians and sponges usually predominate. Some of the crabs which are smaller and conspicuously coloured usually ascend the mud-flats during high tide to feed on a variety of organisms and return with the receding tide to bury themselves in the shallow water and muddy shores (Plate 4.2). Other crabs commonly found along the mud banks are *Cardisoma*, *Samatium*, *Ocypode*, *Sesarma*, *Sesuvium*, *Drippe*, *Metopograpsus* and *Uca*. Many of these have developed curious adaptations to thrive on mangrove mud-flats, sandy beaches, or the shallow coastal waters. For instance, the right claw of the *Uca*, especially of the male, is very much enlarged and highly coloured, perhaps to display strength in defence and to attract the females. A number of species of the family Portunidae, though they swim freely off the coast, are found occasionally on mud-flats brought in by the incoming tides. The swimming crab, *Neptunus pelagicus* and, occasionally the Portunid, *Scylla serrata*, are seen swimming in shallow waters at the margins between the muddy shore and the sea. A variety of insects, especially the fireflies, swarm the mangrove trees quite commonly and their illuminous function can be commonly noticed from dusk to dawn. Some colonial insects, especially the ants of the genus, *Oecophylla*, which are voracious eaters, often cause considerable damage to most cash crops such as coconuts, cashew nut, clove, pepper, and mango trees along the coasts of Ceylon, India and many islands of the Indian Ocean. The mosquitoes of the genus *Anopheles* are widespread all over South-east Asian, African and South American coasts. The oysters, *Crassostrea cucullata* and *C. rhizophora*, colonise in particular the portions of the prop roots which lie just below the high water marks. Many molluscs, especially the *Crassostrea paraibanensis*, prefer to bury themselves in the muddy swamps close to the mouths of estuaries. Among the gastropods, *Ellobium, Haminera, Terebralia, Teloscopium* (Fig. 4.12), and *Cerinthium* are most common in the tropical mangrove muddy flats. The mud crabs (Plate 4.2) remain buried in the mud-flats under low tides.

Fish: a variety of fish, large in numbers and species, are found in mangrove ecosystems. Over 300 species have been listed for mangrove habitats of Asia, and 100–200 species for the other major mangrove regions (IUNC, 1983). Among fish, the most common ones are of the Mugilidae and Gobidae families, especially *Mugil cephalus, M. brasilinsis* and *M. wagensis*, which often enter the river mouths and estuaries to feed on planktonic algae and weeds and when sexually mature they return to sea to breed. Within the swampy water edges, the mudskipper, *Periophthalmus vulgaris, P. koelreuteri* and *Boleophthalmus dentatus* and *B. viridis*, can be frequently seen. These are some of the most curious creatures commonly encountered along mud-flats fringed with shrubby or medium sized mangroves. They are somewhat amphibious, agile and aggressive predatory fish, which can abandon water for long hours and live of the top or the prop roots or other similar objects well above the water, but with the slightest noise they jump into the water and swim away with considerable speed. They represent some of the best adaptations for the mangrove environment. Their large bulbous stalked eyes protrude laterally from the top of the head and are well adapted for wide angle vision both in water and in air. When they swim just below the surface of sallow water the stalked eyes probably act as a 'periscope'. During their aerial respiration, the opercular folds above the gills act as air chambers. Several reptiles, particularly weakly poisonous

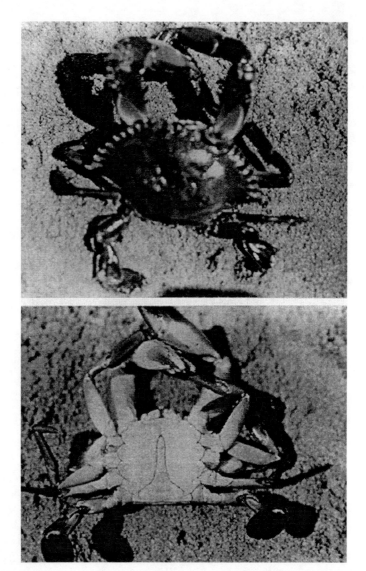

Plate 4.2   The mud crab, often found buried in the mud-flats among the
                   mangrove swamps of Brazil.   The carapace is broad and
                   armed with sharply cut teeth and the celipedes are enlarged.
                   It is an edible crab commonly exploited by the poor coastal
                   populations during low tide.   (A) Dorsal view (top), (B) Vental
                   view (bottom).

snakes of the genus, *Boiga*, of the family Colubridae, component of marine edible turtles, large monitor lizards, and occasionally some marine crocodiles venture to shelter and build nests at the outer fringe of the mangroves to lay eggs and to feed on fish, including mudskippers, crabs, prawns, birds, and smaller mammals in the mangrove swamps. A number of species of birds, including herons and storks, often wade through the mangrove swamps and filter the muddy water between their beaks in search of food. Most cormorants, *Phalacrocorax*, of the family Phalacrocoracidae are commonly encountered all over the Indian Ocean and African coasts. The kingfishers, *Ceryle rudis*, are widespread in most of the tropical Afro-Asian mangroves. Among the mammals which commonly invade the river estuaries or lagoons are the mantees, *Trichechus spp.* and *Dugong dugon*; they probably feed on the mangrove buds and leaves and frequently get stranded in the lagoons and the estuarine coasts of south-east Asia and South American Atlantic coasts. Of the well-known Macaque monkeys of the Tropical south-east Asia and Africa, the Genus, *Macaca*, with elongated tail, is carnivorous. They frequently stroll in small groups, consisting of twenty to thirty of both sexes and of all ages, along the mangrove coasts and hunt for the mud crabs as they much relish the favoured crab meat. Small groups of flying-foxes of the genus, *Pteropus*, often frequent the mangrove swamps in Ceylon and Australia.

4.12   Corals and Coral Reefs

Corals have been known to man for thousands of years because of their richness, beauty and colour, and the fish and many other organisms that live among the coral reefs. Corals and reef-building (scleractinian) corals, are widely distributed in the Indo-Pacific tropical waters, more so than in the tropical Atlantic. The reef building corals are significantly absent close to much of the coastal parts of West Africa, and west of South America; the obvious reason for the absence of the reefs seems to be the changes in temperature due to subsurface cold currents. Difference in latitude, geographical range and extreme environmental conditions seem to have a great impact on the coral growth. Within the warm tropical belt, corals extend from the Red Sea to the Galapagos. The Great Barrier Reef in Australia, the fringing type in the southern part of the Indian Ocean,

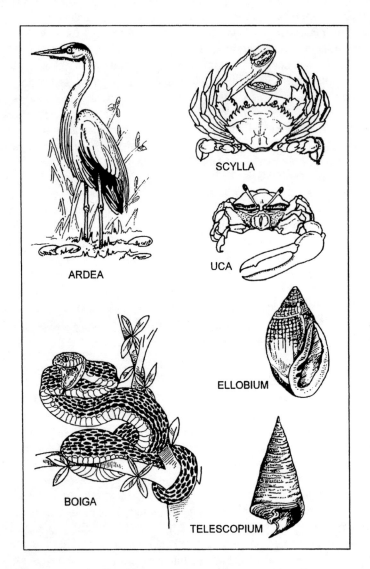

Figure 4.12   The different curious creatures which are most abundant in the tropical mangrove environment. These have remarkable adaptations for survival in mangrove habitats.

including a number of islands such as Ceylon, Andaman, and the Maldives, the Gulf of Kutch, the Red Sea, and Nicobar Islands, and the atoll forming coral around the Maldive and Laccadive Islands are well-known. A number of atoll-type corals are commonly found around the waters of numerous Pacific islands. For a full discussion see Chapter Five on corals.

4.13 Coastal Characteristics

The total length of the global coastline seems to exceed 586,153 km, the shelf to 200 metres depth is some 21,426.5 km^2 and the exclusive economic zone accounts for 94,466.1 km^2 whilst the coastal population is predicted to be $99,855 \times 10^6$ in the year 2000 (UN, 1981). The study of coastal characteristic is not always easy in the Tropics because of the difficulty in gaining access and because of the coastal complexity and the greatest impact by man and nature. Time-scales involving significant geological and evolutionary changes stretch from a few years to many thousands. Most coasts in the maritime Tropics are directly exposed to severe interaction of natural calamities from the air, earth and sea. The fringes of most coasts are constantly eroded by heavy rainfalls, tidal flows, wave actions, and so the topographic features change rapidly. The dimensions of the intertidal zone, together with their endemic floral and faunal organisms, also move beyond their boundaries. Some of the available parameters pertaining to coastline tropical countries are listed in Table 4.2. The world is continuously evolving. In 1950 the world comprised of only 82 countries. Currently, the world comprises of some 191 independent states and 58 dependencies; in 1993 Eritrea became the most recent addition. Antarctica remains the only unclaimed land of the earth (Dorling Kindersley, 1995). Like the expanding trend of each country in the world, the population of the world has increased rapidly; more so within the tropical and subtropical regions of the world.

Table 4.2   Coastal Parameters of the Tropical and Subtropical Regions
(0°–30° north and south of equator) of the World

COUNTRIES AND MAJOR ISLANDS	POPULATION	COAST LENGTH	EXCLUSIVE ECONOMIC ZONE	LAND AREA
ATLANTIC OCEAN				
WEST AFRICA				
W. Sahara				
Morocco	3000	895	278,100	446,600

Mauritania	2.1000	360	154,300	1,030,000
Senegal	7.7000	241	205,700	196,200
Gambia	0.9010	38	19,500	11,000
Guinea-Bissau	1.0000	–	150,500	36,100
Guinea	6.1000	190	71,000	245,900
Sierra Leone	4.4000	219	155,700	71,500
Liberia	2.8000	290	229,700	111,400
Ivory Coast	12.9000	274	104,600	322,500
Ghana	16.0000	285	218,100	238,500
Togo	3.8000	26	1,000	56,000
Benin	4.9000	–	27,100	112,600
Nigeria	8.3000	415	210,900	923,800
Cameroon	12.2000	187	15,400	475,400
Equatorial-Guinea	0.4000	184	283,200	28,100
Gabon	1.2000	399	213,600	267,600
Congo	2.4000	106	24,700	342,000
Angola	9.9000	806	506,100	1,246,700
Namibia	1.5000	804	–	–
S. Africa+	37.4000	1,462	1,101,600	1,222,200

INDIAN OCEAN
EAST AFRICA

Egypt (Red Sea)	54.8000	–	1,73,500	1,001,400
Sudan (Red Sea)	28.7000	387	91,600	2,505,800
Eritrea (Red Sea)	3.5000	–	–	–
Djibouti (Gulf of Persia)	0.5000	–	6,200	22,000
Somalia (Horn of Africa)	9.2000	1,596	782,800	637,700
Kenya	25.2000	247	118,000	582,600
Tanzania	27.8000	669	223,200	945,000
Mozambique	16.1000	1,352	562,000	783,000
Togo	3.8000	26	1,000	56,000
S. Africa+	–	–	–	–
Comoros Islands	0.6000	211	228,400	2,200
Madagascar Islands	12.8000	2,155	1292,000	587,000
Mauritius Islands	1.1000	87	1183,000	1,900
Seychelles Islands	0.0680	–	729,700	0,400
Mahé Island	–	–	–	–
Zanzibar Island Tza	0.579	–	–	–

INDIAN OCEAN
WEST ASIA (MIDDLE EAST) — 18,000 (mainland + 14,000 islands km)

Saudi Arabia	15.9000	1,316	186,200	2,149,700
Yemen	12.5000	–	–	–
Socotra Yemen	–	–	–	–
Oman	1.6000	-	561,700	212,200
Djibouti	0.5000	-	6,200	22,200
Dubai	–	–	–	–
Abu Dabi	–	–	–	–
U. Arab Emirates	0.5620	1,307	59,300	83,600
Qatar	0.5000	204	24,000	11,000
Kuwait	2.1000	115	12,000	17,800
Iraq	19.3000	10	0,700	434,900
Iran	61.6000	990	155,700	1648,000
Pakistan	124.8000	750	318,500	803,900

INDIAN OCEAN				
NORTH ASIA				
China+**	1,2000.0000	3,962	1355,800	9,997,000
INDIAN OCEAN				
SOUTH ASIA				
India	879.5000	2,759	2,014,100	3,280,500
Sri Lanka	17.7000	1,343	517,400	657,000
(Ceylon)				
Bangladesh	119.3000	580	76.800	144.000
Burma	43.7000	1,230	509.500	678.000
Malaysia	18.8000	4,675	475.600	329.700
Singapore	2.9000	193	0.300	0.600
Thailand	50.1000	1,299	324.700	514.000
Cambodia	8.8000	-	55,600	181.000
Vietnam	67.8000	1,247	722.100	332.600
Hong Kong	5.9000	733	-	-
Taiwan	0.8000	-	-	-
(Formosa)				
Philippines	65.2000	6,997	1,890.700	300,000
Indonesia	191.2000	19,784	5,408.600	1,904.300
Papua New Guinea+	4.1000	15,750	7,006.500	7686.900
INDIAN OCEAN				
ISLANDS				
Andaman Island Ind.				
Nicobar Island Ind.				
Brunei	0,3000	161.000	-	-
Maldives	0.2230	644.000	959.100	-
Rodigues Island	0.3275			
Cocos Island	0.0007	-	-	-
Réunion Fr.	0.5978	-	-	-
Christmas Island	0.0013	-	-	-
Chargos Islands (Arch) UK	0.0034	-	-	-
Lacadive Islands Ind.	-	-	-	-
(Lakshadweep)				
Minico Islands Ind.				
ATLANTIC OCEAN				
NORTH AMERICA				
USA+	255.2000	11,650.000	7,825.000	9,372.000
Mexico+	8.2000	4,848.000	2,851.200	1,972.500
Bermuda UK	0.0585	-	-	-
ATLANTIC OCEAN				
CARIBBEAN SEA				
CENTRAL AMERICA				
Anguilla	0.0090	-	-	
Cayman Islands UK	0.2554	-	-	-
Nassau				
Bahamas Is, US	0.3000	759.200		
Virgin Islands UK	0.1019	-	-	-
Cuba	9.5000	747.000	362.800	114.500
Belize	0.1940	-	-	-
Honduras	5.5000	74.,000	200.900	112.200
Nicaragua+	4.0000	445.000	159.800	130.000

Costa Rica+	3.2000	446.000	258.900	50.000
Panama+	2.5000	993.000	306.000	75.600
Columbia+	33.4000	1,022.000	603.200	1,138.900
Haiti	6.8000	584.000	160.500	27.800
Puerto Rico	3,2000	–	–	–
Ari Bau Neth.	–	–	–	–
Antilles Neth.	–	–	–	–
Antigua & Barbuda	0.0640	1,920	–	–
Aruba Neth.	0.0624	–	–	–
Dominican	0.0830	325.000		
Jamaica	2.5000	280.000	297.600	11.000
Barbados	0.064,0	55.000	167.300	0.400
Grenada	0.0840	–	–	
Trinidad and Tobago	1.3000	254.000	76.800	5.100

ATLANTIC OCEAN
SOUTH AMERICA
Colombia+

Venezuela	20,2000	1,081.000	363.800	912.000
Guyana UK	0.8000	232.000	130.200	215.000
Suriname	0.0425	196.000	101.000	163.300
Guiana Fr.	0.1149	–	–	–
Brazil	153.2000	6,500.000	3,168.400	8,512.000
Uruguay	3.1000	305.000	119.300	177.500

PACIFIC OCEAN
WEST COAST
Tropical NORTH AMERICA
USA+
Hawaii Islands
Mexico+
SOUTH AMERICA

Guatemala	9.7000	178.000	99.1000	108.900
El Salvador	5.4000	164.000	91.9000	21.000
Nicaragua+	4.0000			
Costa Rica+	3.2000			
Panama W+	2.5000			
Colombia+	33.4000			
Ecuador	11.1000	458.000	1,159.000	283.600
Peru	22.5000	1,258.000	786.600	1,285.000
Chile	13.6000	2,882.000	2,288.200	756.900
Hawaii Islands				
New Hebrides				
New Caledonia Fr.	0.1642			
North Marshall Islands US	0.0454			
Pitcairn Islands, UK	0.0001			
Norfolk Is. Aus.	0.0019			
Nieu NZ	0.0224			
Wallis and Futuna	0.0137			
Marana Is.				
Revilla Giggedo Is.				
Galapagos Is.				
Islas Juan Fernandez Chile				

143

PACIFIC OCEAN
AUSTRALIA and OCEANIA

Australia+	17.6000	7.069	700.600	7,686.900
Papua New Guinea+	4.1000	5.152	–	461.700
Hawaii US				
Cooks Island	0.018,6	–	–	–
Micronesia (Caroline Is)	0.1010	–	–	–
Marshall Islands US	0.4335			
Tonga	0.0940	419	–	–
Fiji	0.7000	1,129	1,134.700	18.300
Samoa US	0.0593			
Western Samoa	0.1690	40	96.000	–
Kiribati	0.0730			
Palau US	0.0164			
Tuvalu	0.0091	24 km		
Vanuatu	0.1630			
Polynesia (Tahiti) Fr.	0.1999			

Population based on UN *Demographic Year Book*, 1990. The countries delimited by two oceans are marked with + and data on coastal lengths and EEZ are based on *The Times Atlas of Oceans*, Couper ed., 1983. Copyright with the kind permission of Times Books Ltd.

References

Alongi. D., 'The ecology of tropical soft-bottom benthic ecosystems', *Ocenogr*, Mar. Annu. Rev., 28, 1990, pp.381–496

Atkins, W. R. G., 'The hydrogen concentration of sea water in its relation to photosynthetic changes', J. Mar. biol. Ass. (UK), 13, 1925, pp.93–118

Aubert de la Rue, E., Boudiere, R., and Harry, G. P., *The Tropics*, Eeffer and Simons, 1957, p.375

Bascom, W., *Scientific American*, 2 (August), 1959, pp.74–78

Bridges, C. R., 'Ecophysiology of intertidal fish', in *Fish Ecophysiology* (ed. Rankin. C., et al.), London, Chapman and Hall, 1993, pp.375–400

Brown H. W., and Fisher, A. V., Philippines mangrove swamps, in minor ducts Philippines forests, (ed. W. H. Brown), 1, 1920, pp.9–125

Corps of Engineers, US Army, Committee on Tidal hydraulics, 1950. Evaluation of present state of knowledge of factors affecting tidal hydraulics and related phenomena Report no.1, February 1950

Defant. A., *Physical Oceanography*, New York, Pergamon Press, 1961, p.729

Dorling Kindersley, *World Reference Atlas*, Dorling Kindersley Ltd., London, 1994, pp.733

Edwing, M., et al., 'Sediments distribution in the Indian Ocean', *Deep Sea Res.*, 16, 1969, pp.231–248

Ekman, T., *Zoogeography of the Sea*, London, Sidgwick and Jakson, 1953, pp.417

Flint, R. F., 'The position of the sea level in a glacial age', a paper given at INQUA Congr, VII, Paris, 30th August – 5th September, 1969

Hedgpeth, J. W., 'Ecological aspects of the Laguna Madre, a hypersaline estuary', in *Estuaries* (ed. G. H. Lauff), Publication 83, American Association for Advancement of Science, Washington, DC, 1967, pp.408–419

Gessner, F. and Hammer, L., 'La produccion primada em la Laguna Unare', XII Convencion Anual Assoc., Venezuela, Av. Cienc., 1962

IUNC, 'Global status of mangrove ecosystems', Commission on ecology papers (eds Saenger, P., Heged, E. J. and Daves J. D. S.), International Union for Conservation of Nature and Natural Resources, Gland, Switzerland, 1983, p.88

Lacerda, L. D., and Diop, E. S., 'Conservation and sustainable utlization of mangrove forests in Latin America and Africa regions', International Society for mangrove ecosystems, 1993

McEvedy, C. and R. Jones, *1978 Atlas of World Population History*, London, Penguin Books Ltd., p.368

Macnae, W., 'A general account of the fauna and flora of mangrove swamps and forests in the Indo-West-Pacific Region', Adv. Mar. Biol. 6, 1968, pp.73–273

Milward, E. N., 'Mangrove dependent biota', in *Mangrove ecosystem in Australia: structure, function and management*, (ed. Clough, B. F.) Canberra, ANU Press, 1982, pp.121–140

Navarro, N. J., *Marine Diatoms associated with mangrove prop roots in the Indian River, Florida*, USA, J. Cramer, Fl. Vaduz, 1982, p.151

Odum, W. E., and Hall, E. J., 'Mangrove forests and aquatic productivity', in *Compiling land and water systems* (ed. Haseler, A. D.) Ecological Studies, vol. X, Berlin, Springer-Verlag, 1975, pp.139–163

Reyes, G., 'Diatomes litorales de la Familia, Naviculaceae de laguana la Restingha, Isla Margarita, Venezuela', Cumna, Bol. Inst. Oceangr. Univ. Oriente, 14 (2), 1975, pp.199–225

Ramamirtham, C. P., 'On the relative (geostrophic) currents in the south-eastern Arabia sea', J. Mar. Biol. Ass. India, 8 (2) 1966, pp.236–243

145

Seshappa, G and Jeyaraman, R., 'Observation on the composition of bottom muds in relation to the phosphate cycle in the inshore waters of Malabar coast', Proc. Indian Acad. Sci., B 43, 1956, pp.288-301

Sharma, G. S., 'Thermocline as an indicator of upwelling', J. Mar. Biol. Ass. India, 8 (1), 1966, pp.8-19

Sidhu, S. S., 'Studies on the mangroves', Proceedings of Indian Academy of Science, 33 (8), 1963, pp.129-136

Singarajah, K. V., 'Hydrographic conditions, composition and distribution of plankton in relation to potential resources of Paraiba River Estuary', Rev. Nordest. Biol., 1(1), 1978, pp.125-144

Sukardjo, S., 'The present status of the mangrove forest ecosystem of Segara Annkan, Cilacap, Java.', Tropical Rainforest: Leeds Symposium, 53-67., mean monthly and extreme discharges 1972-1975, Stud. Rep. Hydrol., 5, 3, 1984, p.104

Sawyer, J. S., 'Large Scale Disturbances of Equatorial Atmosphere', Mer. Mag., 99, 1970, pp. 1-9

Swell, R. B. S., 'The free swimming planktonic copepods; systematic account', Sci. Rep. John Murrey Expedition (Bdti. Mus. Nat. Hi.), 9, 1948, pp.1-303

Twiley, R. R., Chen, R. H. and Hargis, T., 'Carbon sinks in mangrove and their implications to carbon budget of tropical coastal ecosystems', Water, Air and Soil Pollution 64, 1992, pp.265-288

UNEP, 'Assessment and monitoring of climatic change impacts on mangrove ecosystems', UNEP Regional Reports and Studies no.54, 62, 1994

UNESCO., 'The coastal ecosystems of West Africa: coastal lagoons, estuaries and mangroves', UNESCO, Rep. Mar. Sci, no.17, 60 1981

UNESCO., 'Discharge of the selected rivers of the world', Studies and reports in hydrology, 1979, p.104

United Nations Office for Oceans, (unpublished data) June 1989, New York

Vemberg, B., and Vemberg, W. B., 'Ecological affinities of Brazil and the South-eastern United States of America. A study in comparative physiological ecology', Seminaria de Biologia Marinha, Sao Paulo, dezembro, 1975, pp.69-74

United Nations., Demographic Year Book, New York, 1990, p.1020

Weyl, P. K., Oceanography. An introduction to the marine environment, New York, John Wiley and Sons, Inc. New, Inc., 1970, p.535

'World Resources 1994-1995', a report by the World Resources Institute in collaboration with the United Nations environment programme and the United Nations development programme, New York, Oxford, Oxford University Press, 1994, pp.400

Chapter Five
Corals and Coral Reefs in the Tropics

5.1   Unravelling the Beauty of Nature

Some of the most beautiful and curious groups of organisms are the corals. Their colours are contrastingly vivid. They are interesting because of their richness, variety, and the splendour of their intricate system. They have the ability to form extensive coral reefs in the vast tropical seas, where the water is warm with almost equal diurnal rhythms of light and darkness and rise and fall of the tide corals exist. Corals play an important role in encrusting calcareous algae. They contribute to building islands of archipelagos. Some of the best reef building corals are found in relatively shallow warm tropical waters with temperatures between 21°C–26°C within latitudes 30° north and 30° south of the Equator, but they are normally absent where cold subsurface currents circulate. Corals cover some 2×10^8 km^2 of the tropical seas. Most coral genera and species are widespread along the Indo-Pacific and Brazilian coast of the Atlantic and the Caribbean. The west coast of Africa has relatively less coral growth than on the east coast. Coral reefs are unbelievably rich in a variety of life, forms, and colour besides the corals themselves. The coral reefs and the partially submerged coral pools under low tide provide both shelter and food for an infinite number of sedentary animals such as sponges, sea anemones, sea urchins, starfish, sea cucumbers, bivalves, gastropod molluscs, crustaceans, bryozoans, and several varieties of species of fish, including sharks and moray eels with some of the most bewildering adaptations (Plate 5.1).

A variety of pelagic larvae of diverse array of benthic animals find corals as suitable substrate to settle down. In addition, a number of marine algae, especially of the family Dinophyceae, grow either independently or symbiotically with corals, thus enriching the reef-economy and the subtlety of the coral ecosystem. Almost all reef-building corals contain zooxanthellae. The common scleractinian

Plate 5.1   Some of the fishes, including a moray eel, with most
                   bewildering colour and adaptation inhabiting the coral reef.

corals have remarkable adaptability; and not only provide rich habitats for many resident fish, but also an equilibrium habitat conditions for the co-existence of many fish, in spite of their competition for space, food and aggressive behaviour.

5.2 Origin and Evolution of Corals

The origin of coral reef is doubtful, but several lines of evidence indicate that the first sponge-like corals, the Archaecyathids, became fairly well established during the mid-Ordovician period, nearly 450 million years ago. As these Archaecyathids became extinct, the true 'Rugose' corals of the order Tetracorals began to replace them, and the last of the Rugose also became extinct some time during the end of Triassic, about 200 million years ago and they, in turn, were replaced by reef-building Scleractinian (modern) corals. The Scleractinian corals appeared to have flourished along the shores of the ancient Tethys Sea of warm waters and later drifted eastwards during the continental drift and finally found their way into warm shallow waters where they began to diversify during the Jurassic period and spread out on a global scale in the tropical seas. Many of them have now become extinct, and only about 2,500 species are living today. At the end of the Cretaceous, like the rest of the marine flora and fauna, there was a decline in the coral growth due to major adverse changes in the climatic conditions which existed for many millions of years. The continental drift continued and the American land masses became separated, even though the Tethys Sea existed, there were no land barriers for the evolving organisms in the tropical waters, and there was no ice during both the Mesozoic and the greater part of the Cenozoic, until the end of the mid-Miozene. However, some corals, for example, the blue coral *Heliopora coerulea*, survived and are still found in abundance in the Indo-Pacific regions, although they dominated the fossils of the Tethys Sea of the Cretaceous period. With the advent of more equable environmental conditions, the corals once again began to diversify and spread out. The Tethys Sea began to shrink, giving way to the expansion of the Mediterranean during the Miocene. The Antarctic ice-caps began to be formed and during the Pliocene the Isthmus once again closed, thus causing land barriers between the Atlantic and Pacific Oceans. Plistocene, about 1,800,000 years ago, marked the ice ages; with each successive glaciation, the sea level fell due to formation of the ice-caps. The last glaciation caused a fall of nearly 100 metres lower than its present level and affected the distribution of corals severely. With a subsequent rather more stable climate, with well oxygenated, effective water movements and adequate wave actions which reduced calcareous deposits of sediments, corals became differentiated adaptively into:

(1) Hermatypic: an adaptation of symbiotic association with unialgal organisms.

(2) Ahermatypic: corals which do not depend on the symbiotic association and hence can grow even in relatively deeper waters where light is at premium.

5.3 Taxonomy of Corals

Corals are exclusively marine and polypoid coelenterates of the class Anthozoa of the phylum Cnidaria. With the exception of a few solitary corals of the genus, *Astrangia*, most corals grow in colonies along the seaward slopes of warm waters not exceeding 60 metres depth, but corals have been found at depth of up to 100 metres in the Caribbean Sea. The individual corals are called polyps, each capable of secreting its own slimy calcareous substance which later becomes hardened. This complex cupped exoskeleton is also called corallite and several of them join together and contribute to build the coral reef. The gorgonian corals are soft and do not secrete the exoskeleton. The class Anthozoa consists typically of polypoids, but the medusoid stage is distinctly absent. The class is subdivided into two subclasses:

(1) Subclass – Alcyonaria (Octocoralina) based on the structural organisation. Polyps with pinnate tentacles, colonial and capable of secreting exoskeleton. This subclass, according to Hyman (1940), includes some six orders:

(a) Stolonifera – polypoids, but basally connected by stolon.

(b) Telestacea – colonies of polyps.

(c) Alcyonacea – soft corals.

(d) Coenothecalia – blue corrals.

(e) Gorganacea – consists of some of the most attractive gorganians, sea fans, sea feathers, and horny-corals, including the precious pink corals, Corallium.

(f) Pennatulacea – sea pens.

(2) Subclass – Zoantharia (Hexacorallia) True corals, may be solitary or colonial, tentacles non-pennate and usually in multiples of six. This subclass consists of five orders of which one order includes the sea anemones.

(a) Actiniaria – sea anemones.

(b) Madreporaria – true corals, with some exceptions, the vast majority are colonial.

(c) Zoanthidea – solitary or colonial and lack pedal disc.

(d) Antipatharia – the black corals.

(e) Ceriantharia – solitary with elongated bodies.

5.4 Scleractinia – Reef Building Corals

Alternatively, because of the enormous range, it is essential to know whether the corals are solitary or colonial, hermatypic or ahermatypic, modern or fossil type, and the structural variations of the skeletons of a colony, the corallum, or of individual polyp, the corallite, since the task of taxonomists to identify corals readily and with reliability is considerably difficult. However, Vernon (1986) provides a major 'family-level' classification for the Scleractinian corals.

Pocilloporidae
Astrocoeniidae
Acroporidae

Poritidae

Siderastreidae
Agariciidae
Fungiidae

Oculinidae

Pectiniidae
Mussidae
Meandrinidae
Trachyphyliidae
Faviidae
Merulinidae
Caryophylliidae
Dendrophylliidae

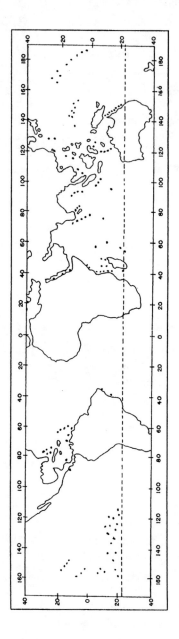

Figure 5.1   Global distribution of corals within the relatively shallow warm tropical waters.

5.5 Range and Distribution of Corals

Through time and space, many corals have become displaced from their own geographical localities, whilst some corals have become geographical variants. However, most genera and species of corals are widely distributed, but limited by only latitudinal and geographical range, ocean circulation, wave action, light, sediments influx, severity of tropical cyclones and temperature. In most tropical coastal waters, from the Red Sea to east of Hawaii, East Africa, along the north-eastern coast of Brazil, Caribbean, the Gulf of Guinea, the Philippines and Australia, especially close to the islands of the Seychelles, Gulf of Kutch, Madagascar, Mauritius, Lakadive, Ceylon, the Maldives-Lakshadweep Archipelagoes, Andaman, Nicobar, Malacca, Solomon, Vanuatu, Marshall Island, New Caledonea, Caroline Island, Bermuda, Clipperton, Galapagos, Cocos, Tonga, Samoa, Revilla Gigegod, and in several other islands many genera and species of corals are well represented. The global distribution of the corals is shown in Fig. 5.1.

Several species of some of the most beautiful genera are found in the Persian Gulf area, but the number of genera is restricted due to extreme environmental conditions such as variations in temperature from 16°C to 40°C and salinity up to 48‰ (Govindha Pillai and Patel, 1988). Except the Gulf of Guinea and Cape Verde, along the west coast of Africa, the corals are generally poorly represented (see Fig. 5.1). Light, both the composition and intensity, is an important determinant of the depths of the reef, and the slope to which the hermatypic corals can extend. The hermatypic corals are usually absent in the Mediterranean Sea. On the other hand, the distribution of ahermatypic corals is not restricted by light or temperature. Some of the interesting reef building corals are Scleractina. The commonly found corals are shown in Fig. 5.2.

REEF-BUILDING CORALS
Montipora explanata
Coscinaraea monile
Favia speciosa
Favites complanata
Gonifora planuiata
Porites luttea
Acropora palifera

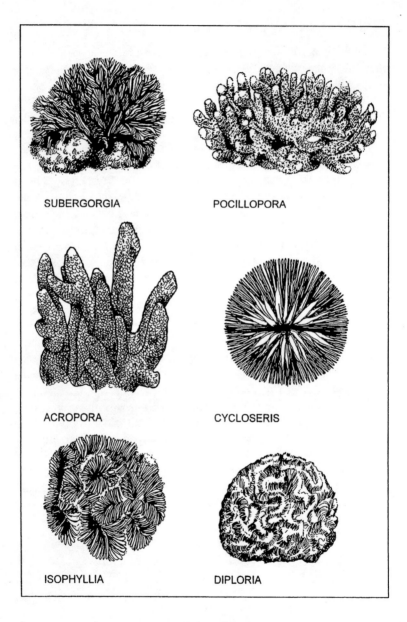

SUBERGORGIA

POCILLOPORA

ACROPORA

CYCLOSERIS

ISOPHYLLIA

DIPLORIA

Figure 5.2   Different genera of the corals commonly found in the tropical and subtropical shallow waters.

Porites lobata
Cycioseris marginate
Fungia repanda
Montastrea annuligera
NON-REEF BUILDING
CORALS
Turbinaria crater
Balanophyllia
Dendrophyllia minuscule

5.6 Coral Biology

The success of the widespread growth of modern corals in tropical coastal waters has been mainly due to the establishment of an intimate and subtle mutually beneficial symbiotic association between the coral polyps and the unialgae called zooxanthaellae, mostly of the planktonic brown algae, such as dianoflagellates (see Chapter Eight on plankton). Not all corals contain the zooxanthellae. These are really algal cells of the species, *Gymnodinium microadriaticum*, which can also exist quite independently of the corals. During symbiotic existence, as the algal cells enter into the tissue of the coral polyps, they lose their two long flagellae, and are able to use the CO_2 to carry out photosynthesis; the zooxanthellae derive some benefits from the corals (see below), but the hermatypic corals do depend on them for organic nutrients, oxygen and slimy exuded matter for reef-building, while the ahermatypic group does not need to associate with the algae (see below).

5.7 Symbiotic Relationship of Corals

As mentioned earlier, the coral polyps are usually small and of variable size and almost reminiscent of sea anemones. They are sedentary and multicellular animals leading a solitary or colonial form of life. In common with many other coelenterates, the corals have near cylindrical bodies with hollow gut cavities; an oral and an aboral end. The oral end is provided with a crown of tentacles which possess nematocysts. The tentacles surround an opening which serves as both mouth and anus. The gut is a simple cavity with a number of vertical and radially disposed mesenteric folds directed towards the centre. The body wall consists of three distinct layers: the ectoderm, the mesoglea, and the endoderm.

Corals are notable for their symbiotic relationship with a variety of planktonic unicellular algae, particularly dianoflagellates. When the unicellular algae grow in association with reef building corals, they lose their flagellates and the cell membranes and become embedded within the wall of the host; rebuilding in brightly coloured spherical bodies, and hence called zooxanthellae. Coral symbiosis is one of the major systems operating in coral reef building. Although corals are basically carnivorous, this symbiotic relationship provides a relatively large proportion of the food and oxygen which accelerate the growth of the corals, and in addition, some defensive mechanisms. Of the organic materials synthesised by the algae, nearly 80% diffuse out of the zooxanthellae to be utilised by the corals. On the other hand, the zooxanthellae derive some of the essential nutrients such as CO_2, PO_4, NO_3, NO_2, SO_2, NH_4 from the excretory products of coral hosts. Zooxanthellae are not the only symbionts which grow among coral reefs. Many other species of brown, red, and green algae, especially *Halimeda*, also grow in great abundance. The latter contains a toxic substance, diterpenoid halimedathal, which when liberated in the sea water seems to deter the predators, particularly the damselfish which harvest the coral polyps so heavily. Not all corals have a symbiotic relationship with zooxanthellae. Those which have are called hermatypic reef building corals; while others which do not have are called ahermatypic, for example, *Tubastrea, Dendrophyllia,* and *Balanophyllia.*

5.8   Reproduction

The growth of scleractine is by asexual multiplication of polyps, although sexual reproduction is common to all corals. Asexual polyp duplication can result in the formation of new colonies. The gonads are formed on the mesenteric wall. The gonads on maturity release the sperms into the sea water which are then drawn by the ciliate action into the coelentron of another individual and then fertilisation is completed. The sexual reproduction often results in the prolific liberation of 'planula' larvae. The planula larvae, when released, are small and provided with cilia; they usually swim towards light and then swim down until they find suitable substrate either close to their parent colonies. Alternatively they drift and lead a relatively brief planktonic existence before settling down on a suitable substrate to metamorphose into a new polyp and build new colonies. Although most coral polyps are hermaphrodites, some have separate sexes. When hermaphrodites, the eggs ripen relatively more rapidly than the sperms and the sperms are released separately to ensure fertilisation. The fertilisation may be internal. Many reproduce all year round, whilst some reproduce only once a year. When sexes are separate, every polyp in the colony is either male or female.

5.9 Behaviour

Corals are carnivorous. They usually feed on a variety of planktonic organisms such as copepods, shrimps, eggs and larvae of other invertebrates and fish. The tentacles are armed with nematocysts which can paralyse the zooplankton and direct the prey through the mouth into the stomach of the polyp where the fragmented food is digested. Some aggressive behaviour has often been reported in corals. When competition for food and space becomes greater, some corals, particularly *Euphyllia* and *Goniophora*, are said to be antagonistic. They extend their tentacles and attack or extrude mesenteric filaments and digest the parts of neighbouring individuals. The corals, especially the larvae, during dispersal, are faced with a number of predators; the chief being the starfish, *Acanthestar*, the gastropod *Drupella*, the mussel, *Lithophaga lessepsiana*, and the worm, *Spirobranchas giantiis*.

5.10 Coral Reefs

Coral reefs are the hard calcarious complicated framework of limestones formed by living coral polyps. They are also often in association with a variety of coraline algae through the incorporation of the chemical components, especially $Ca(CO)_3$, detritus, and the other marine sedimental deposits of argonite and magnesium calcite derived from the corals themselves, and some molluscs, echinoderms, brachiapoda, foraminefera and unicelular algae of the family coccolithophorae. The biological process of reef building usually involves the ion exchange mechanisms of both active and passive transports. These processes are controlled by the amount of solar energy available and the concentration of an enzyme, the carbonic anhydrous. Under optimal conditions, the rate of calcification increases, resulting in the rapid growth of coral reefs. Coral reefs rise from the sea floor and grow upwards to the sea level of low tide limit. The principal reef-building corals are the Madreporarian (Scleractinian) colonial corals of the class zoantharia (see below). Corals are quite successful in reef building in those tropical regions where the sea is warm and shallow. On the basis of their ability to build coral reefs the corals are grouped into:

(1) The Hermatypic – reef building corals.

(2) The Ahermatypic – non reef building corals.

The hermatypic corals essentially establish a symbiotic association with a number of coraline algae, notably the zooxanthellae and the ahermatypic corals that distinctly lack them. However, the reef building corals are carnivorous.

Darwin postulated the hypothesis that in the development of a fringing reef around a foundation such as a volcanic island, the fringing reef becomes a barrier reef and finally an atoll by organic growth and slow subsidence of the island. But, since Darwin's time, many subsequent investigators either support or provide alternative hypotheses based on the glacial control theory (Daly, 1910).

However, on the basis of chemical nature and the mechanisms of their formation, the coral reefs can be recognised as falling into three distinct types:

(1) The fringing reef.

(2) The barrier reef.

(3) The atoll reef.

Fringing Reef: this was considered by Darwin as the basic reef type. It is a relatively narrow ridge which extends either continuously or discontinuously, and usually in shallow waters parallel to the coasts. The fringing reef is separated from the shores by a narrow stretch of water with varying depths. The fringing reef provides a rich variety of habitats for many animals. The upper surface of the reef may be exposed under low tides; storms and heavy tropical rains can be particularly destructive. The destructive effect is more common in the north Atlantic, especially off the coast of Jamaica, but is relatively scanty in the Indo-Pacific waters.

Barrier Reef: though resembling the fringing reef in its basic characteristics, it grows a little further from the coast somewhat asymmetrically, steeply oceanward and is of greater dimensions. The water that stretches out between the coast and the massive barrier reef forms a lagoon and is often over 30 metres deep. The well-known Great Barrier Reef off the north-eastern coast of Australia, the barrier reef off the north-west coast of Fiji, the Tagula barrier reef, Borneo barrier reef, and the Caledonia and Palau Island barrier reefs are a few good examples. Recent satellite studies have provided much information on the Greater Barrier Reef. Apparently, the present form of it is a continuation of the deep substrate which existed during the glacial times, about 6,000 years ago, when the sea level was about 100 metres below the present level. The Great Barrier Reef extends from the Gulf of Papua to the Tropic of Capricorn, and beyond. It consists of a complex series of reefs, and covers some 230,000 km^2 and occupies about 8% of the continental shelf. Others are found off the Bahamas and the tropical western Atlantic.

Atoll Reefs: an atoll is nearly a ring-shaped reef, i.e. a group of coral reefs formed in a complete or incomplete circle and thus enclosing a body of water with a diameter of about 30 km. It is thought to have formed on the top of small flat surfaced volcanic elevations. As the corals grew, the bottom subsided at the same rate as the growth of the coral reefs and hence the theory of subsidence as proposed by Darwin (1842). Atoll reefs are common in the Indian Ocean, especially around Ceylon, the Maldives, Nicobar, and various other islands, and in the Pacific around the north-east of Australia, Kiribati, Gilbert, and many other islands and in the China Sea.

5.11   Coral Reef Ecosystem

Because coral reefs provide a most suitable substrate to many sedentary organisms, food and shelter for mobile and active predators, there is an intricate relationship between the diverse groups of animals and the coral reef environment. Coral reefs are some of the most productive sites within the marine environment. This is highly significant and the overall productivity far exceeds 1,200 gC m^{-2} y^{-1} per year in a relatively nutrient depleted area. The enhanced productivity in a relatively impoverished environment appears largely due to an increased surface area of the coral tissue and, consequently, an increased amount of zooxanthellae or other related filamentous or unicellular photosynthetic algae. The blue-green algae, in particular *Calothrix crustacean*, help to fix the free atmospheric nitrogen which, in turn, is consumed by certain first trophic herbivorous organisms, including fish, and their excretory products find their way over the coral reefs by circulation. Other essential nutrients such as phosphates are also obtained through the recycling process. The corals secrete large quantities of lime deposits which are added to the process of reef-building. The coral ecosystem has been considerably damaged by global warming. In the recent conference held at Berlin (Germany) in March 1995, with one hundred and ten large and small nations, substantial evidence has emerged concerning the adverse effects of corals, especially many tropical islands, including the Maldive Islands in the Indian Ocean. 'Cynide fishing' for live-fish supply to very large commercial restaurants and export to Hong Kong, the Philippines and Indonesia, has brought about a devastating effect on the coral growth, and their natural habitats have already been destroyed. There are other potential threats to reef ecosystems by human activities, especially tourism, and it is also thought that the terrigenous input of silt has increased and that the corals are less able to clear the silt load. Mass mortality of living corals has been reported. The living polyps are eaten by the starfish, *Acanthaster planci*, resulting in a reduction in coral reef formation. Coral reef productivity is very high – 300-500 g $Cm^{-2} y^{-1}$ – when compared with surrounding waters which have an average productivity of 20-40 g C $m^{-2} y^{-1}$.

5.12 Economic Importance of Coral Reefs

Economically, coral reefs can provide excellent tourist attractions, for underwater photography, bird watching, angling, and trawling for prawns. Corals also are of industrial use for building materials. The reef can be cut into rectangular blocks, 'free stones', which can be sun dried and used in the construction of houses in the Tropics. Owing to the porous nature of the coral stones, they have a very cooling effect. Coral stones are also used in roads and air field constructions. Coral lime stones are burnt down to fine powder form and used as valuable cement. There are many species of fish that live on coral reefs. Fish of economic value include many species of sharks, rays, panulirid lobsters and crabs, and a variety of bony fish, small and large, of Indo-Pacific areas. Others are very colourful and of ornamental values (see Chapter Nine on necton, including fish). Cephalapods, sea urchins, sea slugs and holothurians are fished for Chinese delicacy, in addition to trepangs, oysters, pearl oysters, turtles, and many species of sponges. Coral reefs can also be hazardous to navigation and a vast number of ships have been wrecked in unmarked areas, especially in the China Sea and Coral Sea and some parts of the Great Barrier Reefs of Australia.

5.13 Potential Hazards to Coral Reefs

(1) Being sedentary organisms of tropical warm waters, they grow well at a mean optimal temperature of 23–28°C. With the exception of a very few hermatypic corals, a fall in temperature below 18°C usually affects the coral growth.

(2) Reefs can be considerably reduced at sites where a fresh supply of nutrients is lacking.

(3) At times, when phytoplankton blooms red tides can inhibit coral growth, though temporarily.

(4) Wave action due to wind-driven forces, such as hurricanes, can wreck coral colonies and cause considerable damage.

(5) Efflux of freshwater from river mouths and silt generally inhibits coral growth.

(6) Suitable substrates and optimal depth can favour coral growth.

(7) Man-made ecological disasters such as radioactive wastes and nuclear tests can wipe out coral growth.

References

Barnes, D. J., (ed.) *Perspectives on coral reefs*, Canberra, Brain Clauston Publishers, 1983, p.277

Bernard, H. M., 'The genus Turbinada, the genus Astreporea' (Catalogue of the Madreoporarian corals in the British Museum) *Nat. Hist.*, 2, 1896, pp.1–106

Darwin, C. R., *The structure and distribution of coral reefs*, Berkeley, University of California Press, 1942, p.214

Daly, R. A., 'Pleistocene glaciation and the coral reef problem', *Amer. Jour. Sci.* (ser. 4), 30, 1910, pp.297–308

Hyman, L. H., *Protozoa through ctenophore*, New York, London, McGraw Hill Book Company, 1940, p.726

Goreau, T. F., 'The Ecology of Jamaican coral reefs', 1. Species composition and zonation, *Ecology*, 40, 1957, pp.67–90

Goreau, T. F. and Goreau, N. T., 'Life in the Sea', in *Scientific American*, San Francisco, W. H. Freeman and Company, 1980,pp.130–140

Govindaha Pillai, C. S. and M. I. Patel, 'Scleractinian corals from the Gulf of Cutch', J. Mar. Biol. Ass. India, 30 (1 and 2), 1988, pp.54, 74

Jackson, J. B. and Huges, T. P., Adaptive strategies of coral reef invertebrates, *American Scientist*, 73, 1985, pp.265–274

Newell, N. D., 'The evolution of reefs', *Scientific American*, 1972, pp.901

Odum, H. T. and Odum, E. P., 'Trophic structure and productivity of windward coral reef community of Eniwetok Atoll', *Ecological monographs*, 25, 1955, pp.381–410.

Reader's Digest Book, (ed. Mead and Beakett,), *The Great Barrier Reef*, Reader's Digest Publishers, Sydney, 1984, p.384

Rosen, B. R., 'Reef coral biogeography and climate through the late Cenozoic: just islands in the sun or critical patterns of islands', in *Fossil and Climate*, (ed. Benchley), New York, Wiley, 1984, pp.201–259

Russel-Hunter, W. D., *Life of Invertebrates*, New York, MacMillan Publishing Co. Inc., London, Collier MacMillan Publishers, 1979, p.650

Sale, P. F., 'Co-existence of coral reef fish – a lottery for living space', *Env. Biol. Fish.*, 3, 1978, pp.85–102

Scheer, G. 'The distribution of reef corals in Indian Ocean with a historical review of investigation', *Deep Sea Research*, Pt. A 31, (6–8), 1905, pp.885–900

Veron, J. E. N., *Corals of Australia and the Indo-Pacific*, NSW Australia, Angus and Robertson Publishers, 1986, p.644

Wilkinsion, C. R., 'Nutrient translocation from symbiotic cyanobacteria to coral reef sponges', in (ed. C. Levi and N. Boury-Esnault), *Biologiedesspongiaires*, Colloques Internationaux du Centre National de la Recherche Scientifique, no291, 1979, pp.373–380

Young, C. M., 'The biology of coral reefs', *Advances in Marine Biology*, 1, 1963, pp.209–260

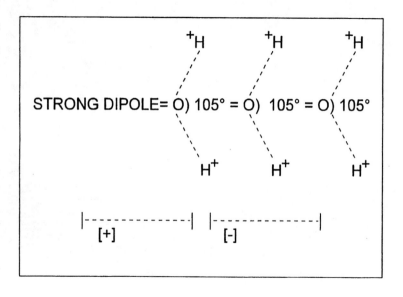

Figure 6.1   In an electrical field, water molecules line up with their negative ends facing positive electrode and their positive ends to the negative electrode.

Chapter Six
Sea Water

6.1 Chemical Nature of Sea Water

In view of understanding more detailed tropical marine environment, the knowledge of sea water is extremely important, particularly in relation to biological processing. All marine organisms need to adjust through evolution to adapt to the interactions of subtle range of physical, chemical, geological and biological conditions. The sea has remained a suitable environment to a diverse groups of organisms, since the beginning of life on this planet. Sea water contains two major components – the water and the salt dissolved in it. Much of the nature of the chemical composition of sea water is determined by the dissolved substances of terrestrial origin. The great rivers' run-off and coastal erosion are more common and widespread to the Tropics than to temperate regions. Heat is exchanged between the sea and the air. The heating of the surface water in Tropics often tends to form stratification. The sporadic seasonal upwellings in many isolated areas within the Tropics can increase not only the productivity, but also can vary the concentration of normally available essential nutrients of phosphates and nitrates following the phytoplankton bloom. In this section some essential features, as they pertain to the tropical and subtropical regions, are emphasised.

Sea water is a complex aqueous solution of salts in certain natural proportions. Due to the continuous flux from the land to the sea, it carries products of the earth's crust. The sea water contains almost all the known natural elements of the biosphere dissolved in it. Table 6.1 lists the known dissolved chemical species of sea water. It also contains a certain amount of organic matter derived essentially from decomposition and subsequent dissociation of the dead organisms and from the excretory products of the living plants and animals. These may be found in dissolved form or in particulate colloidal suspension. The minor elements, though minute in concentrations, are also important as they are reactive in organic processing in the sea. Before proceeding with the discussion of sea water, it may be helpful to know a little more about the nature of water as a solvent for most ionic substances. Water is a covalently bonded substance containing two hydrogen-oxygen bonds. The asymmetrical structural arrangement of the water molecule plays a unique role in all living systems. The three atoms of the water molecule – one anionic $O^=$ and two cationic H^+– remain together sustaining an angle of about 105° between the bonds and resulting in a strong dipolar nature as can be seen in Fig. 6.1.

Table 6.1 All the Known Natural Chemical Elements of the Sea
(Atomic masses in this table are based on the atomic mass of carbon-12 being exactly 12.)

Name	Symbol	Atomic Number	Atomic† Mass	Name	Symbol	Atomic Number	Atomic Mass
Actinium	Ac	89	(227)	Gadolinium	Gd	64	157.2
Aluminium	Al	13	27.0	Gallium	Ga	31	69.7
Americium	Am	95	(243)	Germanium	Ge	32	72.6
Antimony	Sb	51	121.8	Gold	Au	79	197.0
Argon	Ar	18	39.9	Hafnium	Hf	72	178.5
Arsenic	As	33	74.9	Helium	He	2	4.00
Astatine	At	85	(210)	Holmium	Ho	67	164.9
Barium	Ba	56	137.3	Hydrogen	H	1	1.008
Berkelium	Bk	97	(247)	Indium	In	49	114.8
Beryllium	Be	4	9.01	Iodine	I	53	126.9
Bismuth	Bi	83	209.0	Iridium	Ir	77	192.2
Boron	B	5	10.8	Iron	Fe	26	55.8
Bromine	Br	35	79.9	Krypton	Kr	36	83.8
Cadmium	Cd	48	112.4	Lanthanum	La	57	138.9
Calcium	Ca	20	40.1	Lawrencium	Lr	103	(256)
Californium	Cf	98	(251)	Lead	Pb	82	207.2
Carbon	C	6	12.01	Lithium	Li	3	6.94
Cerium	Ce	58	140.1	Lutetium	Lu	71	175.0
Cesium	Cs	55	132.9	Magnesium	Mg	12	24.3
Chlorine	Cl	17	35.5	Manganese	Mn	25	54.9
Chromium	Cr	24	52.0	Mendelevium	Md	101	(258)
Cobalt	Co	27	58.9	Mercury	Hg	80	200.6
Copper	Cu	29	63.5	Molybdenum	MO	42	95.9
Curium	Cm	96	(247)	Neodymium	Nd	60	144.2
Dysprosium	Dy	66	162.5	Neon	Ne	10	20.2
Einsteinium	Es	99	(254)	Neptunium	Np	93	(237) -
Erbium	Er	68	167.3	Nickel	Ni	28	58.7
Europium	Eu	63	152.0	Niobium	Nb	41	92.9
Fermium	Fm	100	(257)	Nitrogen	N	7	14.01
Fluorine	F	9	19.0	Nobelium	No	102	(255)
Francium	Fr	87	(223)	Osmium	Os	76	190.2

Oxygen	O	8	16.00	Sulphur	S	16	32.1
Palladium	Pd	46	106.4	Tantalum	Ta	73	180.9
Phosphorus	P	15	31.0	Technetium	Tc	43	(97)
Platinum	Pt	78	195.1	Tellurium	Te	52	127.6
Plutonium	Pu	94	(244)	Terbium	Tb	65	158.9
Polonium	Po	84	(210)	Thallium'	Ti	81	204.4
Potassium	K	-19	39,1	Thorium	Th	90	232.0
Praseodymium	Pr	59	140.9	Thulium	Tm	69	168.9
Promethium	Pm	61	(145)	Tin	Sn	50	118.7
Protactinium	Pa	91	(231)	Titanium	Ti	22	47.9
Radium	Ra	88	(226)	Tungsten	W	74	183.9
Radon	Rn	86	(222)	Unnilhexium	Unh	106	(263)
Rhenium	Re	75	186.2	Unnilpentium	Unp	105	(262)
Rhodium	Rh	45	102.9	Unnilquadium	Unq	104	(261)
Rubidium	Rb	37	85.5	Unnilseptium	Uns	107	(267)
Ruthenium	Ru	44	101.1	Uranium	U	92	1238.0
Samarium	Sm	62	150.4	Vanadium	V	23	50.9
Scandium	Sc	21	45.0	Xenon	Xe	54	131.3
Selenium	Se	34	79.0	Ytterbium	Yb	70	173.0
Silicon	Si	14	28.1	Yttrium	Y	39	88.9
Silver	Ag	47	107.9	Zinc	Zn	30	65.4
Sodium	Na	11	23.0	Zirconium	Zr	40	91.2
Strontium	Sr	38	87.6				

† Numbers in parentheses give the mass number of the stable isotope. More precise values of atomic masses are given in Appendix I.

Because of this dielectric property, H_2O molecules tend to cluster around a charged particle. The distinct effect is that a significant force develops that causes the attraction of molecules of water. Thus water, with its high specific heat, provides an excellent heat source, and the high values of latent heat during fusion and evaporation, make both the freezing and evaporation more difficult. Hydrogen bonding is also responsible for the physical property of unusual expansion as water freezes. Water is slightly compressible, but more so at great pressures. This can affect the density. The maximum density of water reaches $+4°C$ and therefore ice is less dense than water and hence it floats. Some of the other physical properties displayed by water other than compressibility are: thermal conductivity, surface tension, viscosity, light absorption, sound absorption, raising of boiling point, lowering freezing point, etc. As a solvent, water is capable of dissolving salts and absorbing the atmospheric gases, especially CO_2 which, in the Tropics, becomes supersaturated as lime and can reach nearly 300%. This is essential for coral growth and foraminefera. Increase in salt contents in water will also contribute to the osmotic pressure might limit the capacity of living organisms that inhabit the marine and other salt water environments. Oceans of the world contain over 5×10^{16} metric tons of salt. Chemically, on average, the natural sea water is composed of about 96.50% of pure water and the remaining 3.50% is represented by inorganic salts, organic products, and dissolved gases. In this small amount of salt from sea water, at least ten known major ions: cations – sodium, magnesium, calcium, potassium, and strongium; and anions – chloride, bromide, sulphate, bicarbonate, and carbonate, together with undissociated boric acid, are represented, thus making up 99.95% of the total salt in solution. The principal constituents of natural sea water are listed in Table 6.2. The most remarkable property of sea water is the relative proportions of the major constituents which are independent of salinity, irrespective of geographical variations. The analysis of a collection of 77 classical samples of the *Challenger* expedition (Ditmar, 1884) forms the basis of this relative proportion of the major constituents of the sea water; i.e. although the concentration of the individual ions varies quite considerably in different parts of the sea, their proportions relating to one another remain remarkably constant, especially in offshore waters, over a long span of geological time.

Table 6.2   Major Salt Contents of Sea Water

IONS/MOLECULES	CONCENTRATION‰	TOTAL SALT %
CATIONS:		
Na^+	10.547	30.60
Mg^{++}	1.272	3.69
Ca^{++}	0.400	1.16
K^+	0.379	1.10
Sr^{++}	0.012	0.03
ANIONS:		
Cl^-	18.980	55.07
SO_4^{--}	2.646	7.68
HCO_3^-	0.141	0.41
Br_3^-	0.065	0.19
HBO_3^*	0.024	0.07
TOTAL	34.465	99.95

* Boric acid is undissociated

On the contrary, the important minor constituents show marked variations and they hardly comply with the law of relative proportions because of the selective preference for them by both plants, especially phytoplankton, and animals.

Besides these major and minor constituents, there are many other organic substances known to have been detected in the normal sea water. Some of these are toxic 'ectocrine' metabolites, organic acids, carbohydrates, proteins, amino acids, vitamins, hormones, hydrocarbons, sterols, lipids and enzymes.

6.2   Salinity, Chlorinity and Chlorosity

The total salt content of sea water, which is significantly variable, is generally understood as salinity.   Conventionally, the salinity is defined as the total salt content in grams dissolved in 1 kilogram of sea water, and is universally expressed as g/kg and the symbol for parts per thousand is ‰.   Theoretically speaking, this definition assumes that all carbonates and organic matter have been oxidised and bromides and iodides have been converted to chlorides when the salts have been dried to constant weight at 480°C.

Chlorinity was defined as the mass of chlorine in grams equivalent to the total halogens contained in 1kg of sea water.   The total mass of halogens can be estimated, with considerable precision, by precipitating silver halides.   However, since first defining chlorinity, the accepted atomic weights of chlorine and silver have changed somewhat and chlorinity has been revised in terms of the amount of silver precipitated and to make chlorine independent of any changes in the atomic weights.   Therefore, chlorinity is redefined as the mass in grams of pure silver necessary to precipitate the total halogens in 0.3285234 kg of sea water.   Occasionally, the total concentrations of halides in sea water are expressed as chlorosity.   It is defined as the number of grams of chloride plus chloride equivalent to the other halides contained in one litre of sea water at 20°C.   Because of the relative proportions of the major ionic species in natural sea water being constant, Knudsen et al., (1902), in examining the relationship between density, salinity and chlorinity in a number of sea water samples of geographically different localities, found a linear relationship between salinity and chlorinity.   From the measurement of chlorinity, the salinity can be calculated.   The empirical relationship between salinity (S) and chlorinity (Cl) is that salinity equals 0.03

more than 1.805 times the chlorinity, and the equation is written as follows:

$$S\%o = 0.03 + (1.805 \times Cl\%o)$$

However, for most practical purposes, it is desirable to determine the salinity of sea water directly, and the details of the procedure are treated in the Appendix I. Since sea water contains ions, it is also a good conductor of electricity. When salinities are determined by the electrical conductivity method, an international convention is now adopted and the chlorinity-salinity relation is formulated as:

$$S\%o = (1.806655 \times Cl)$$

6.3 Importance of the Chemical Evolution

One of the basic requirements of life having evolved on earth was the availability of a liquid medium in the form of sea water. Water, being a 'universal solvent', is capable of interacting with a number of other molecules surrounding it, and is rarely found in nature in a 'pure' state. This capacity probably resulted in the complex process of evolution. Life on earth is inevitably bound to water. All living matter contains water. It is the major component of all living organisms. To survive, every organism needs water. Our own body contains nearly 60% of water by weight; and that of a jelly fish constitutes some 98% of water. Water is an indispensable source of hydrogen ions and the transference of which generates ATP in all living systems. Fortunately, water is a renewable resource and, like many other renewable resources of biological origins, it can be recycled. This recycling process in nature started in the sea itself and has gone on for many millions of years profoundly modifying the surface of the earth. Marine organisms exist in a salt water environment, and there is a remarkable similarity in chemical composition between sea water and the body fluids of extremely diverse life forms. The constant interactions between the organisms and the non living environment constitutes an ecosystem. In nature, water is continually recycled. As explained in the outset of this chapter, this process involves heat balance, particularly along tropical and subtropical regions, between 0° and 35° north and south of the Equator. As a result, the distribution of plants and animals is largely determined by the changes that are associated with the marine environment. However, in the tropical and subtropical seas, except in areas of upwelling, the vertical mixing of the water masses is limited and enormous quantities of nutrients remain at great depths compared with higher latitudes.

6.4 Compositon of Sea Water and Origin of Salts

The sea is infinitely vast. The huge body of water is in a state of continuous motion. As mentioned previously, several salts were already dissolved in the primeval sea as a result of the condensation of the primitive atmosphere. In addition, due to the violent volcanic activities of the solid earth, many of the known elements found in the earth's crust would have been continually added to enrich the salt contents of the sea water. For many of the biological purposes this total salt content of the sea is important.

6.5 Characteristics of Tropical and Subtropical Sea Water and Distribution of Salinity

Seas are never at rest. Many factors operate to set the sea water in constant motion. Some of the processes which influence the spatial and temporal variations in tropical sea water characteristics are: intense volcanic activities, transport of sediments and dissolved materials, advective flow, diffusion, heat transfer, evaporation, precipitation, temperature, salinity, density, monsoons, circulation and, in some cases, horizontal stratification of sea water and bottom topography of the sea.

6.6 Salinity Variation

Salinity varies quite considerably regionally and, to a moderate extent, seasonally. The variation of salinity is due to a number of factors such as climatic changes, freshwater run-off, river influx, influence of increased or decreased precipitation or evaporation, surface mixing or subsurface circulation, freezing or melting of ice, isolation by land barriers and differences in latitudes. The normal salinity of most oceanic waters is about 34.70‰. The average salinity in the tropical sea waters is about 36‰, except in the Bay of Bengal where the salinity is about 30‰ because of the often unpredictable monsoons and sporadic influx of the river Ganges; other areas where salinities

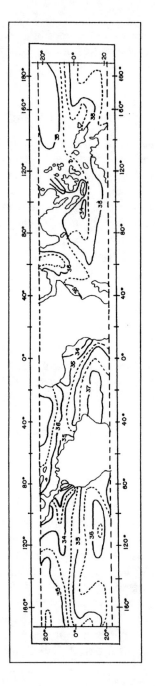

Figure 6.2   The distribution of surface salinity within the tropical belt.

are frequently high, reaching 36–37‰, are the north-eastern coast of Brazil and the Sargasso Sea, where precipitation is relatively low and evaporation is high. Some areas in the world have two other extremes. Abnormally high salinities are found, for example, in the Red Sea and the Persian Gulf (see Table 6.3 below), where evaporation is greatest especially when the sea is surrounded by semiarid land masses and circulation is restricted with open sea, reaching above 41‰. On the other hand, in the Arctic and Antarctic waters, where there is a considerable influx of freshwater from the melting ice and from rivers, continual snow and precipitation, the salinity is surprisingly low – about 28‰. For similar reasons, the water in the Baltic Sea is nearly brackish with salinity often reduced to 7‰. Fluctuations of salinity also occur in coastal waters, sheltered bays, estuaries, lagoons, large harbours, ports and river mouths and deltas compared with open sea where variation is so little. Depth-related salinity variation too may often occur; at the bottom depressions of the Red Sea, where the mineral salts are directly derived from the underlying crustal layer, the salinity is exceptionally higher, often reaching up to 265‰. The higher salinity is usually accompanied by a higher temperature of up to 60°C. Usually the surface salinity is higher than at great depths as seen in polar regions where the much diluted cold surface water may be carried to deeper layers. Depth-related salinity variations can be frequently observed in large tropical river mouths or estuaries where the sea water enters under tidal flow, forming vertical as well as horizontal gradients. The distribution of the normal range of surface salinity within the tropical belt is shown in Fig. 6.2.

6.7 Salinity and Temperature

Although the mean salinity and the temperature of the global sea are around 34.71‰ and 3.5°C respectively, maximum variations often occur within the tropical and subtropical regions. Within these low latitudinal zones, the lowest values of salinity occur along the equatorial belt (see Fig. 6.2), but increase to a maximum of 37‰, particularly in the Atlantic (north-eastern coast of Brazil) though lower in the Indian Ocean where the salinity varies from 33 to 36‰ on either side of the equator; being highest in the Red Sea exceeding 40‰ and the Pacific equatorial belt shows relatively less than the

Atlantic but higher than the Indian Ocean. The mean value of the surface salinity for the whole tropical and subtropical zone between 30° south and north is about 34.61‰. Some of the major factors which influence the salinity variations in the tropical and subtropical zones are high evaporation, precipitation, turbulence, high salt solution in the bottom layers, especially along the Gulf of Eden and the Suez Canal, coastal effluence and vertical mixing. The variations of salinity and temperature in geographically different tropical marine environment are summarised in Table 6.3.

Table 6.3   The Average Salinity and Temperature within the Tropical and Subtropical Marine Environment

Major Oceans and Tropical and Subtropical Geographical Locations	Salinity ‰	Average Temperature °C	Depth m
MAJOR OCEANS			
Atlantic	35.50	17.80	0–200
Indian	34.76	24.57	
Pacific	34.62	20.83	
MINOR SEAS			
Andaman sea	33.50	24.00	
Arabian sea	36.50	24.00	
Gulf of Oman	37.82	30.85	
Persian Gulf	42.00	32.70	
Red sea	40.00	32.05	
Mediterranean	38.50	12.00	
Gulf of Mannar	33.25	24.40	
Gulf of California	35.56	23.10	
Marina Trench	34.67	02.48	10,000

6.8   Oceanic Boundaries within the Tropics and Subtropics

As pointed out in Chapter Four, the oceanic boundaries are difficult to define because of the continuous stretch of water masses. However, the existence of continents, islands and the bottom submarine ridges may guide to divide the major oceans to some extent, even though only arbitrarily. For example, a line parallel to meridian about 20°E and passing through Cape Agulhas may be considered as the boundary between the Atlantic and Indian Oceans; the meridian through the South Cape and Tasmania of about 147°E may be recognised as the boundary between the Indian and Pacific Oceans; and the boundary between the Pacific and Atlantic may be formed by the line going from Cape Horn through the south Shetland Islands and Bering Straits. Overall, the Pacific is the largest in size, next the Atlantic, and the last the Indian Ocean. The dimensions of the major oceans are shown in Table 6.4. The limits of the tropical and subtropical have already been defined in Chapter Four.

Table 6.4 Dimensions of the major oceans

Oceans	Surface areas		Average	
	km^2	%	Depth (m)	Volume (km^3)
Atlantic	80,784,050	22.37	3,678	297,123,730
Pacific	162,615,370	45.03	4,059	660,055,770
Indian	72,261,459	20.01	3,952	285,772,800
Antarctic	31,670,814	8.77	3,720	117,815,430
Arctic	13,795,041	3.82	1,523	21,009,848
TOTAL	361,126,730	100.00	3,386	1,381,777,600

References

Defant, A., *Physical Oceanography*, New York, Pergamon Press, 1961, vol. 1, p.729

Ditmar, W., 'Report on researches into the composition of ocean waters collected by *HMS Challenger*', Challenger Rept. Physics and Chem., 1, 1884, pp.1–251

Dorgain, M. M., and Mofta, M., 'Environmental conditions and phytoplankton distribution in the Arabian Gulf and Gulf of Oman', September 1986 Symposium, J. Mar. Biol. Ass. India. 31 (1 and 2), 1989, pp.36–53

Harvey, H. W., *The chemistry and fertility of sea waters*, Cambridge University Press, 1963, p.240

Knudsen, M. (ed.), Hydrological Tables, GCE GAD, Copenhagen, 1991, p.104

Johnston, R., 'On salinity and its estimation', *Oceanogr*, Mar. Biol. Rev. (ed. H. Barnes), 7, 1969, pp.31–48

Martin, D. F., *Marine Chemistry*, New York, Marcel Dekker, INC., 1968, vol. 1, p.280

Riley, J. P. and Chester, R., *Introduction to marine chemistry*, London and New York, Academic Press, 1971, p.65

Sverdrup, H. U., Johnson, M. W. and Fleming, R. H., *The oceans, their physics, chemistry, and general biology*, Bombay, London, New York, Asian Student Edition, Publishing House, 1961, p.1087

Chapter Seven
MarineResources

7.1   Potential Tropical Resources

Oceans cover vast areas and their boundaries are arbitrary, and access to marine resources is not monopolised by any single nation. However, because of advanced technology, which only a few rich nations can afford, these nations derive a fair share of the resources. The tropical countries used the sea for generations, mainly for fishing. The coastal nations are now becoming increasingly aware of the need to exploit other available natural resources of the sea. As already noted, the potential of the sea is enormous and the sea has been envisioned as an inexhaustible repository of resources. The vast marine items of food, energy, and minerals are derived from both the living and non-living resources of the sea.

7.2   Living Resources

The living resources broadly include plankton, pelagic, nectonic, and demersal fish. Apart from the other major groups of benthic shellfish, edible echinoderms, reptiles and mammals, including the largest whales there are a greater variety of coral resources since many animals live in association with coral and coral reefs as they provide excellent marine habitats, especially a number of species of coral fish, crustaceans, pearl oysters, molluscs of blue ringed octopus which, together with puffer fish, are used in extracting a costly toxic substance called tetradotoxin used in medical research; and turtles, and the edible, medicinal and non-edible seaweeds.

7.3   Non-Living Resources

On the other hand, the non-living resources consist of the rich reserves of fossil fuels such as coal deposits, petroleum and oil, natural gases, dissolved chemicals such as salts, iodide, bromide; and many other mineral substances mainly sand, silt, gravel, tin, gold and some rare minerals such as platinum, zirconium, gem stones and cerium, chromite, phosphorites, and others. The major offshore oil-producing tropical countries of the world today are: Saudi Arabia, Abu Dhabi, Qatar, Persian the Gulf, Nigeria, Ghana, the Gulf of Guinea, and many nations of Southeast Asia, particularly China and India, Thailand, Indonesia, the Philippines, Brunei, and South American countries of Mexico, Venezuela, Brazil and some Caribbean islands, apart from the tropical USA. Some of these countries have gas and seabed minerals too and they extend their exploration into the exclusive economic zones. The Red Sea area is very rich in mineralised muds. The living resources are renewable, i.e. for example, the species of plants and animals are renewable and are capable of regenerating themselves through reproduction, whereas the non-living resources cannot be replaced once they are used up, except in some cases where recycling is possible. The presence and prospects to retrieve these new minerals, especially gold and nodules of manganese, cobalt, and nickel from the seabeds, have often encouraged and stimulated many tropical coastal nations to resort to more sophisticated technology since these valuable items may subsidise their ambitious economic enterprises. There is a growing awareness of the importance of protecting and preserving mineral resources for the future among the nations. Massive deposits of marine mineral resources such as sulphides have been found in the Pacific, and relatively lesser manganous nodules, especially nickel, copper, and cobalt in Australia, Gabon, South Africa, and Brazil. In addition, many of the marine organisms are used as a source in medical and pharmaceutical industries; this is now a rapidly expanding industry. The puffer fish, porcupine fish, and sun fish, lampreys and trunk fish contain a very powerful neurotoxin called tetrodotoxin ($C_{11}H_{19}O_9N_9$) which is concentrated in their livers, intestines and reproductive organs. This biotoxin is used in neurophysiological research especially, to block the action potentials in nerves, synaptic transmission, and for neuromuscular junction.

7.4   Marine Ecosystem

As already noted, the largest ecosystem on this planet is the biosphere, and the oceans and seas can be considered as the largest of the aquatic ecosystem. The sea water and the sea-floor together comprise the abiotic (non-living) components of the environment. The whole organisms which live in them comprise the biotic (living) components. The natural interactions between the dynamic living and the complex non-living entities, where the energy flow is regulated and the nutrients are recycled by natural forces, are broadly considered as an ecosystem. However, this can be broken down into simpler functional subunits – for example, the marine organisms and the marine environment; the coral communities and the coral reefs; the mangroves and their marshy habitats, the organisms and the intertidal pools; the organisms at the interface between the sea water surface and the atmosphere; the areas where the sea has constant intercourse with the estuaries of the larger river mouths, lagoons, bays, gulfs and the endemic organisms; and the coastal organisms and the intertidal zones, etc. Thus, there is an inseparable and infinite continual exertion of influence between the living organisms and the non-living environment. Diversity of ecosystem (variety of habitats), is important to provide species diversity and for species to exist continuously through the availability of nutrients and other essential materials necessary for life. Globally, the marine habitats' destruction is on the rise, for example, the damage to coral reefs, the inexorable over-exploitation of fish beyond maximum sustainable yield could seriously disrupt the ecosystem and bring disastrous environmental impacts, as many tropical nations depend on fish as their basic source of protein-rich food. The current global estimate of potential sustainable yield of marine fish seems to be between 62 and 87×10^6 metric tons (Goodwin, 1990).

7.5   Marine Biotas and the Major Groups

Biota, the flora and fauna of a region, and their occurrence is limited primarily by light, temperature, available nutrients and other environmental conditions. The marine environment supports a rich variety of organisms of considerable antiquity. As pointed out earlier, life probably originated from the marine environment. Except bacteria and blue-green algae, which derived from the prokaryotes with no nuclear membrane, all the remaining forms of life evolved from eukaryotes with distinct membrane bound nuclei. Among animals, almost all the major phyla, except Amphibia, are represented in the marine environment; the phyla Cnidaria, Brachiapoda, Siphonculata, Chaetognatha and Echinodermata, together with Tunicates and Urochordates of the Protochordata, are exclusively marine.

Marine life is quite diverse and very complex. So are the varied ecosystems of the tropical waters of coastal, offshore, deep sea, coral and mangrove areas. Besides the non-living resources, the sea provides a variety of living organisms. Depending on their habits, the different habitats that they occupy, distribution, and ability to afford an efficient power of swimming, or locomotion by development of the neuromuscular system, the marine organisms can be grouped into three major and important categories of resources:

(1) Plankton.

(2) Necton.

(3) Benthos.

The plankton are the drifting organisms, the majority of which are microscopic; the nectons include the powerful swimmers such as fish, squid and whales; benthos embrace all the engybenthic and benthic organisms, for example, shellfish crustaceans, bivalves and many other bottom living organisms. In this context of diversity of habitats and distribution, the rooted seaweeds may be considered benthic, and the free floating seaweeds pelagic, irrespective of their origin. Since these major groups of biotas constitute the basis of the living resources and an important potential source of protein-rich food of the sea for the tropical regions, they merit individual and detailed discussions and will be dealt with shortly (see Chapters Seven, Eight, Nine and Ten).

References

Balakrishnan, N. N., et al., 'Ecology of Indian estuaries ecology and distribution of benthic macrofauna in the Ashatmudi estuary', Kerela. Mah. Bull. Nati. Inst. *Oceanogr.* 17, 1984, pp.89–11

FAO., (ed. Gulland, J. A.) *The fish resources of the ocean*, England, Fishing News (Books) Ltd., 1971, pp.255

Goodwin, J. R., 'Crisis in the world fisheries: People, Problems and Policies', California, Stanford University Press 1990, pp.155-160

Firth F. E., (ed.), *The encylopedia of marine resources*, New York, Van Nostrand Reinhold Company, 1969, p.740

Lonfhurat, A. R., Proceedings of a symposium on the oceanography and fisheries of the tropical Atlantic, Abjan, Ivory Coast Pads, UNESCO, Pads., 1969, pp.147-168

Moiseev, P. A., 1971. 'The living resources of the world ocean', Israel Programme for scientific translation, Jerusalem, p.334

Quasim, S. N., et al., 'Organic production in a tropical estuary' Proc. Indian Acad. Sci., 59, 1969, pp.51-94

Repetto. R., ed., 'The global possible resources, development, the new century', New Haven and London, Yale University Press, 1984, p.538

Singarajah, K. V., 'Hydrographic conditions, composition and distribution of plankton in relation to potential resources of Paraiba River Estuary', Rev. Nordest Biol., 1 (1), 1978, pp.125-144

Steele, J. H., 'Some problems of the study of marine resources', Spec. Publs. int. Commn. NW. Atlant. Fish., no.6, 1965, pp.463-476

UN., World Wide Resources, New York, Oxford University Press, 1994

Walford, L. A., *Living resources of the sea*, New York, Ronald Press, 1958, p.321

Weeks, L. G., 'The ocean resources', *Offshore*, 1968, pp.28, 39-48, 87-88

Wenk, E., 'Physical resources of the oceans', *Scientific American*, 221, 1969, pp.166-176

'World Resources 1994-1995', A report by the World Resources Institute in collaboration with the United Nations Environment Programme and the United Nations Development Programme, Oxford, New York, University Press, 1994, p.400

Chapter Eight
Plankton, Pleuston and Neuston

8.1 Plankton Concepts and Definitions

Among the biotas, plankton are the most numerous life forms in the sea. Plankton are ubiquitous and they play a very important role in the economy of the sea because their components are the direct source of basic food for most organisms of the major groups of nectons, including the largest baleen whales, engebenthos, benthos, pelagic and many other marine organisms. The study of plankton is called planktology; and one who specialises in planktology is a planktologist. Planktology is a vast and fascinating field of study and much work has been done in temperate waters, most of them in taxonomy, but relatively less is known about plankton in the tropical and subtropical waters, and published data is scanty and the need for more work becomes obvious.

The term plankton (=drifting), though a Greek word, was first applied by Victor Hensen, a German physiologist, in 1887, for those organisms, both plants and animals, which, in nature, passively drift or float or feebly swim and, left without greater control, are at the mercy of winds, waves, and currents. Some are certainly able to swim and are capable of moving vertically, but they are unable to resist the mobility of the water. Though this form of existence seems to be an effective device for dispersal, the mechanism is less economical as many of the plankton may perish or fall an easy prey to other predators long before they reach their destination of habitats or maturity.

8.2 Composition and the Bases for Classifying Plankton into Different Groups and Subgroups

The community of plankton as a whole is known as 'true' or euplankton and consists of representatives from most of the major and minor phyla, including the free floating eggs and early stages of larvae of necton, benthos and engebenthos. Contrastingly, the pseudo-plankton are dead organic-detrital-particulate materials which are frequently found in large amounts in the plankton collections, especially in the neritic waters. The word plankton cognisably implies both singular and plural in its wider usage. However, the planktonic individuals are called planktonts or plankters. Although they are minute in size, not all are microscopic, and there are several which fall out of this range; for example, some of the coelenterates such as *Cyanea* and *Physalia* – Portuguese man-of-war – whose tentacles often extend to over twelve metres, are also plankton, though some authors prefer to include them in the subgroup of pleuston (see below). Since plankton is a collective term for all such floating organisms, many of which are neither clearly plants nor animals, they will be more appropriate to be considered as 'Protists'; alternatively, however, though somewhat arbitrary, the most convenient and readily useful way of grouping plankton is based on the following features: mode of subsistence, size, developmental characteristics, habitats, and spatial and temporal distributions during their life cycles.

Authorities often differ on the question of the assortment of plants and animals into some meaningful and natural groups of even closely related species. This is largely because that pre-Darwin system of classification relied heavily on the fundamental external characteristics and hence the information sought to classify the animals and plants rationally is somewhat incomplete. There are not less than 5 million species of organisms living in the world, but at least 65% of them are in the tropical regions. Nevertheless, since Linneaeus (1745) introduced the binomial system of Latinised generic and specific names, just over 500,000 plants and 1,200,000 animals have been named. Of these, the marine diatoms and copepods alone represent some 12,000 and 10,000 species, respectively. Further advancement in knowledge of the marine realm will undoubtedly add several more new members to this continuing list of flora and fauna. Recent workers, however, seek other information and evidence from such studies of palaeontology, evolution, embryology, cytogenetics, comparative anatomy, biochemistry, physiology, behaviour, psychology, ecology, zoogeography, and even electrophoroses and scanning electron microscopy etc., to reinforce the similarities of characters of a given group of animals and plants. Yet, some are more difficult than others to analyse for patterns of morphological variations in living things, and plankton classification is no exception. For example, most of the planktonic organisms resemble one another, during their early stages of development, and perhaps the safest way of identifying them is to rely on the experience of the individual workers and also, where possible, to rear them in the laboratories until the adult features of most species are clearly noticeable and to make a comparative study of life histories.

8.3 Major Groups of Plankton

Firstly, the plankton can be broadly placed into two major natural groups, namely:

(1) Phytoplankon – Phyto (= plant) plankton.

(2) Zooplankton – Zoo (= animal) plankton.

Within these two great groups, however, many other subgroups based on a number of criteria occur and these subgroups will be considered in some detail now.

8.4 Sub-Groups of Plankton Based on Common Characteristics

Plankton are a whole assemblage of organisms, both plants and animals, related by some common characteristics, but their true systematic positions may differ considerably.

8.4.1 Size Criterion

Although most phytoplankton are microscopic in size, individually they are composed of a single cell or sometimes in loosely bound colonies. Zooplankton is composed of a greater part of micro and macro animal plankton. The plankton may be further subgrouped on the basis of their size as shown in Table 8.1.

Table 8.1 Sub-Group of Plankton by size criterion

SUB-GROUPS	SIZE CRITERIA		
(1) Ultrananoplankton or Picoplankton	>00μ	–	<02μ
(2) Ultraplankton	>02μ	–	<05μ
(3) Nanoplankton	>05μ	–	<20μ
(4) Microplankton	>20μ	–	<02mm
(5) Macroplankton	>02mm	–	<02cm
(6) Megaplankton	>02cm	–	<02m

8.4.2   Developmental Criterion

Depending on whether they spend the whole or part of their life in the pelagic realm, the plankton may also be subgrouped into: holoplankton (= permanently) – for example, the majority of copepods, and many other organisms of the major and minor phyla; meroplankton (= temporarily), for example, a great variety of benthic or nectonic larvae during their different series of developmental stages until they become adults.

8.4.3   Circumstantial Criterion

Some plankton which may be of benthic habitat, but diurnally pelagic are called hypoplankton, while others which are benthic but may be circumstantially swept off the bottom by current or tidal changes into the neritic zone are called tychoplankton.

8.4 4   Forms and Size

During functional evolution, many members of both phyto- and zooplankton have developed a variety of physical devices that help them to float or suspend themselves in water as adaptive mechanisms to the pelagic mode of life. Some possess spherical or round or discoid forms. The latter are more common among the diatoms and are called discoplankton. While others are relatively solid and rod- or cigar-shaped, called rhabdoplankton; and still others, with gelatinous and transparent bodies, are called physoplankton. Examples of these groups are considered towards the end of this chapter.

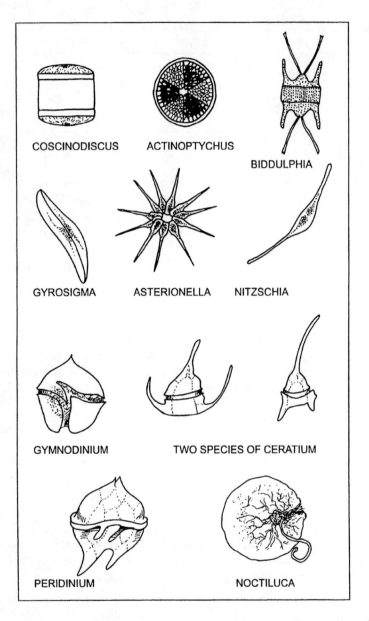

Figure 8.1   Some of the common planktonic species, mostly of
                  phytoplankton, the plant forms.

8.4.5   Regional or Seasonal Criterion

Plankton confined to neritic regions over the continental shelf are called neritic plankton, while those in the oceanic province are called oceanic plankton. However, there is a considerable degree of overlap between the two, though the zone of their separation may be recognised by some indicator species.

8.4.6   Subgroup Based on Depth Criterion

Plankton can also be grouped according to the depth-related zones that they occupy. In their appropriate environmental realms, either during diurnal migration of their holoplanktonic, or meroplanktonic existence, plankton are subgrouped into the following:

(1)   Epiplankton – which live in the epipelagic zone.

(2)   Mesoplankton – plankton of the mesopelagic zone.

(3)   Bathypelagic plankton – plankton of the bathypelagic zone.

(4)   Abyssal pelagic plankton – of abyssal pelagic zone.

(5)   Hadopelagic plankton – of hadopelagic zone (relatively recent discovery).

8.5   Phytoplankton

From a standpoint of the productivity of the sea, the phytoplankton are of immense importance. They are the primary producers and the basic link in the food cycles of all organisms. They alone are capable of elaborating the essential nutritional requirements such as carbohydrates, proteins, fats and minerals (see Chapter Twelve on primary production). The phytoplankton are microscopic, largely unicellular plants (Fig. 8.1), which are freely suspended at different levels of water mainly in the euphotic zone. They dominate virtually every plankton haul taken in the sea, using the finer nets, particularly in the coastal waters where nutrients are available in such abundance.

Although most of the phytoplanktonic organisms are unicellular, their cells contain sufficient amounts of chlorophyll and other pigments and hence they are autotrophic, i.e. they are capable of leading an independent life. Consequently, they are able to derive their energy requirements by photosynthesis, utilising the green pigments, chlorophyll, dissolved nutrients of the surrounding medium, and the free carbon dioxide, and the light energy that penetrates into the sea water. Thus, they are primary producers and constitute the bulk of the plankton. In most photosynthetic cells, the principal organic products of photosynthesis are carbohydrates. The overall reaction is represented as follows:

$$CO_2 + H_2O + light + (Chlorophyll) \Rightarrow (CH_2) + O_2 + H_2O$$

In this case, the principal limiting factor is the depth-specific light that penetrates, to be absorbed by the photosynthetic pigments of the phytoplankton. This limiting factor, however, varies depending on the transparency of the sea water of the regions and seasons, but extends to more than 180–200 metres, particularly in the tropical and the subtropical zones. Owing to the presence of chloroplasts, which contain different pigments, and when phytoplankton blooms, from time to time, in response to changes in local conditions, the surface of the sea often appears to be green, blue, or even occasionally red (see below, Red Tide 8.6.1).

There are some species of phytoplankton which depart from the totally independent forms, in the sense that they are auxotrophic, i.e. they lack the ability to synthesise certain of their own specific organic growth factors such as vitamin B_{12} complexes. Similarly, based on the mode of nutrition, some other species of the phytoplankton are considered heterotrophic. These are usually found in deeper waters, below the euphotic layer, where, besides their need for inorganic substances, including carbon dioxide, nutrients of nitrates, phosphates, sulphates and many other trace elements, they derive a wide range of dissolved or particulate organic metabolites, aside from amino acids and vitamins, from the environment. Despite these differences, the marine phytoplankton produce nearly 40% of the total earth's primary production (Golley, 1972).

Next to phytoplankton, the rest of the plankton consists of zooplankton and are either heterotrophic or holozoic, and for this reason they are consumers. The zooplankton are migratory and can control their levels of vertical distribution diurnally. Between these two major categories of autotrophics and heterotrophics lies a relatively small group of mixed nature of which some are autotrophic, while others are holozoic or heterotrophics as seen in dianoflagellates; these will be discussed in due course. Apart from these common characteristics, the tropical warm water plankton show some unique features inherent to regional differences and seasonal variations imposed by the natural barriers of continents, islands, ocean and seas. However, the distribution of tropical plankton is normally confined to the warm water band of 30°N and 30°S on either side of the equator. Other tropical planktonic characteristics will be discussed in the appropriate sections.

8.5.1   Main Components of Phytoplankton

Among the planktonic algae, nothing is more dominant in numerical abundance and widespread distribution, divergence of form and size, fine sculpture and beauty than the diatoms. Perhaps, for this reason, they were known as the 'jewels of the sea'. Diatoms are also of considerable economic, industrial, and medicinal importance. Many organisms feed directly on diatoms and are the first link in the food chain of the sea. Marine diatoms are the principal producers of the sea and constitute some 20–26% of the world's total primary production. The number of valid marine diatom species has been estimated to be between 10,000 and 12,000 (Hendy, 1964; Van Landingham, 1975), but now it seems that the list exceeds 16,000 species, and all of them are placed into the phylum Bacillariophyta. Because of the extremely minute size and the uniquely distinct characteristic features, the chief algal components of the marine phytoplankton can be divided into the following main divisions:

(1)   Diatoms.

(2)   Dianoflagellates.

(3)   Flagellates.

(4)   Coccolithophores.

(5)   Silicoflagellates.

(6)   Cryptomonads.

8.5.2   Diatoms

Diatoms are algae of the phylum Bacillariophyta. They may be either unicellular or colonial. For the purposes of identification and taxonomy, the structure and the arrangement of the pattern on the siliceous skeletal-wall or frustule are essential. The orderly disposed patterns are mainly of two types: one in which the patterns radiate

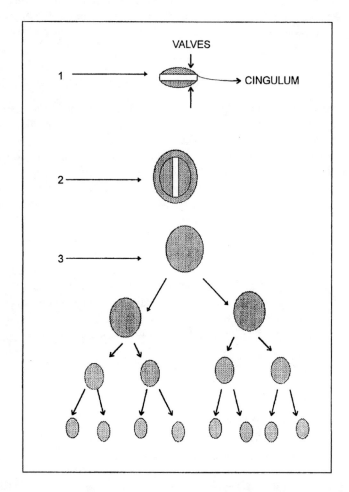

Figure 8.2   Reproduction in diatoms: 1. An adult cell with the cingulum facing the side; 2. The dividing cell where the two halves separate, each with the half-cell contents; and 3. The daughter cells (only one seen) divide and subdivide.

from a central point with reference to the round valve thus looking radially symmetrical, and in the other, the patterns radiate from an axial 'stria' on an elongated valve, hence providing semblance of a bilateral symmetry. These often facilate to place the diatoms into two main orders (see below). Most diatoms are planktonic and usually of the centric type. Others are benthic and pennate type, and the latter especially are of the intertidal zone. Diatoms are autotrophic and the chloroplasts contain chlorophyll a, c, and carotenoid pigments, together with fucoxanthin. The stellate masses of protoplasmic components of the diatoms are enclosed by cell walls, transformed into two rigid halves of silicified shells. The wall is usually heavily impregnated with silica (SiO_2nH_2O). The upper half or epitheca slightly overlaps the lower half or hypotheca, just like a closed pair of minute Petri dishes, where the bottom one fits snugly into the top one. The silicified skeletal halves, with or without the protoplasm, are also called 'frustules'. The periphery of the top and bottom rather flattened and round walls or 'valves' is encircled by the vertical walls, or 'bands'. These are also called cingulums or 'girdles' which help to secure the two halves together (Fig. 8.2).

The planktonic diatoms are non motile and the flagella are absent. Food reserves are in the form of oil droplets which, together with some tiny gas bubbles, secretion of mucus and extrusion of fine protoplasmic processes, and hairs or seta, may contribute substantially to their buoyancy (see below); perhaps a functional adaptation to increase the efficiency of photosynthesis by suspending themselves in the euphotic layer. Diatoms show distinctive regional dominance and the species are widespread in the tropical waters of Malay-Archipelagos, Indo-Pacific; Atlantic warmer waters, between the coasts of Africa and Americas. In brief, a few of the most common neritic species, though some are seasonal, are: *Actynoptypchus undulatus, Asterionella japonica, Biddulphia aurita, Biddulphia sinensis, Coscinodiscus centralis, Chaetoceros affinis, Chaetocerus curvisetum, Ditylum brightwellii, Gramatophora oceanica, Gyrosigma sp., Hemilaulus sinensis, Lauderia annulata, Navicula sp., Navicula lyra, Nitzschia closterium, Planktoniella sol, Pleurosigma naviculum, Rhizosolenia setigera, Skeletonema costatum, Thalassiosira subtlis,* etc.

8.5.3   Reproduction of Diatoms

Reproduction of diatoms are mainly of two types:

(1) Vegetative.

(2) Sexual reproduction.

Vegetative reproduction: prior to cell division the protoplast enlarges and the nucleus divides first mitotically, followed by the division of chloroplast along a transverse plane and parallel to the valve, and the large pyrenoid bodies also divide concurrently. The divided contents of protoplasts separate out as two halves, each carrying with it the hypotheca and epitheca, respectively. A new valve is soon secreted which, as a rule, becomes the hypotheca of each daughter cell, while the maternal half always becomes the epitheca, irrespective of its previous position, thus resulting in one line of descendants in vegetative division, and remaining the same size; whereas the other line becomes progressively smaller with each generation of division. The cell division continues until a minimum cell size is formed, and eventually may give rise to auxospores by the escape of the protoplast from the cell wall. A siliceous new membrane, 'perigonium', is now secreted around the escaped protoplasts. Periodically, however, these restore the rejuvenation and consequently upon germination they assume the normal size.

Sexual Reproduction is rare, and reproductive mechanism has been less understood in the past. With the advent of the electron microscope, the knowledge on the sexual reproductive process of the diatoms has been improved. Sexual reproduction essentially results in gametic fusion and genetic recombination and restoration of a much reduced cell size once again to the original size with vigour and dominance. Sexual reproduction in centric diatoms takes place by oogamy, in which, following meiotic division, the large non-motile female gamete or egg becomes fertilised by small flagellated motile male gamete or sperm.

During gametogenesis, the vegetative cells destined to become sperms divide repeatedly resulting in 2, 4, 8, 16, or 32 'depauperizing' mitosis, a term introduced by Stosch and Drebes (1964) for this type of mitotic division. The daughter cells thus produced remain within the mother frustule until meiosis is completed. By two successive meiotic divisions, accompanied by cytokinesis, each of the spermatocyte transforms into a potential sperm. By this time, the flagella are evident and the four unflagellated sperms thus formed escape from the mother frustule, leaving behind the residual bodies. With successive divisions, the cells diminish in size. For this reason, the developing daughter cells of the same diatom have been mistakenly identified as different species in the past. Periodically, however, spore formation restores the normal size of the diatoms. The concentric diatoms also produce resting spores to tide over the unfavourable conditions such as nutrient depletion, perhaps as a functional adaptation for the survival of the species.

Depending on the presence or absence of cytokinesis, following the second meiotic division, the gametogenesis is recognised as merogenous and hologenous respectively. In all diatoms studied so far the flagellar axis does not conform to the universal pattern of $9+2$, but instead shows a $9+0$ pattern, essentially lacking central filament (Heath and Dadey, 1972).

During the oogenesis, the oogonia develop directly from vegetative cells. The oogonium enlarges and divides twice meiotically, the first being accompanied by cytokinesis, but not the second, and thus resulting in two large spherical ova, while the others transform into degenerative pycnotic types. Other types of oogenesis, where only one egg and pycnotic bodies or one egg and polar bodies form, are also common. The pycnotic or polar bodies usually degenerate eventually.

Fertilisation is completed by the free swimming sperms by fusion of the sperm with oogonium. The entry of the sperm usually takes place at the extremely thin membrane spot where enzyme action can dissolve the membrane. As a rule, when the nuclear fusion is complete, the zygote or 'auxozygote' swells up while still within the fertilisation membrane or outside the membrane. Any subsequent division is by mitosis resulting in auxospores which develop new theca. The auxozygote eventually ruptures and the enlarged cells liberate. Auxospore formation varies from species to species; and involves, besides the normal type described above, pathenogenesis. Alternatively, purely vegetative mode occurs.

Sexual reproduction appears to be inhibited by high light intensity or continued illumination, but slightly reduced salinity seems to enhance the process. Nutritional deficiencies also affect the sexual reproduction (Steele, 1965).

8.5.4　Classification of Diatoms

A systematic classification of diatoms at present is far from satisfactory. Hendey (1964) placed the entire class of Bacillariophyceae into a single order of bacillariales, consisting of 22 families. On the basis of recent studies on discernible morphological details, symmetry, radial or bilateral patterns of sculpture of the cell valves and size, the diatom classification has become more complex. However, in spite of much controversy and incompleteness centring around the taxonomy of diatoms, they are now recognised as two main orders – namely, the centrales and pennales. Of the 22 families, recognised by Hendey as bacillariales, the first ten, in numerical order, correspond to Centrales, and the remaining twelve families to Pennales.

Centrales: this order includes centric diatoms, where the cells are mostly circular in valve view and with ornamentations of pits and striae, radiating from the central space. The cell walls are without raphae or pseudoraphae. The spontaneous movements are not apparent in these diatoms.

Pennales: this order includes pennate diatoms, where the cells are variable in shape from elongate, ova, rectangular, crescent or boat, cigar, sigmoid to wedge shaped. The valve sculpture is oriented in relation to a longitudinal raphae or pseudoraphae on one or both valves resulting in bilateral symmetry.

Diatoms occur from Jurassic to the present-day in great quantity. From a phylogenetic standpoint, the centrales are regarded as more primitive and an older group among the diatoms from which pennales probably have evolved in the late Cretaceous or early Tertiary time (Simonsen, 1979).

8.6　Dianoflagellates

Phylum dianoflagellata are a diverse group of unicellular organisms. Next to diatoms, dianoflagellates are the major and most important group of phytoplankton in the Tropics, in terms of numerical abundance, widespread distribution and beauty. They also serve as the fundamental food resource for many other organisms. Like the diatoms, with some exceptions, the dianoflagellates have well developed protective exoskeletal plates called 'thecae'. These protective coverings are usually composed of a mixture of cellulose and calcium compounds. The anterior region, or epicone of the body is covered by an epitheca, while the posterior part or hypocone is covered by hypotheca. The two are linked at the transverse groove by the girdle. On the basis of the thecae, they are roughly placed into two groups: the unarmoured or 'naked', and the armoured. The dianoflagellates differ from the diatoms in having two motile flagella, one is extensible and trails behind while the other remains usually coiled round the transverse or spiral 'groove'. Dianoflagellates are autotrophic, phagocytic or saprophytic, and a few are parasitic. Many of them have chlorophyll a, and c, in addition to beta carotene. Apart from these, the free living dianoflagellates have a distinct xanthophyll or peridin pigment. While some are heterotrophic and capable of ingesting other cells, many develop a symbiotic relationship with a variety of multicellular organisms such as coelenterates, particularly with corals, to form zooxanthellae (see Chapter Five on Corals and Coral Reefs).

Another interesting phenomenon associated with some of the dianoflagellates is that they are luminous. It is a common experience for those who have been on a boat to collect plankton in tropical waters, particularly during nights, to encounter the 'shine by night' organisms. If one dips his handkerchief on the surface of the sea water, he is bound to collect several hundreds of these organisms which produce bioluminescence. The most popular and well-known examples of Dianoflagellata that cause phosphorescence in the sea are *Noctiluca scintillans, Noctiluca miliaris*, the latter is more widespread even in temperate waters. Other forms which display the 'light by night' are *Ceratium tripose, Ceratium furca, Ceratium fusus, Ceratium massiliense, Ceratium trichoceros, Ceratium vultur, Peridinum depressum, Gonyaulax monocanha, Gonyaulax polyhedre, Gymnodinium flavumetc*, and among still others, *Prorocentrum micans* is cultured for feeding experimental bivalves etc. Dianoflagellates reproduce asexually by binary fission by dividing along the longitudinal axis of the cell. Sexual reproduction has also been reported in some species.

8.6.1 Phenomenon of Red Tide

One of the common and curious phenomena in the coastal areas of tropical seas is the occurrence of red tide, when dianoflagellates, under most favourable conditions, burst out in dense population and discolour vast areas of the sea surface – hence 'red tide'. During such phytoplanktonic blooms, the cells multiply rapidly, and at the same time they release a toxic substance into the surrounding water with drastic effects. In different regions of the world more than two species cause red tides. Within the inshore waters of the warm tropical belt, especially the south-west coast of India (Kerela), the Arabian Sea, south-west Africa, the north-eastern coast of Brazil, Peru, the Gulf of Mexico, southern California, Florida and Japan, dianoflagellates of the genus, *Gymnodinium,* and in other areas *Gonyaulax* dominate the surface plankton during red tide. Major outbreaks frequently occur during certain seasons and soon after heavy rain falls. It has been estimated that the cell concentration of *Gymnodinium* far exceeds 50,000,000 cells per litre during population explosion. The red tides often have a profound detrimental effect on a variety of fish, crustaceans, molluscs, and other marine organisms. They are well associated with mass mortality of fish populations. Many edible fish, especially shellfish, some of which are filter-feeders, fill their guts with these toxic phytoplankton. When these fish, in turn, are eaten by man, the toxic substance accumulates in the gut and causes gastrointestinal upset. The toxic substance eventually enters, through circulation, the central nervous system causing neurotoxic shellfish poisoning and finally becomes fatal. Intoxication from these organisms is known as paralytic shellfish poisoning. This condition is also called ichthyosarcotoxism, or 'cirguatra', a term originally derived from the West Indies. At times, during strong winds, respiratory irritation, sneezing, spasmic coughing, and a burning sensation in the eyes may result from inhibition of the neurotransmission of the diaphragm through the inhalation of the contaminated air from the red tide areas.

8.6.2   'Brown Tide'

Like the red tide, the phenomenon of brown tide also exists to a relatively lesser extent. This seems more common in temperate waters than in warmer tropical regions. However, it may be of considerable interest to observe the similarities and differences, briefly. Recently, a massive phytoplankton bloom in Long Island Marine Bay has been reported to have caused catastrophic effects on the commercially important bivalves, especially the scallop (*Agropectin*), the mussel (*Mytilus edulis*), and the oyster (*Crassostrea virginica*) populations. The brown tide has been associated with a small Chrysophyte microalgal species, *Aureococus anophagefferesns*, of the picoplankton. Currently, there is a general consensus that the nutrient levels on a global scale have been increasing in many marine coastal waters by rivering inputs throughout the world. Consequently, this phytoplankton bloom, with variation in the abundance and composition, appears to occur in direct response to changing environmental factors. NO_3 increased to some 32% during 1981, PO_4 level increased by 34%, and the ratio of N:P:S and Si:P in the sea was altered. During bloom, these micro-organisms have been reported to exceed 10×10^6/ml, resulting as the principal factor for the failure of the oyster and other related bivalves industry.

8.7   Coccolithophora

These are phytoplanktons, also called coccoliths, and are found in great abundance in the tropical and subtropical waters. They have variously shaped scale-like coverings formed of calcium carbonate; and are provided with two flagella. The fossils are important as indicators of the petroleum deposits. The species, *Coccolithus pelagicus*, is distributed in oceanic waters while many others are found in neritic waters.

8.8   Silicoflagellates

Silicoflagellates are unicellular planktonic organisms with a stellate tubular siliceous skeleton. The skeleton is usually covered by protoplasmic materials. A single flagellum is present. Relatively only a few living species exist; the genus *Dictyocha* is widely distributed. The fossil records are traceable as far back as the Upper Cretaceous period.

8.9   Cryptomonads

These are small, often oval shaped planktonic organisms. They have variously coloured chromoplasts and are typically biflagellate with the flagella rising from the gullet-like groves. They may be halophytic, holozoic or even saprophytic. Some marine species have been

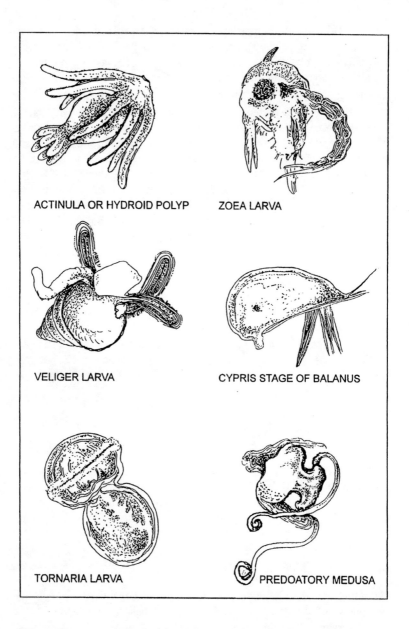

ACTINULA OR HYDROID POLYP   ZOEA LARVA

VELIGER LARVA   CYPRIS STAGE OF BALANUS

TORNARIA LARVA   PREDOATORY MEDUSA

Figure 8.3   Planktonic larvae often found floating with permanent plankton.

reported as symbiotic or parasitic on radiolarians. The genus *Cryptomonas* includes mostly chlorphyll-bearing species.

8.10 Phytoplankton Growth and the Growth Factors in the Tropics and Subtropics

In the tropical regions, where the solar energy is fairly constant throughout the year and there is relatively little seasonal variation of heat intensity, the phytoplankton growth continues all the year round. High production of the phytoplankton has been shown (Nielson, et al., 1957) in the shallower tropical coastal regions. Phytoplankton is generally abundant in the tropical and the subtropical zones, particularly more so near the coast within the neritic province than the oceanic province. The total primary production has been estimated by a number of sources to be 13.3 mg $C/cm^2/hr$ in the neritic province and 3.6 mg $C/cm^2/hr$ in the open ocean. Some of the factors affecting phytoplankton growth are:

(1) Concentration of nutrients.

(2) Illumination.

(3) Density stratification of water.

(4) Grazing by zooplankton.

(5) Turbulence.

(6) Mixing of water masses.

The availability of the nutrients depends on the regenerative and replenishing processes by current and coastal effluents. From a nutritional point, besides CO_2, nitrogen (as nitrate and ammonia), phosphate, and silicates, together with some trace elements and vitamins are of basic importance. It has been reported (Ryther and Dunstan, 1971 and other sources from kinetic experiments) that in the ocean, where the nutrients are relatively less, plankton have adapted to take up nutrients more efficiently. On the other hand, the inshore or neritic regions where the waters are rich in nutrients, plankton have developed the ability to take the dissolved nutrients less effectively, but in quantitative terms they take more nutrients.

Different regions have vastly different magnitudes of plankton production. The most highly productive regions in the Indian Ocean are those influenced by monsoons in the north-western part, the Bay of

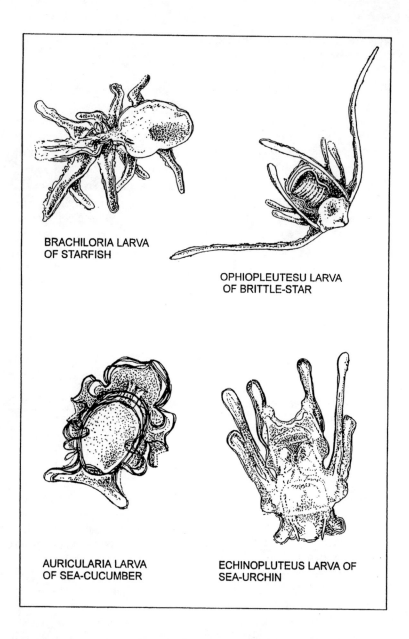

BRACHILORIA LARVA
OF STARFISH

OPHIOPLEUTESU LARVA
OF BRITTLE-STAR

AURICULARIA LARVA
OF SEA-CUCUMBER

ECHINOPLUTEUS LARVA OF
SEA-URCHIN

Figure 8.4   The larvae of echinoderms which lead meroplanktonic
existence until they settle down to the bottom.

Bengal, and waters of the Indonesian archipelagos. In regions of upwelling and of equatorial divergences, a high level primary production occurs. Because vast tropical oceanic areas have low primary productivity, there is relatively lower concentration of fish and invertebrates in these regions.

8.11   Zooplankton

Zooplankton are animal plankton; the second major group of plankton. Many of the zooplankton communities depend on the phytoplankton for their food. Next to phtytoplankton, zooplankton constitute the largest part of the planktonic community in the sea, and play a very significant role as both consumers of phytoplankton, and an important food link in the second trophic level. With some exceptions, the majority of zooplankton are herbivorous and microscopic within the range of nanomicroplankton size, (see Table 8.1 above). As already observed, zooplankton can be divided into two groups: temporary or meroplankton, which consist mainly of planktonic eggs and diverse forms of larvae of the pelagic necton, engybenthos and benthos; permanent or holoplankton, which live their complete life cycles in a planktonic state (Figs 8.3, 8.4, and 8.5).

Permanent plankton include representatives from most phyla of the animal kingdom. Among the protista, foraminifers and radiolarians are the most abundant and worldwide in distribution. The other dominant species well represented in the tropical warm waters are a variety of crustaceans. The important fraction of crustaceous zooplankton are the copepods. They are holoplankton and most of them play a significant role at second trophic level as herbivores. Copepods of the calanoida: *Calanus, Paracalanus, Calocalanus, Eucalanus, Centrophagus, Peleronemma, Temora, and Acatia;* and the harpacticoida: *Euterpina, Macrostela, Microsetella;* and cyclopoida: *Oithonia, Corycaeus, Sapphirina, and Copilia;* and the many other varied forms of larval stages of crustaceans such as prawns, shrimps, mysids, crabs, lobsters; flagellated larvae of sponges, hydroids, siphonophores and ptenophores; chaetognaths or 'arrow worms', trochophore larvae of annelids, especially *Tomopteris* of polychaetes, the trochophore and veliger larvae of mollusc, especially of bivales and the exclusively marine and most characteristic auricularia, bipinnaria, and plutei larvae of echninodermata, and tonaria larvae of

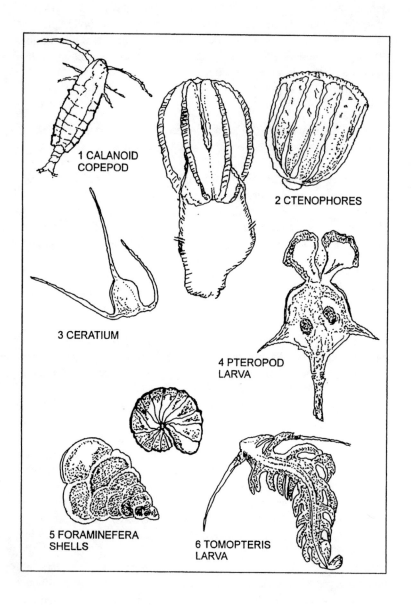

Figure 8.5   Some of the holoplankton or permanent plankton.

hemichordata are the most dominant and widely distributed members of the zooplankton. The copepods are also an important group of 'herbivorous grazers' and effectively control the phytoplankton population. The copepods form nearly 80% of the biomass among the herbivorous zooplankton. The environmental factors of salinity, depth and proximity of shore are important variables to the distribution of copepods. Most of the tropical copepods can tolerate the variations of salinity between 34‰ and 36‰, and temperature between 18°C and 32°C.

8.11.1   Foraminefera

Many foraminefera are planktonic protozoa, some are epizooic or live on sand and others are heterotrophic. Some are benthic and occupy characteristically different depths from tidal pools to the deep sea floor. Some benthonic foraminefera are of epiphytic nature dwelling on a variety of algal substrates of the infralittoral belt. The foramineferans have shells which are perforated. The shells are made up of organic substances reinforced with inorganic materials, especially calcium carbonate. The shells are often called tests which may be single or multi chambered and tend to be spiral (see Fig. 8.5). The cytoplasmic projections through the pores of the tests are used for a variety of functions such as locomotion, feeding and in the incorporation of inorganic materials in the construction of their tests. They have a complex life cycle with an alternation of generations involving haploid and diploid division. Some of the common examples are *Globigerina*, and *Elphidium*. The foraminefera are indicators of oil deposits. Their ooze and skeletal substances have contributed quite substantially to form the bulk of bottom sediments throughout the world. Because of their wide distribution, abundance and evolutionary changes the foramineferans offer an excellent record of the geological time since the Triassic Period, about 220 million years ago.

8.11.2   Radiolarians

Radiolarians are unicellular marine pelagic organisms found either singly or in colonies. The cells are usually spherical and covered by tiny perforated shells composed mainly of silica, calcium, strontium and sulphate. The protoplasm is generally differentiated into a central capsule enclosing nuclei and a peripheral capsule. The protoplasm extends from the inner capsule through the extra cellular capsule forming fine radiating pseudopodia as locomotory organelles. The radiolarians are holozoic or symbiotic often forming zooxanthellae. Reserve foods include oil droplets. Most common radiolarians are of the genera, *Thalasiocola, Collozoum,* and *Sphaerozoum.* The reproduction is asexual by binary fission, followed by sporulation into flagellated isogamates. The shells contribute to the bottom ooze.

8.11.3   Ciliates

The ciliates are planktonic protozoans represented mainly by tintinids; some of the important genera are those of *Tintinnopsis, Favella, Dictyocysta, Metacylis,* and several others.   Generally, both agglomerated (Tintinopsis-like) and non-agglomerated (Favella-like) species are more dominant in the neritic waters than in oceanic waters.

8.11.4   Coelenterates

Coelenterates, especially siphonophores, jelly fish, and the 'comb jellies' or ctenophores or sea walnuts are common both in tropical as well as in temperate waters.   The planula and actinula larvae of hydrozoa are commonly encountered in plankton collection.   Among the several ubiquitous tropical and subtropical species are the larvae of anthomedusae, *Bougainvillia,* leptomedusa larvae of *Obelia,* and the trachymedusae of *Clytia.*   The hydromedusae, though exclusively carnivorous, are well represented in the neritic waters of Indo-Pacific regions, especially tropical Australia, Thailand, Indonesia, along the coast of India, Arabian sea, east coast of Africa and Atlantic waters.

8.11.5   Chaetognaths

Chaetognaths are holoplanktonic and often known as the marine 'arrow worms'.   Although they are exclusively marine and worldwide in distribution, they constitute an important group of epiplankton virtually spreading out in great numbers from the littoral to oceanic waters of all the seas of equatorial Tropics, and subtropics from the surface to depths down to 1,000 metres and below.   Chaetognaths are bilaterally symmetrical and with elongated bodies, the overall lengths range from 30–100 millimetres. The body has a well formed head at the anterior end bearing a pair of eyes, a slender trunk, and a broad tail. Of the nearly seven genera in the tropical waters, the common genus, *Sagitta*, is more advanced and provided with two pairs of lateral fins, and adequately represented by several species. *Sagitta enflata, Sagitta pulcra, Sagitta decipens, Sagitta pacifica*, and some others are the common species in the epipelagic plankton of Indo-Pacific Oceans, but these species are significantly absent in the tropical Atlantic waters. The reason for this difference seems to lie in the fact that when the water masses flowing from the Pacific into the Bay of Bengal through the Malacca Strait, during north-east and south-west monsoons, the mixing process is reversed (Srinivasan, 1981). The other genera show only a single pair of lateral fins and are represented by the species, *Pterosagitta, Krohnitta, Heterokronia, Bathyspadella, Eukrohnia* and *Spadella*; the last is typically a benthic species dwelling usually attached by its tail to some substrate of rocks or algae. The chaetognaths are some voracious predatory zooplanktons preying principally on other planktonic organisms such as copepod, fish larvae, tunicate, and medusae larvae. They are extremely sensitive to changes in environmental conditions, including sensitivity to hydrostatic pressure (Singarajah, 1966). Chaetognaths are often considered as the biological indicators of oceanic waters of different origins.

8.11.6   Molluscs

Of the molluscs, the larvae of the various bivalves, especially of the oysters, pectens, mussels, clams, and many other gastropod larvae are meroplankton, while a variety of other molluscan larvae, and the wing-snails or pteropoda are holoplankton. The pelagic cephalopod larvae can be regarded as meroplankton.

8.11.7   Crustaceans

The most dominant, both in terms of numbers and species, are the crustaceans. Planktonic crustaceans are well represented by a great variety of larvae, juveniles and adults. The copepods constitute some 12% of the zooplankton and occupy an important systematic division among crustaceans. There are more than 1,000 species of planktonic copepods. Many more are benthic or parasitic copepods. They are economically important because the pelagic fishery, particularly herrings, depend on them. The euphasids, *Euphausia superba*, are the principal food of the baleen whales, especially in the Antarctic where they feed, but resort to the Tropics to breed. The zooplankton generally show the most variety in tropical and subtropical waters. The most numerous being crustaceans, coelenterates, molluscs, tunicates, fish larvae and eggs of a variety of organisms. The zooplankton concentration varies with diurnal and seasonal migrations which extend through several hundreds of metres of descent during dawn and ascent during dusk. The magnitude of diurnal changes in the quantitative and qualitative distribution of zooplankton varies in relation to physio-chemical environmental parameters. With some important exceptions, zooplankton generally increase during the night. Apart from some ecological factors, light seems to be the most important factor influencing the diurnal changes in the occurrence and abundance of the zooplankton. From a practical point of view it is important to know, in quantitative and qualitative analysis of plankton population, the differences due to night and day collections to avoid the effects of vertical migration, since many zooplankton, including larvae, undertake vertical movements (see below).

8.11.8   Ichthyoplankton

This is a vast topic and will be discussed only briefly. The fish larvae are collectively called ichthyoplankton. As already noted above, the meroplankton consists of a majority of larvae of true plankton themselves, benthos and necton. Depending on the habitats, feeding habits, and their duration of planktonic existence, the larvae can be of two types: planktrophic larvae, which spend more time in pelagic life; lecithotropic larvae which have highly proteinous yolk sacs and spend relatively less time as plankton. The abundance and distribution of fish eggs and fish larvae in the tropical and subtropical waters are reflected in the pre- and post-monsoonal periods of spawning seasons.

8.12   Diel Vertical Migration

An extensive list of literature is available on this topic, but many pertain to temperate waters related to vertical, horizontal, and seasonal movements of plankton and fish with the basic purposes of feeding and spawning. Most species of zooplankton are able to bring into effect extensive vertical migration. This complex migratory behaviour of marine plankton may differ to some degree between different species, but the pattern of behaviour is remarkably consistent. Although light seems to be the most important factor which governs the diurnal variations of zooplankton, the effect of the water movements associated with the tidal flows and seasonal changes in the environmental conditions can also influence the abundance and distribution of zooplankters. They are triggered to concentrate near the surface just after dusk and just before dawn. Most zooplankters migrate vertically daily. As noted above, the tendency to migrate diurnally is associated with a number of intrinsic and extrinsic factors. The migratory behaviour varies with feeding and breeding stimuli, developmental and functional stages, seasonal and regional changes, especially latitudinal, and the physico-chemical conditions of the sea. Most planktonic animals ascend from the deeper layers toward the surface at dusk and spend much of the night in the upper layers, either to graze on the phytoplankton or in pursuit of prey for predation and only to descend again to the levels where optical stimulus is minimum. A variety of zooplanktonic groups migrate diurnally. Clearly, copepods comprise an important component of epiplankters that regularly undertake diurnal vertical migration. They can swim down to depths of over 800–1,000 m. For example, a calanoid copepod can swim upwards at a speed of 15–20 metres per hour. Other common groups of planktonic organisms that habitually migrate are the chaetognaths, siphonophores, ctenophores, decapod larvae, *Lucifers*, polychaetes, appendicularians, fish larvae, and a variety of other planktonic animals (Fig. 8.6). The reasons for their vertical migration have puzzled scientists for many centuries. Many intrinsic and extrinsic factors, such as physiological state, light, pH, temperature, salinity, hydrostatic pressure, movements of bodies of water and gravity, are thought to be involved. However, some of the adaptive advantages of this behaviour seem briefly to be as follows:

(1) Escape from predation.

(2) Conserve energy by descending into deeper and cooler waters by lowering the metabolic rate.

(3) They are often carried away by currents into phytoplankton-rich patches of waters.

(4) To retreat from any deleterious effect in the sunlit zone.

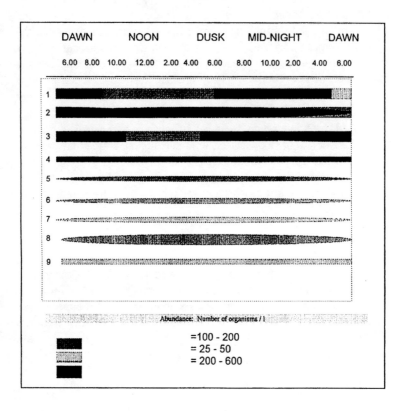

Figure 8.6   Diel (Vertical) migration of major groups of zooplankton. Data based on abundance of groups of zooplankton on an average of twelve monthly collection (January – December 1978). 1. Copepods, 2. Crustacean larvae, 3. Chaetognaths, 4. Ctenophores, 5. Polychaetes, 6. Amphipoda & Isopoda, 7. Cladocerans, 8. Appendicularia and 9. Fish larvae.

(5) The fecundity in some planktonic animals seems to be enhanced by their stay in deeper layers during their descent.

(6) Some planktonic species, for example, *Acetes,* which live in neritic waters, where there is poor visibility, use chemosensory clues to track down and catch falling food (Hammer, 1977).

Great patches of plankton often form 'deep scattering layers' because of their sound-reflective properties which can be detected by an echo-sounder. Zooplankton constitutes the major food for many fish. Herring larvae depend largely on copepods of *Calanus* type; the horse mackerel, anchovy, sardine, saury and squids also depend on zooplankton and consume in their large numbers. Zooplankton productivity in the Indian Ocean is relatively low, though the planktonic biomass increases appreciably in areas of upwellings and divergences and reaches peak density in such areas as the Gulf of Aden, the Seychelles, off Ceylon, Java, the Bay of Bengal, western Australia and the west coast of Africa. Other very productive upwelling areas are off the coasts of Peru and California. Clearly, depth-related standing crop or biomass of zooplankton varies in the tropical belt of the Atlantic, and the Indian and Pacific Oceans. There appears to be some relationship between the distribution of zooplankton and light intensity. The maximum biomass of zooplankton is found in the layers between the surface and 25 metres; less down to 200 metres, and least down to depths of 500 metres and below.

8.13 Some Adaptative Mechanisms of the Planktonic Organisms in the Tropical Waters

Size and shape: the universal phenomenon of the reduction in size and great variety of forms of plankton, particularly for the pelagic mode of existence, are well recognised. In the Tropics, generally, the surface water is warmer and the density is relatively less than in temperate waters. However, with the exception of macroplankton and pleustons (see below), most planktonic organisms are smaller in size and can only be seen under a microscope. On the other hand, the variation in excessively elaborate shape is well reflected in almost all tropical planktonic organisms. The smaller size is a functional adaptation

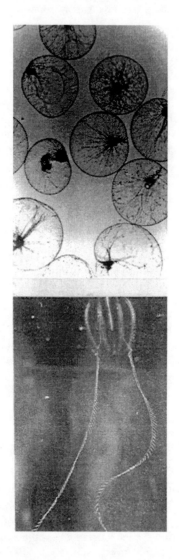

Plate 8.1   A. Ctenphore, *Pleurobrachia pileus*, floating mouth upwards and the tentacles extended; living (by electronic flash); B. *Noctiluca miliaris*; living (by microphotography).

primarily to increase the surface area relative to their volume to reduce density and diminish the frictional resistance and to keep the organisms more buoyant or in suspension within the range of optimal depth in warmer waters. The smaller size in most phytoplankton also enables them to absorb nutrients and enhance cell division. Obviously, many of the smaller, especially phytoplankton, are disk shaped, and hence are called discoplankton, while most zooplankton are spherical in shape; some of these differences among planktonic species in size and shape are associated with the regional and seasonal conditions of the sea that they occupy. Among the diatoms, *Chaetoceros, Ditylum brightwell;* and the dianoflagellates: *Noctiluca miliaris, Ceratium tripose, Ceratium furca, Ceratium fusus, Ceratium massiliense, Ceratium trichoceros, Ceratium vultur, Peridinum depressum, Gonyaulax monocanha, Gonyaulax polyhedra, Gymnodinium flavum,* etc. and eggs and larvae of a variety of zooplankton such as actinotrocha, cyponautes, polychaetes, crustaceans, especially copepods, and the leaf-like, flattened *Phyllosoma* larvae of scyllaridae and paniluridae, and nauplius and zoea larvae of a number of other crustaceans, gastrapoda, echinoderms, and fish etc., use different devices and mechanisms for floatation or to make them buoyant. In most plankton, oil droplets and gas vesicles are often formed within the protoplasm. In tropical waters, where the surface water layer density is relatively less (average 1.022) than that of higher latitudes, the temperature of surface water varies from 26°C to 32°C, the viscosity decreases and the tendency to sink is even greater. Many pelagic organisms have developed neutral buoyancy, the common examples are the euryhaline (salinity 28-36‰) Chaetognath species, *Sagitta enflata, Sagitta bedoti, Saggita bipunctata,* and many other transparent and gelatinous animals. The density may be regulated by ionic exchange in some animals. The dianoflagellate *Noctiluca miliaris* (regarded as zooplankton) and the ctenophore, *Pieurobrachia,* show this phenomenon. *Noctiluca* being small and spherical, its density can be determined from the rising rate, a movement independent of the movement of tentacle (Harvey, 1917), and the values can be substituted for calculation in the formula for Stokes's Law.

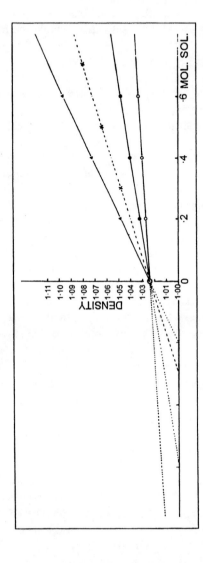

Figure 8.7 Incremental effects of varying salt concentrations, using water as solvent. 1. Sea water; 2. NH_4 Cl; 3. NaCl; 5. NH_2SO_4 (isotonic NaCl, NH_4 Cl); and 6. Na_2SO_4.

$$V = \frac{2g[d_1 - d_2]r^2}{\mu^9}$$

V = terminal velocity.
g = acceleration due to gravity.
d_1 = density of the sphere (*Noctiluca*).
d_2 = density of sea water where planktonic organisms were collected.
r = radius of the sphere.
μ = viscosity of sea water.

Evidence from recent research shows that some of these organisms – for example, *Noctiluca miliaris,* and the ctenophore, *Pleurobrachia pileus* (Plate 8.1) – float by buoyancy mechanisms by regulating the ionic composition of their body fluids. Concentrations of ammonia and sulphate ions were comparatively high in *Noctiluca* but low in *Pleurobrachia.*

The buoyancy mechanism in the former is largely attributable to a high concentration of ammonia, presumably replacing sodium, and surprisingly, to a lesser extent, to the heavier divalent sulphate being at 71% of its concentration in sea water. In the latter, perhaps, the buoyancy mechanisms rely almost exclusively on the sulphate ions, being at only 32% of their concentration in sea water (Fig. 8.7), for gas bubbles and fat seem to be absent but ammonia is relatively sparse (Singarajah, 1979).

Many planktonic organisms have developed chitinous exoskeletal outgrowths of setae, bristles, hairs, spines, labial protuberances, etc. and are called chaetoplankton. In some phytoplankton, and most larval forms of zooplankton, the elongated and profusely branched appendages support the floatation mechanisms – for example, the spiny outgrowths are common among the genera of the diatoms *Chaetoceros, Biddulphia,* and *Ditylum.* Among the zooplankton, the pelagic larvae at different developmental stages of polychaete worms and a variety of crustaceans, especially the calanoid copepods, for example, the tropical species, *Calocalanus pavo* (Fig. 8.8A), have developed very extensive caudal furca and profusely branched antennules and antenna; the very widely distributed phyllosoma larvae of several genera and species of both palinurids and scyllarids of the Tropics have developed extensive appendages to float (Fig. 8.8B); a

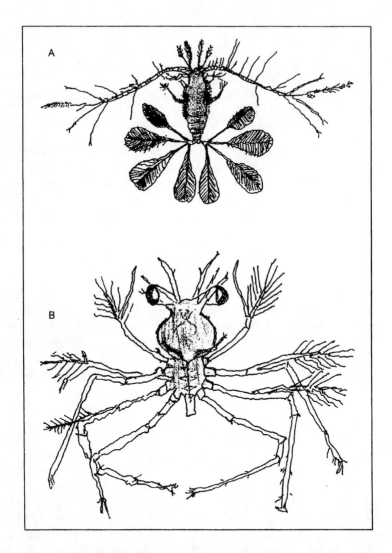

Figure 8.8   The copepod A. *Calocalanus pavo*, and B. the phyllosoma larva of the lobster, *Panulirus* sp, with their appendages modified to adapt for floating mechanism in the tropical and sub-tropical waters.

great many larvae of colonial or polymorphic coelenterates, ctenophores, crustaceans, molluscs, protochordate, especially salps, and appendicularians, are characteristically transparent and gelatinous and are able to float by reducing the friction of water.

Copilia, a tropical pelagic copepod, is extremely transparent; *Velella*, the sail-by-the-wind siphnophore polymorphic hydrozoan, with pneumatophores of a chitinous disc raised vertically to form a sail, and the pleaustinic nudibranch opisthobranch snail, *Glaucia atlanticum*, drift upside down on the water surface. Some plankton are vesicular with bladder-like transparent bodies, for example, *Pleurobrachia and Beroe*. The tropical waters are relatively clearer than those of temperate waters, and plankton have evolved some adaptive mechanisms for both floatation and transparency, together with some opportunistic colouration. *Sagitta, Spadella*, and *Krohina* may be considered as chaetoplanktons. They are some of the permanent and cosmopolitan members of the plankton and widely distributed in tropical waters. They are typically transparent, having a solid elongated body with well developed head, trunk and tail, and lateral and caudal fins. They are provided with two conspicuous eyes, and their teeth are chitinous. They are usually active swimmers and often dart in the surface waters and undertake vertical migration. They are extremely sensitive to changes in hydrostatic pressure (Fig. 8.9) changes in the ambient conditions (Singarajah, 1966) and some baroreceptors sense depth regulation (Singarajah, 1991 a, b).

8.14   Pleuston

Pleuston (= float) are planktonic organisms which occupy characteristically the interface between air and sea surface and have developed special gas sacs and increased the surface area of contact with water as floatation mechanisms. Their distribution is largely determined by the direction of the wind. Some of the well recognised examples are the representatives of coelenterates, molluscs and arthropods – for example:

Physalia – a polymorphic siphonophore; its thin membranous elongated gas filled vesicles, which act as floats, and the genera have extensible tentacles, some up to 12 metres long and lined with poisonous 'stinging cells' powerful enough to paralyse a human. Though seasonal, they get stranded in thousands on the tropical

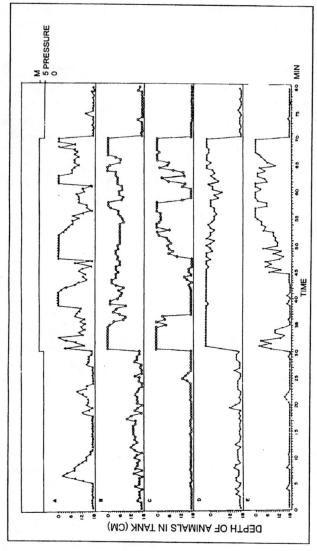

Figure 8.9 Responses of individual species to sustained pressure increases equivalent to 5m of sea water lasting 40 minutes. (A) Calanus copepod; (B) Zoea larva of Porcellana; (C) Megalopa larva of Porcellana; (D) Zoea larva of Carcinus; and (E) Megalopa of Carcinus.

beaches often mixed with seaweeds and at times are a menace to sea bathers. The tentacles are used to capture and paralyse the fish larvae as food. The gonozooids are specialised gonad cells, a variation of polymorphism, and produce sperms and eggs.

Porpita – hydrozoan polymorphic colony found floating mostly on the surface layer of the sea exposed to air. The functional individual cells are called zooids. The colony as a whole is disc-shaped and contains gas chambers which help to float. The marginal zooids are formed into tentacles to capture food.

Velella – (by-the-wind-sailor) hydrozoa, usually has an oval gas filled sail which supports and keeps the animal on the sea surface. Members of the dianoflagellates often form a symbiotic association with them.

Tomopteris – is a common pelagic polychaete worm which has developed the parapodia into a wing-like structure, an adaptation for planktonic existence.

Janthina – a gastropod mollusc, is devoid of eyes, and often floats with shell hanging, and when it is opportune it associates with the *Velella* (Fig. 8.10).

Thalia – the common tropical salp, often found floating either solitarily or in smaller chains, are usually transparent and slimy.

Pyrosoma – closely related to the family thaliacea, lives in thick 'tests' in the pelagic zone in colonial form, and can display powerful luminescence.

Glaucus – as noted above, this nudibranch gastropod usually lives in water, hanging upside down. It is predatory and gulps much air; the air bubbles help it float in the surface layer. The blue colour is also an adaptation to provide camouflage from predatory sea birds.

Halobates – the true sea insects, several species are found in tropical waters of the Indo-Pacific, but they are rare in the tropical Atlantic. They have long appendages, usually kept folded.

Lepas – stalked barnacles attached to floating objects, commonly found in oceanic tropical waters, they hang by elongated stalks or peduncles; 3–4 individuals stick together and buoy themselves.

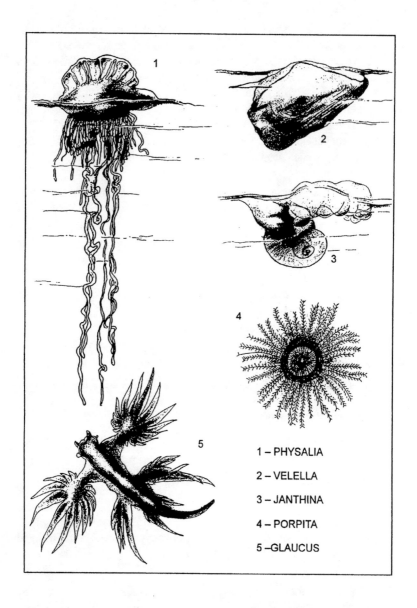

1 – PHYSALIA

2 – VELELLA

3 – JANTHINA

4 – PORPITA

5 –GLAUCUS

Figure 8.10   Some of the well recognized representatives of Pleuston.
These are also regarded as megaplankton by some.

Creseis virgula, (sea butterfly) – the planktonic sea snail, has a special adaptation of muscular wings to float, and secretes a mucus-like substance which helps to entangle the phytoplankton as food.

Sargassam – the large free floating algae of the Tropics; begins as attached thallus but soon becomes afloat. The branchlets give rise to leaf-like blades with modified stalks on which several bladders are borne to make the branchlets buoyant. This is the seaweed that has given the well-known Sargasso Sea its name; this and a number of other related multicellular seaweeds are also included in pleuston.

8.15   Neuston

Neuston: the surface micro-organisms, a mixture of heterogenous micro flora and fauna, predominantly the bacterial population, some protozoans, and marine viruses, which live in the uppermost surface layer of the sea forming a film about 1–2 centimetres thick. This group of organisms that straddles on the thin air-sea interface of the sea is also collectively called fematoplankton. These are capable of breaking down fats, carbohydrates and proteins. The marine algal species, *Ochromonous,* is often considered as neuston. In some of the tropical coastal areas, the bacteria in the sulphur cycle become more active in the salt, brackish, and mangrove environments where sulphur compounds such as sulphates, sulphides, colloidal sulphur and pyrites are rich. The main nonbiotic constituents of neuston include a variety of detritus materials, surface-active substance, and other substances of biochemical nature. Under natural undisturbed conditions, particularly when brackish waters of mangrove swamps are subject to tidal flushes, daily wave and wind actions, and the bacterial processing by the rapid oxidising bacteria, for example, *Thiobacfflus dentrificans,* tend to adjust a relative equilibrium between oxidising bacteria and sulphur-reducing agents. These natural forms of equilibria are often upset by continuous human intervention for major issues of socio-economic, fisheries and pollution. Generally, the latitudinal variation of biomass of neuston occurs throughout the Tropics and subtropics, but the concentration of the constituents was reported to be higher during the night than during the day.

220

8.16 Nutritional Aspects of Plankton

Plankton, particularly zooplankton, have been considered as a potential source of human food. A few countries like the USSR and Japan have been exploiting the krill resource in the Antarctic for some years. Apart from their calorific values of protein, carbohydrates and fats, the crustacean euphosids (krill) in particular are claimed to have high concentrations of vitamins. These also form the natural food of the great baleen whales (see Chapter Nine on whales) and many sea fisheries depend on plankton resources.

References

Anderson, D. M, White, A. W., and Baden, D. G., 'Toxic Dinoflagellates', proceedings of the Third International Conference on Toxic Dinoflagellates, St Andrews, New Brunswick, Canada; Elsevier, New York, 1985, p.561

Cheng, L., 'Marine Pleuston – Animals at the sea-air surface', *Oceanogr. Mar. Biol. Ann. Rev.*, 13, 1975 , pp.181–212

David, P. M., 'Illustrations of oceanic neuston', Symp. Zool. Soc. Lond., 19, 1967, pp.211–213

Denton, E. J. and Shaw, T. I., 'The buoyancy of gelatinous marine animals', Proc. Physiol. Sc. J. Physiol., 161, 1961, pp.14–15

Dorgham, M. M., et al., 'Environmental conditions and photoplankton distribution in the Arabian Gulf and Gulf Oman', J. Mar. Biol. Ass. India, 31 (1 and 2) 1989, pp.36–53

Fraser, J., *Nature Adrift*, London, Foulis and Co Ltd., 1962, p.178

Goiley, 'Energy flux in ecosystems' *Ecosystems, Structure, and Function* (ed. A. Weians), Corvalles, Oregon State University Press, 1972, pp.69–88

Goswamy S. C., et al., 'Zooplankton production along central west coast of India', Proceedings of the symposium on warm zooplankton, Natl. inst. oceanogr., Goa, 1977, pp.337–353

Hamner, W. M., 'Observations at sea of live, tropical zooplankton' Proceedings of the Symposium, Warm water Zooplankton, 1977, pp.284–296

Harvey, E. B., 'A physical study of specific gravity of luminescence to *Noctiluca* to special reference to anesthesia', Camegie-inst. Publ., 1917, pp.27–253

Hendy, *An introductory account of the smaller algae of the British Coastal waters*, London, Her Majesty Stationary Office, 1964, p.317 + Plates XLV

Nielson, E. S., and Jensen, E. J., 'Primary oceaning production. The autotropic production of organic matter in the oceans', Galathea Report, 1957, vol. I

Omod, M., 'Distribution of warm water epiplanktonic shrimps of the genera *Lucifer* and *Acetes* (Macura, Penaeidea, Sergestidae)' Proceedings of the symposium on warm water zooplankton, Nati. Inst. Oceanogr., Goa, 1977, pp.1-20

Round, F. E., Crawford, R. M., and Mann, D. G., *The diatoms. Biology, and Morphology of the Genera*, Cambridge University Press, 1990, p.747

Rounsefell, G. A., and Nelson, W. R., 'Redtide research summarised to 1964, including an annotated bibliography', US Fish Wildl. Serv., Spec. Sci. Rep. Fish 535, 1966, pp.1-85

Ryther, J. H., and Dunstan, W. M., 'Nitrogen, phosphorus and eutrophication in the coastal marine environment', *Science*, 171, 1971, pp.1008-1012.

Sastry, A. N., 'Pelagic Larval ecology and development', *The Biology of Crustacean* (eds J. Vemberg and W. B. Vemberg), New York, London, Academic Press, 1983, pp.213-282

Simonsen, R., 'The diatom system: Ideas on phylogeny', Bacillada, 2, 1979, pp.9-71

Singarajah, K. V., 'Pressure sensitivity of the Chaetognath, *Sagitta setosa*' J. Comp. Biochem. Physiol., 19, 1966, pp.475-478

Singarajah, K. V., Moyse, J., and Knight-Jones, E. W., 'The effect of feeding upon the phototactic behaviour of cirripede nauplii' J. Exp. Mar. Biol. *Ecol.*, 1, 1967, pp.114-153

Singarajah, K. V., 'Escape reaction of zooplankton: effects of light and turbulence' *J. Mar. Biol. Ass.* UK, 55, 1975, pp.627-639

Singarajah, K. V., 'Escape reactions of zooplankton: the avoidance of a pressure siphoning tube', *J. Exp. Mar. Biol. Ecol.*, 3, 1969, pp.171-178

Singarajah, K. V., 'Ionic regulation as buoyancy mechanism in *Noctuluca miliaris and Pleurobrachia pileus*', Rev. Brasil., 39 (1), 1979, pp.53-65

Singarajah, K. V., 1991a, 'Behaviour of *Pleurobrachia pileus* to changes of hydrostatic pressure and the possible location of baroreceptors', *Mar. Behav. Physiol.*, 19, pp.45-59

222

Singarajah, K. V., 1991b. Responses of Zooplankton to changes in hydrostatic pressure, *J. Mar. Biol. Ass.* India, 33, (1 and 2), pp. 317-334

Smith, G. M., *Cyrptogamic Botany, algae and fungi*, International students' edition, Kogahusha Co. Ltd, Tokyo, 1955, vol. I, p.546

Specter, D. L., *Dianoflagellates*, Academic Press, INC, 1984, p.442

Srinivasan, M., 'Chaetognatha from northern Arabian Sea collected during the cruses of *INS Darshk*' *J. Mar Biol. Ass.*, India, 23, (1 and 2), 1981, pp.151-160

Stosch, H. A. Von and Drebes, *Entwicklungesche Untersuchungen an zentrihen daatomen IV. Die Planktodiatomee* Stephnopyis turns - *ihre Behandlung und Entwicklungsgesshichte*, Heigol. Wiss. Meerresunters, 11, 1964, pp.209-257

Subrimanyan, R., 'Studies on phytoplankton of the west coast of India' Proc. Indian Acad. Sci., 50 (3), Sec. 5, 1959, pp.113-252

Van Landingham, *Catalogue of the fossil and recent genera and species of diatoms and their synonyms*, 8 vols, Vaduz J. Cramer, 1967-1979, p.4654

Wickstead, J. H., *An Introduction to the Study of Tropical Plankton*, London, Hutchinson and Co. Ltd., 1965, p.160

Wimpenny, R. S., *The Plankton of the Sea*, London, Faber and Faber Ltd., 1966, p.426

Wood, E. J. F., *The Living Ocean, Marine microbiology*, London, Croom Helm, 1974, p.146

Zaitsev, Yu. P., Marine neustonology, *Nauk Dumka*, 1970, pp.264 (in Russian Israel Programme for Scientific Translation.)

Chapter Nine

Necton, and the Major Components of Tropical Fish Resources

9.1   Concepts and the Major Constituents

The term necton was first applied by Haeckel in 1890. He recognised that during the ontogeny of an organism it may have both planktonic eggs and larvae and nectonic adults. Necton are relatively large, both invertebrate and vertebrate, animals which are capable of rapid and sustained swimming over considerable distances. Generally, they have developed a very efficient neuro-muscular system for powerful locomotion. The functional morphology of the body contributes substantially to the nectonic mode of life. Nectonic composition ranges from invertebrates such as crustaceans, small and large cephalopods and many of the chordates especially a rich variety of fish, marine reptiles, and among the mammals (seals, porpoises and whales). The whales are the largest of necton which graze on plankton and smaller necton. Under the broad heading of necton, only some of the economically important aspects of fish fauna will be considered here. These will include mainly fish, crustaceans and cephalopods which play a decisive role in the neritic, oceanic and benthic composition as well as the bulk of the marine fisheries. The distribution of many species is restricted by a number of abiotic as well as biotic factors such as constraints of basic nutrients for phytoplankton growth, food, competition, reproductive potential and behaviour. The largest species of the necton, the whales, will be treated separately because of the enormous commercial importance (see Chapter Ten on whales).

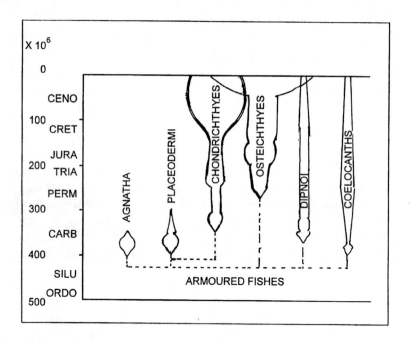

Figure 9.1   Evolution of fishes.

9.2 Occurrences of Fish in the Ordovician Period

Fish are aquatic vertebrates of considerable antiquity and diversity, and predominate the nectonic constituents. Although the origin of fish remains obscure, they probably originated during the Ordovician period, about 4.75×10^6 years ago. By Silurian time, about 4.80×10^6 years ago, fish with thick bony protective armour were well developed, and since then have evolved into a most successful group and radiated widely. Both the cartilaginous and bony fish evolved from these armoured fish before the end of the Devonian period Fig. 9.1.

Fish are cold blooded; with the exception of lung fish, they breathe through gills; they range both in size and shape, are exquisitely suited for a totally aquatic environment. The body may be spindle shaped, rounded or cylindrical and is usually covered with scales, alternatively the skin may be smooth and slippery; they swim with their limbs which have become modified into lobular fins. The fins may be paired or unpaired. The paired fins are the pectorals and pelvics; the unpaired fins are the dorsal-median fins, the anal and the caudal fins are differently shaped in different groups of fish. A comparison of representatives of the two major groups is as shown in Fig. 9.2.

9.3 Taxonomy of Fish, the Major Groups and Classes

Although taxon implies an evolutionary line where members are descended from a common ancestry, experts differ quite considerably in their approach to classifying fish on their evolutionary and phylogenetic relationship, the latter often being difficult to establish. The classification of fish has always been controversial from classes to species. However, it seems logical to arrange fish according to their diverse characteristics and evolutionary sequences, though some are primitive while others are more genealogically advanced. Despite many conflicting views, there is a greater unity of opinion among various researchers that fish certainly would have evolved from some forms of acraniates during the Cambrian period (Young, 1956). Several different classifications are employed by different workers, each system differing due to its own merits. Some of the major groups are diverse and include:

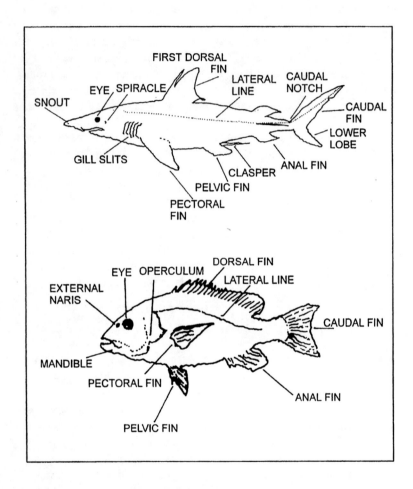

Figure 9.2   Comparison of the limbs which are modified into fins in the representatives of the two major groups of fish.   (A). Chondrichthyes; and (B). Osteichthyes.

(1)   Cyclostomata (lampreys and hagfish).

(2)   Chondrichthyes (sharks, rays and chimaeras).

(3)   Osteichthyes (bony fish).

(4)   Dipnoi (lung fish).

(5)   Crossopterygii (lobed fin fish).

The last group is thought to have been extinct for millions of years but a *Coelacanth* was caught off South Africa in 1938 which is now regarded as a 'living fossil'. Although a complete list of fish of the Tropics and subtropics, on the basis of the geographical or regional limits and distribution of species with sufficient accuracy and in considerable detail based on available knowledge, is desirable, it falls outside the scope of this book. Therefore, only a basic schematised or outline classification, particularly of some living and commercially most important species endemic to tropical and subtropical zones, will be given. In spite of the fact that many of the fish families are well represented throughout the Tropics and subtropics, often the consistency of the scientific name of the species can differ, i.e. one and the same species in one region may be named or renamed by a different specific name in another place – for example, the common usage of the specific name, *Mugil cephalus*. These controversies need to be resolved by taxonomists and by the traditional practice of the localities.

Approximately 21,000 species of fish have been recognised (Nelson, 1976), of which the vast majority (about 65%) are of marine environment and many species, including the coral-reef fish, are of tropical waters, particularly of south-east Asia, along the Red Sea, the Indian Ocean, Papua New Giunea, Australia, Polynesia, the West Indies, the north-eastern part of Brazil, Peru (see below), the North American coasts and the west coast of Africa. Many live exclusively in marine habitats, especially on continental shelves, but some often invade freshwater streams either to spawn or for nutritional purposes, and hence are of diadromous nature. Those fish which habitually ascend up stream of freshwater specifically for spawning, while spending the rest of their life in marine habitat, are called anadromous – for example salmon, herrings and sardines. Others, which regularly descend into oceans for spawning, but spend most of their time in freshwater, are called catadromous, e.g. eels. Their functional adaptation allow them to live in salt water and freshwater or *vice versa.* Among the many other marine species, some are native to coastal waters while others may live in oceanic waters, or the tropical mangrove swamps; still others may well be adapted for coral reefs with most exciting colours, sizes and shapes, ranging from the tiny gobies, (Syngnathidae), some of which are blind and live in coral holes and crevices, slim pipes and sea horses, little puffers and damselfish, the ferocious moray eels and sharks, both small and large, and the largest marlins, besides a variety of crustaceans and the molluscan bivalves, gastropods, cephalopods and echinoderms.

9.4   Phylogenic Relationship of Pisces

Fish are the most diversified and abundant aquatic vertebrates of the phylum Chordata. The fish were formerly considered as a single class known as 'pisces', but, based on several recent studies of the relationships of phylogenic (evolutionary history) and evolutionary characteristics, many distinct divisions, superclasses, classes, orders, suborders and families are now established by different icthyologic experts on taxonomy. The study of fish science is called ichthyology and one who specialises in it is an ichthyologist; and the purpose of ichthyologists is enormous. They try to draw conclusions based on the comparative nature of affinities and the possible lines of lineages or descents of the different classes of fish. It has been observed (Longhurst and Pauly, 1987), that the diversity and abundance of warm water fish species, especially in the Indo-Pacific tropical regions are greater than those of the Atlantic; and the reason for this is twofold: firstly, the geographical features and climatic variability, and secondly, the associated natural barriers for the dispersal of pelagic larvae (McManus, 1985). The richest areas of fish resources of the world are the Indo-Pacific regions, especially Indonesia and the Philippines, India, and off the coast of Peru. But, in the Atlantic, except West Africa around New Guinea and the Caribbean Islands, fish population is much less varied. The deep sea fish are more abundant and permanent residents in the tropical and subtropical waters where sunlight penetrates deep waters.

9.5   Distinguishing Characters Between Chondrichthyes (Cartilagenous) and Osteichthyes (Bony) Fish)

Fish are the forerunners and the oldest of the vertebrates. In relation to their mode of life in the warm water tropical environment, fish have evolved some remarkable structural modifications, including the internal structures. On the basis of the presence of a skeleton of cartilaginous or bony nature, fish are placed in two major natural groups – the elasmobranchs and teleosts (see below).

9.5.1   Elasmobranchs

The group includes sharks, skates, sawfish, and ray-fish. These are usually characterised by five platelike gills which are not covered by an operculum; a single nostril on either side; a skeleton composed of cartilage; fairly smooth skin, although in some types it is covered with small tooth-like placoid scales; an asymmetrical caudal fin, where the dorsal lobe is much larger than the ventral and hence heterocercal. The sexes are separate and the males have a pair of claspers which are modified from the pelvic fins and used in copulation. Most species of sharks are viviparous while others, whose females lay eggs in tough horny egg-cases, also called 'mermaid-purse', are of differing lengths and shapes. A shark can produce about a hundred offspring at a time. They are voracious predators living on fish and cephalopods. The sharks range in size from a few centimetres to about 18 metres. The majority of sharks and rays are marine fish, but may invade the river estuaries of lower salinity where the salinity may fluctuate with the tidal rhythms. Usually all elasmobranchs have high content of urea in their blood. Almost all species are found in tropical seas; they are frequently taken in large numbers both in pelagic and benthic fisheries. Some species of the family Pristidae, *Pristis perroiteti, Pristis zysron,* whose rostrum is elongated and armed with denticles, have been known since Cretaceous (Young, 1962). Other related species often visit the brackish waters of the Indo-Pacific regions, especially Borneo, Indonesia, the Philippines, Australia, East and West Africa, Brazil, and the Gulf of Mexico. Recently, it has been reported that the filter feeding, 'megamouth' sharks, about 5 metres long, were taken from deeper waters which

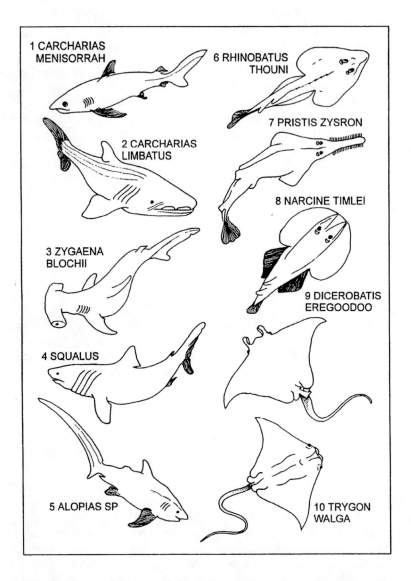

1 CARCHARIAS MENISORRAH

6 RHINOBATUS THOUNI

7 PRISTIS ZYSRON

2 CARCHARIAS LIMBATUS

8 NARCINE TIMLEI

3 ZYGAENA BLOCHII

9 DICEROBATIS EREGOODOO

4 SQUALUS

5 ALOPIAS SP

10 TRYGON WALGA

Figure 9.3   Some of the common sharks and skates of the tropics; the sharks are omnivorous and some can attack humans. They are predatory, especially on small schools of fish.

were previously unknown. Some commonest sharks and rays that inhabit tropical and subtropical waters are shown in Figures 9.3 and 9.4.

9.5.2   Teleostei

The 'modern' teleosts are the most diverse and dominant among fish and constitute about 90% of the living fish species today. About 60% of them are marine and more than half the number of species occupy the oceans and seas of the Tropics and subtropics. Teleostean or bony fish consist of a skeleton which is composed of ossified bones; gill slits covered by a bony flap of operculum (see Fig. 9.2); a body generally covered with different types of scales, although in a few cases the scales may be absent. The caudal fin is usually symmetrical and differently shaped. Most bony fish are oviparous and fertilisation is effected in the water, but a few are viviparous and fertilisation is internal. Teleosts are adapted to a wide variety of habitats of rocky marine shores, mud-flats, coral reefs, offshore waters, deep-waters, pelagic realms and bottoms of diverse conditions. As already noted, some fish migrate between the seas and freshwaters through estuaries. Despite the vast spatial and temporal dispersion of fish within the tropical and subtropical belt, the important and commonly exploited species of bony fish are shown in Figures 9.5 to 9.7.

9.5.3   Outline Classification of Fish

Kingdom: Animalia
Subkingdom: Eumetazoa
Phylum: Chordata
Subphylum: Vertebrata (Craniata)
(1) Superclass: Agnatha (without jaws)
   Class: Cephalaspidomorphi
   Order: Cyclostomata
   Suborder: Petromyzontidae
   Suborder: Mixinoidae
(2) Superclass: Gnathostomata (with jaws)
   Class: Placadermi (extinct)
   Subclass: Chondrichthyes (cartilaginous)
   Class: Elasmobranchii
   Order: Selachii

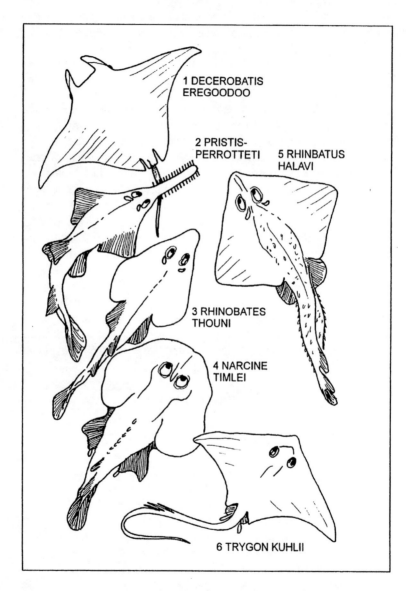

1 DECEROBATIS EREGOODOO

2 PRISTIS-PERROTTETI

5 RHINBATUS HALAVI

3 RHINOBATES THOUNI

4 NARCINE TIMLEI

6 TRYGON KUHLII

Figure 9.4   Some common species of ray fishes are of commercial importance. They are widely distributed in all oceans and seas. Some ray fishes are dangerous because of their poisonous spines at the base of their tail.

Suborder:	Heterodontoidea
Suborder:	Hexanchoidea
Suborder:	Galeoidea (sharks)
Suborder:	Squaloidea
Suborder:	Batoidea (rays and skates)
Subclass:	Bryodonti (= Holocephali)
(3) Class:	Osteichthyes (Teleostomi)
Subclass:	Actinopterygii
Infraclass:	Holostei
Infraclass:	Teleostei
Superorder:	Elophomorpha
Superorder:	Osteoglossomorpha
Superorder:	Clupeomorpha
Superorder:	Ostriophysi
Superorder:	Protacanthopterygii
Superorder:	Paracanthopterygii
Superorder:	Antherinomorpha
Superorder:	Acanthopterygii

Oceans and seas contain nearly 90% of the living biomass of the world, of which fish resources and edible seaweeds constitute a considerably significant proportion. Marine seaweed flora have a tremendous potential for food production but their distribution is restricted to the intertidal zone; they are usually attached to rocks, sandy bottoms, and other suitable substrates; and some float on the surface in open oceanic waters (see Chapter Fourteen). In most tropical coastal countries, marine fishery is not only a major national industry but it is also the livelihood of many local populations. Fish have long since been considered as their main source of food and for generations they depended on the sea for their food. Because of the lack of precise knowledge of fishing grounds or banks, fishery parameters, limited expertise, relatively high investments on heavy crafts, gears and tackle, the exploitation of the fish resources has been confined to more of the inshore waters, whose near-surface does not exceed between 40 metres and 100 metres. Large-scale enterprise for these resources on an individual basis, like in the Western world, is rare and much of the exploitation by trolling for the school of pelagic fish and trawling for bottom fish has been monopolised by the government or co-operative and commercial agencies. Basically, the stocks of fish are related to different habitats and depths. Many seek

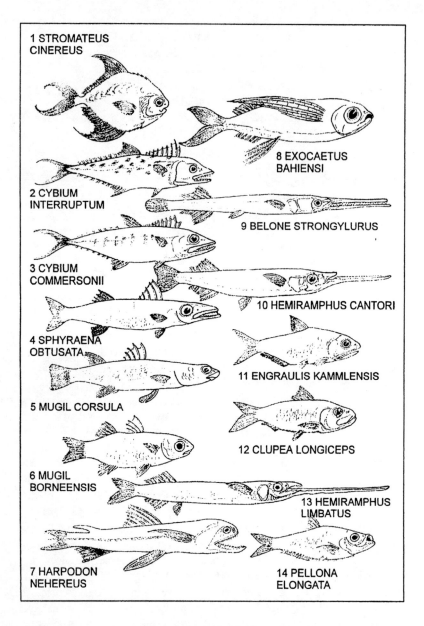

1 STROMATEUS CINEREUS

2 CYBIUM INTERRUPTUM

3 CYBIUM COMMERSONII

4 SPHYRAENA OBTUSATA

5 MUGIL CORSULA

6 MUGIL BORNEENSIS

7 HARPODON NEHEREUS

8 EXOCAETUS BAHIENSI

9 BELONE STRONGYLURUS

10 HEMIRAMPHUS CANTORI

11 ENGRAULIS KAMMLENSIS

12 CLUPEA LONGICEPS

13 HEMIRAMPHUS LIMBATUS

14 PELLONA ELONGATA

Figure 9.5   Some of the tropical pelagic fishes of considerable commerical importance.

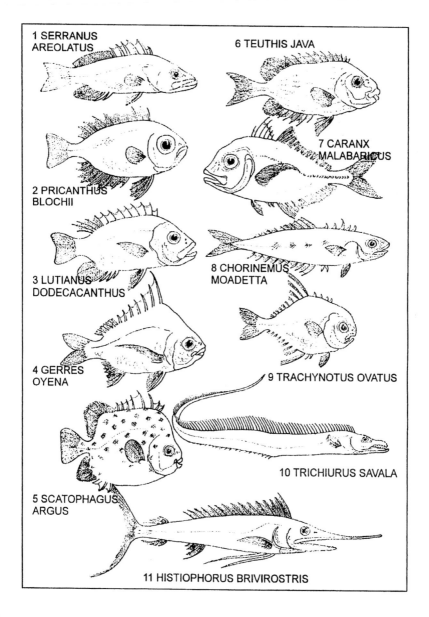

Figure 9.6   Commercially important fishes of inshore and offshore
              waters.

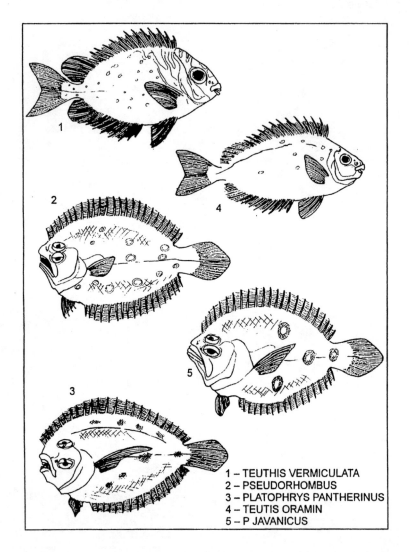

1 – TEUTHIS VERMICULATA
2 – PSEUDORHOMBUS
3 – PLATOPHRYS PANTHERINUS
4 – TEUTIS ORAMIN
5 – P JAVANICUS

Figure 9.7   Demersal teleost fishes of commercial importance in the tropics and subtropics. Some of them bury themselves in the sandy bottoms.

to take the larger sized species such as marlin, sail fish, sier fish and yellow fin tuna, but these are of deep waters, close to the edge of continental shelves. The distribution of fish is greatly influenced in the tropical waters by depth, water movements, salinity, concentration of oxygen, abundance of nutrients and productivity of plankton as food, as well as many other factors. The most important marine species taken from different depth zones can be of three types: pelagic, demersal or benthic. On the other hand, the exploitation of the relatively untapped demersal fishery resources of the deep sea zones (depth below: 100, 100–200 metres, and 200–500 metres) in the Tropics by indigenous people has only just begun. However, some estimates of the catches and abundance of species of the depth-related fishing zone are shown in Fig. 9.8. Some of the commercially important fish species of deep sea in the Tropics are: flat fish, sharks, rays, crustaceans and cephalopods, which are mostly exported. On the other hand, for a few other technologically advanced nations, sea food is a matter of taste as seen in the USA, where only about 7% of the world's population consumes almost 12% of the global marine fishery products. The Scandinavian countries, especially Norway, Denmark and Iceland, also fall into this category of great consumers of fish and fish products.

9.6.1 Fish as Food and Fisheries

The demand for marine fish as food is ever increasing. The coastline of China alone consists of 18,000 kilometres, including its several islands. Marine waters stretching across tropical, subtropical and some temperate zones contain some 1,500 species of which 300 have economic value. The average annual catch of marine species amounts to about 254.63×10^6 (World Resources, 1994). India has a coastline of 5,550 kilometres and an exclusive economic zone covering some $2,014.9 \times 10^6$ km^2. The Indo-Pacific tropical region is characterised by a great number and an abundance of species. The Atlantic has a relatively long coastline and there is a dominant fishery on either side. *Food*: Many species of fish can be effectively used as food. Food and fisheries have subjective and objective connotations. Fishery denotes the actual catch of all types of living resources from the sea, including fish, clams, mussels, oysters, gastropods, abalones, cephalopods, sea cucumbers, crustaceans – lobsters, prawns, shrimps crabs – and the

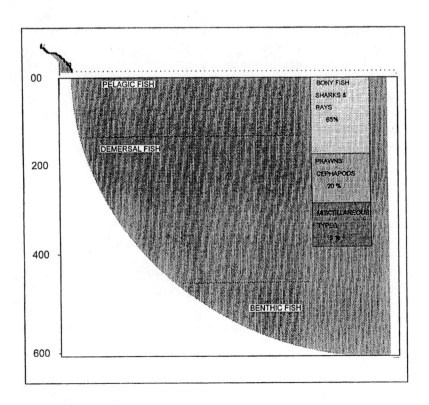

Figure 9.8   Annual average catch of demersal fish components, expressed as a percentage of landings by small trawlers. Data based on 6 years observations from Porto Cabedelo, Brazil. Depth in m.

harvesting of the edible seaweeds. The culture of prawns and bivalves was known to have been well-developed in the Far East from ancient times. Turtles are mostly oceanic, but they come to sandy beaches regularly to lay their eggs; the green turtles are considered to be an endangered species as they have been hunted for their eggs and meat. Whales, porpoises, and seals are also considered by some nations as useful marine products. The development of marine fish and the growth of fish catch throughout the world has been intensified over the past several decades. The total world fish catch of marine fishery for the year 1966, was some 59×10^6 tons. Of this, about 87.80% was fish, 7.12% were invertebrates, 3.73% whales, and 1.35% algae. Since the Second World War, the marine harvest has been steadily increasing and, according to FAO statistics, the maximum world's production is approximately about 700,000,000 metric tons; this is excluding China, North Korea and North Vietnam as they do not supply fishery data to FAO. This figure corresponds to some 10% of animal protein required by the global needs. Within the tropical zone, particularly the Asiatic, African and Latin American countries, sea food has been an indispensable and relatively cheaper source of protein. Despite the demand for home consumption, the sea food is an important economy for some countries by way of exportation, earning valuable foreign exchange and offsetting the balance of payment. Based on the recent statistics (FAO, 1984), the total catch of marine fish resources taken far exceeds some 70×10^6 metric tonnes although the exploitation could be increased. Currently, it has increased to a sustainable yield (see Fig. 9.12). The world catches of marine fish resources are summarised in Table 9.1.

Table 9.1   Total Marine Catch of Fish, including Shellfish,
From the Major Waters of the World (M.T)

	1976	1977	1978	1979	1980	1981	1982
WORLD TOTAL	62,462,500	61,361,300	63,157,700	63,966,100	64,695,900	66,819,000	68,197000
TROPICAL AREAS*							
ATLANTIC OCEAN							
Western Central 31 –	1,574,800	1,417,400	183,400	1,800,800	1,791,300	189,800	2,131,300
Eastern Central 34 –	3,625,600	3,742,100	323,500	2,752,500	3,439,600	3,217,700	3,195,900
South-west 41 –	818,400	1,049,800	12,80,700	1,504,700	1,273,500	1,253,000	1,541,500
INDIAN OCEAN							
Western 51 –	1,977,500	2,083,600	2,053,200	2,022,300	2,091,000	2,007,400	2,021,900
Eastern 57 –	1,139,000	1,354,200	1,383,800	1,437,200	1,458,900	1,512,700	1,540,100
PACIFIC OCEAN							
Western Central 71 –	5,077,800	5,743,900	5,910,500	5,735,100	5,808,800	6,108,600	5,938,500
Eastern Central 77 –	1,519,300	1,720,700	1,811,700	2,004,400	2,422,100	2,607,400	2,370,100
TROPICAL AREAS (excluding area – 47)			Catch as % of total				
	26.00	28.06	27.75	26.97	28.90	28.52	27.98
BY OCEANS			Catch as % of total				
Atlantic Ocean	42.56	41.78	40.15	38.67	38.64	37.31	36.17
Indian Ocean	4.99	5.60	5.44	5.41	5.49	5.27	5.22
Pacific Ocean	52.35	51.98	53.77	55.2	54.96	56.57	57.66
Southern Oceans	0.96	0.64	0.64	0.93	0.96	0.85	0.95

Based on FAO, 1984, according to latest published revision.

The sea food resources provide some of the excellent forms of high quality protein to humans. The fish, shellfish, and molluscs constitute some 10% protein by body weight, while the contents of fat and fish oil account for some 5% by weight. The fats and oils extracted from fish species are also equally important to human diet as they are of a much higher nutritional value. The average dietary requirement per person per day has been estimated to be 2,500 calories with 80 g of protein of which 18 g of them are of animal origin (Table 9.2).

Table 9.2 Energy and Protein Requirement per Person per Day (according to UN Recommendation)

Daily intake	Man	Woman
Body weight (kg)	65.00	55.00
Energy (kj)	12,750.00	9,218.00
Per kilo of body weight		
Energy	193.00	168.00
Protein	0.82	0.75

Widespread protein deficiency and nutritional imbalance are commonly recognised among a number of tropical countries. Their diet includes only an insufficient proportion of animal proteins. Countries like India and Pakistan consumed 6–7 grams of animal protein. In a few nations, like India, the existence of protein deficiency is largely due to religious beliefs, traditions, culture, and social status, although abundant fish supply is possible. In many other coastal nations of Africa, Brazil, Venezuela, Panama, the Caribbean, Indonesia, the Philippines, Malaysia, Vietnam, Taiwan, China, etc., protein deficiency is still predominant.

9.6.2   Principal Tropical Marine Fisheries

The marine fishing zones of the world (FAO, 1982, 1992) have been divided into nineteen fishing areas and many of the central areas fall into the tropical and subtropical regions. These areas are shown in Fig. 9.9.

The tropical marine fisheries, particularly the exploitation of the different genera and species of pelagic, demersal, and bottom-dwelling resources, are of considerable importance. It is difficult to deal adequately with all the potential resources of the tropical seas as their catch statistics are not quantified with sufficient accuracy. Statistics are incomplete on many commercial fisheries throughout the tropical regions, though some are available through the national agencies and the FAO year books. However, some of the most abundant and commercially most valuable species of the pelagic, demersal and benthic fish resources are listed below (Table 9.3). The demersal species often show great taxonomic diversity and habitats, and are able to adapt to bentho-pelagic conditions.

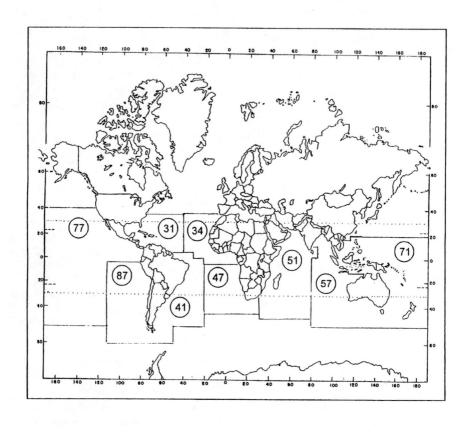

Figure 9.9   Major fishing area within tropical and subtropical belt.

Table 9.3 Some Commercially Important Tropical Fish Resources

General name	Class, Order, or Family	Genus (several) Species	Depth-zone* Pelagic (1) Near-surface (2) Demersal (3) Bottom-dwelling (4)
TELEOSTEI FISH			
CLUPEFORMES			
Anchovies	Engraulidae	*Engraulis*	Inshore 1-2 schooling
Herrings	Clupeidae	*Clupea*	Inshore 1-2 schooling
	Clupeidae	*Sardinella*	Inshore 1-2 schooling
	Clupeidae	*Sprattus*	Inshore 1-2 schooling
	Chirocentridae	*Chirocentrus*	Inshore 1-2 schooling
PERCIFORMES			
Groupers	Serranidae	*Serranus*	In-off shores 1, 2-3
	Centrolophidae	*Centrolophus*	In-off shore 1, 2-3
	Carangids	Carangidae	Caranx Inshore 1-2
Perches	Lutjanidae	*Lutjanus*	schooling
	Gerriade	*Gerres*	schooling
	Sciaenidae	*Johnius*	Inshore 1-2
	Carangidae	*Trachinautus*	
	Priacanthidae	*Priacanthus*	In-off shore 1, 2-3
Mackerel	Scombridae	*Scarnberomorus*	Offshore 1-2-2
	Scombridae	*Scomber*	schooling
	Scombridae	*Rastrelliger*	Inshore 1-2 schooling
Tuna	Scombridae	*Thunnus*	SCHOOLING
	Scombridae	*Euthynnus*	Offshore 1-2-2
Skipjack			
Tuna	Scombridae	*Katsuwonus*	Offshore 1-2-2
Swordfish	Ziphiidae	*Xiphias*	In-offshore 1, 2-
Pomferets	Stromateidae	*Pampus*	In-offshore 1, 2-3 schooling
	Stromateidae	*Stromateus*	
	Acanthuridae	*Acanthurus*	In-offshore 1, 2-3
Silverbellies	Leiognathidae	*Leiognathus*	In-offshore 1, 2-3
	Leiognathidae	*Naso*	In-offshore 1, 2-3
Ribbonfish	Trichiuridae	*Trichiurus*	In-offshore 1, 2-3
Bream	Nemipteridae	*Nemipterus*	In-offshore 1, 2-3
BELONISFORMES			
	Belonidae	*Belone*	In-offshore 1, 2-3
	Hemiramphidae	*Hemiramphus*	In-offshore 1, 2-3
Flying fish	Exocoetidae	*Exocoetus*	Inshore 1, 2 schooling
MUGILIFORMES			
Mullets	Mugilidae	*Mugil*	Inshore 1, 2
	Sphyraenidae	*Sphyraena*	Inshore 1, 2-
MYCTOPHIFORMES			
Lizardfish	Myctophoridae	*Surida*	In-offshore 1, 2-3
	Chloropthalmidae	*Chloropthalmus*	In-offshore 1, 2-3

SCOPELIFORMES			
Catfish	Ariidae	*Galeichthys*	
	Tachysuridae	*Tachysurus*	In-offshore 1,2-3

ELASMOBRANCHS
PLEUROTREMATA

Sharks	Galaiodae	*Scoliodon*
	Pristidae	*Pristis*
	Sphyrniade	*Sphyrna*

RAJIFORMES

Rayfish	Rajidae	*Raja*
	Dasyatidae	*Dasyatis*
	Mylobatidae	*Mylobatis*
	Mobulidae	*Manta*
	Mobulidae	*Mobula*

SHELLFISH
BIVALVES

Oysters	Crassostreidae	*Crassostrea rhizophora*
		Crassostrea paraibenensis
		Crassostrea cucullatus
		Pinctada vulgaris
	Mytilidae	*Mytilus viridis*
		Perna perna
	Pectinadae	*Pecten*
	Donacidae	*Donax*

DECAPODA

Prawns	Peneidae	*Penaeus indicus* (90 m)
	Peneidae	*Penaeus carinatus* (40 m) neritic
	Peneidae	*Penaeus brasilenesis* (60 m)*
	Peneidae	*Penaeus notialis* (50 m)
	Peneidae	*Penaeus monodon* (Giant tiger prawn)
	Peneidae	*Penaeus schmitti*
Deep sea	Peneidae	*Metapaenus monoceros* (40 m) neritic
	Peneidae	*Parapenaeopsis stylifera* (non-migratory)
	Non-Peneidae Aristaeidae	
		Aristeus (4 large deep sea)
Prawns		*Heterocarpus* (4 large deep sea)
		Parapandalus (4 deep sea)
		Plesionika (4 deep sea)
Shrimp		*Acetes spp.*
Lobsters	Palinuridae	*Palinunis poliphagus*
	Palinuridae	*Palinurus laevicauda*
	Palinuridae	*Palinurus gracilis*
	Palinuridae	*Palinurus omatus*
	Palinuridae	*Palinurus regius*
	Palinuridae	*Puerules angulatus* (4 deep sea)
	Scyllaridae	*Thenus ariantalis*
	Scyllaridae	*Scyllarides brasiliensis*
	Nephropidae	*Metanephrops andammanicus*
Crabs	Decapoda	*Scylla* (4 deep sea neritic)

		Neptunus (4 deep sea neritic)
	Portunidae	Portunus
Cephalopds	(Cephalapoda)	
Squids	Loliginidae	Loligo (4 deep sea)
	Sepiidae	Sepia (4 deep sea)

Others are of the genera *Centropus, Nemipterus, Priacanthus,* etc.; among the prawns the genera *Parapandalus, Metapenaeopsis;* the deep sea lobsters *Puerulus* and *Pampus* occupy coastal tropical and subtropical waters up to a depth of 100 metres; distribution extends from Tawain to the Iranian Gulf.

9.6.3 Pelagic Fish and Schooling

The pelagic fish of the warm water belt has depth limits and can occupy from the surface to depths down to 200 metres, but some pelagic species can descend down to 500 metres. One of the curious phenomenon of the fish is their strong schooling behaviour, but despite various explanations the subject is little understood. Most of the pelagic fish such as herrings, mackerel, sprats, tuna, swordfish, and many others tend to swim in enormous groups or 'schools'. But they probably form schooling as a sense of security, to adjust their body temperature, in search of food, and secure places to lay their eggs (see also Chapter Sixteen). Many become predators on the long migratory routes. The prawns and lobsters also migrate seasonally.

9.6.4 Subsistence Fisheries of Developing Countries of Tropics and Subtropics

The clupeoids are the most abundant and popular fish of the tropical pelagic waters. There are several genera, but to cite a few representatives: sardines – *Sardinella, Clupea, Dussumeria,* and anchovies – *Thrissocles, Engraulis* and *Hilsa ilisha.* These are often caught on commercial scales all along the offshore waters of Taiwan, the Philippines, Java, the Bali Straits, the Malay Peninsula, the Andaman Islands, Ceylon, the east and west coasts of India, Pakistan, the Gulf of Iran, Arabia, the west coast of Africa, Brazil and Peru, as they provide both food and oil. They inhabit the coastal waters of most tropical regions, including the coast of California and Florida. The Peruvian anchovy fishery is well-known (see below). The other by-products derived from these fisheries are phosphate and nitrate-rich guano and fishmeal.

Oil Sardine, *Sardinella longiceps,* are caught along the coast of India, Ceylon and Pakistan on a large scale, exceeding 280,000 tons per year, and a number of subspecies, *Sardinella aurita,* are caught along the Atlantic coast of Africa, Brazil, Venezuela, Argentina, Florida and other warmer areas.

Mackerel are also an important commercial fish of the most tropical waters of Australia, India, South Africa, and the North Atlantic coast of Brazil. The mackerel *Scomber* of the Indian Ocean is considered synonymous with *Rastrelliger kanagurata* (Russel). The horse mackerel are caught on a commercial scale off the coast of Taiwan, Vietnam, East China Sea, the Philippines, Malaysia, Indonesia, India and Ceylon. The *Scomber japonicus* and *Trachurus symmetricus,* are caught in large numbers along the coastal waters of California. Of the family chirocentridae, the *Chirocentrus dorab* and *Chirocentrus nudus* are very popular. The length range of adults varies between 220 and 715 millimetres, and the males are relatively slender and weigh more than the females. Observations on stomach content revealed that they generally feed diurnally more during the daytime than the night on a variety of planktonic organisms; especially, crustaceans: copepods, larvae of *Lucifer,* zoea and megalopa larvae of other crustaceans; the fish larvae of *Stolephorus,* and occasionally on segments of seaweeds.

Tuna occur mainly all along the equatorial tropical and sometimes in subtropical waters of the Indian, Atlantic, and Pacific oceans. Tuna, swordfish, skipjack, garoupa, pomprets, flying fish, red snappers, and others are caught in large numbers by a number of tropical countries, including Japan, USA, France, Spain, and Peru. Tuna fishery, for different species, has now been developed and they are caught in open oceanic waters of the Pacific, Atlantic and Indian Oceans. Just over 1.2×10^6 tons of tuna have been caught annually. The largest catch is confined to Pacific waters while in the Indian and Atlantic Oceans they are caught in relatively fewer numbers. Tuna are mainly pelagic predatory fish and they migrate very long distances. They are large and fast swimmers. They are eurythermal and can tolerate a relatively wide range of temperature difference of between 10 and 20°C. They are mostly sought by local fishing fleets because of their attractive size and weight; each fish can grow up to more than 2 metres in length and weighs about 700 kg, and is economically a most valuable fish. Some of the most popular and important species are the bluefin tuna, *Thunnus thynnus,* of the Atlantic and the yellowfin tuna, *Thunus albacores, Euthunus alletteratus,* of Indo-Pacific regions of the Tropics. They usually spawn in coastal waters and their larvae develop in surface waters of the Pacific, Indian and Atlantic Oceans. They are caught commercially in large numbers all along the tropical waters of the Indo-Pacific, especially Hawaii. Tuna fishery is an essential national industry of the Maldive Islands. Nearly 80–90% of the total catch consists of tuna species of skipjack, yellowfin tuna and frigate mackerel and this is about 2.6% of the world's total. Tuna have been exported as dried 'Maldive fish' for centuries. Other important and larger pelagic fishery include swordfish, sailfish, marlin, and tarpon (see Table 9.31).

9.6.5   Tropical and Subtropical Upwellings

Upwelling areas are those where nutrient-rich waters from colder bottom layers rise up and mix with the relatively warmer waters of the surface layers. The major upwellings are located off the west coast of North and South America, Africa and off the coast of the Arabian Sea and Agua Frio in Brazil (see below). It has been estimated that the world's oceans cover only about 0.1% of upwelling areas, yet they contribute nearly 50% of the world fish catch (Ryther, 1969). The 'classical' upwellings in the Tropics and subtropics are relatively less; a few are permanent, while others are temporary, specific to seasons and last for shorter lengths of time. Many of these areas are of great interest because of their association with the prevailing wind patterns and wind-drift currents, and continental configuration. Upwellings can occur either near the costs or in the open ocean. Generally, most upwellings are found on the western coasts of the continents in low latitudes (Fig. 9.10), but in other upwelling regions climatological differences may be involved.

Some of the most important upwelling areas are those off the coasts of Peru, Southern California, the Atlantic coast of Morocco, the coasts of south-western Africa, and the Gulf of Persia. As a result of the NAGA Expedition new potential upwelling areas have been found along the coasts of South-east Asia, for example, the Gulf of

Figure 9.10

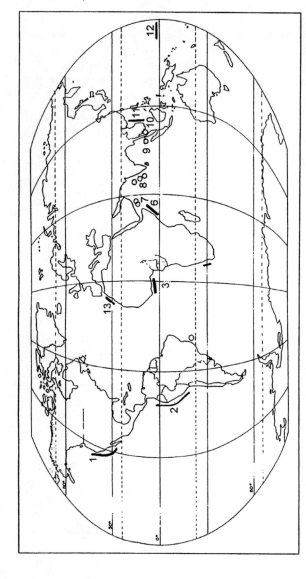

The distribution of upwelling areas within the tropical and subtropical belt. The elongated solid lines indicate the permanent nature of the upwellings while the open circles represent temporary upwellings, either during winter or summer. Upwellings are highly productive areas mainly located off the west coast of (1) California; (2) Peru-Chilli; (3) off the west coast of Africa, New Guinea, (4) off Southeast Africa; and (5) Cabo Frio on the Atlantic coast of Brazil (6) off the coast of Somalia, (7) Saudi Arabia, (8) Arabian Sea; (9) off the west coast of India – Cochin; (10) Gulf of Thailand, (11) off the west coast of Vietnam in the South China Sea; and (12) west Pacific close to equator and (13) off the west coast of Morocco.

Thailand, and off the coast of Vietnam. Mainly the regions and prevailing current systems, rather than entirely wind, are responsible for the upwellings of the south-west coast of India (especially Calicut, Cochin, Quilon), the Gulf of Mannar, the Arabian Sea off the coasts of India, and many other still new potential upwelling areas which have recently been discovered; and these have scarcely been utilised. These areas during upwelling usually bring about nutrient-rich waters to the euphotic layer, depending on marine meteorological conditions, resulting in enhanced primary productivity, concentrations of zooplankton which, in turn, attract a variety of commercial species of fish such as sardine, mackerel and tuna, and hence the establishment of the important commercial fisheries. However, these productive areas are largely determined by seasonally varying currents and the dominant wind systems (See Fig. 13.3).

9.6.6   Upwelling System off the Coast of Peru

The marine areas off the coast of Peru are considered to be one of the richest marine fishing areas of the world. Prior to 1950, the anchovy fishery off the coast of Peru was smaller and remained almost as an untapped resource. By 1953, because of the economic potential in the fish oil and fish meal for export, anchovy fishery increased rapidly in response to growing demands, and soon Peru became the top fishing nation in the world. Because of the Humbolt current, and the richness of the nutrients, the productivity of both phytoplankton and zooplankton increased. Consequently, the pelagic anchovies were found in abundance and the fishery was intensified. During the good fishing years (1969–1970), Peru took more than 11 million metric tons of anchovy; the sustainable quantity of anchovy to be caught seems less than 9.5 million per year. However due to 'El Nino' and other factors, the anchovy fishery has declined: 2.5 million metric tons of anchovies were purse-seined, followed by a further decline, and during 1977–1987 the catch averaged only 1.2 million metric tons per year (Caviedes et al., 1992); the reason for the decline seems to be over-fishing.

9.6.7   Gulf of Guinea – Senegal Coastal Pelagic Fishery

The coastal area of the Gulf of Guinea and Senegal, up to six miles and beyond, is a fishing area exclusively reserved for traditional fishing, using simple crafts and tackle of both pelagic and benthic species. Some of the pelagic species include *Sardinella maderensis, Ethmalosa fimbriata, Caranax rhonchus*, and *Pomadas spp.*

9.7   Benthic and Demersal Fish

Benthivorous fish such as flat fish of the bothidae and solidae families (see Fig. 8.7) are caught in large numbers in coastal waters at depths of up to 60 metres. Several species of sharks, rays and skates are also caught on a large scale for food and other uses. The ribbon fish, *Trichiurus,* is seasonally caught off the coasts of Burma, Bangladesh, Ceylon, South India, the Philippines, Indonesia, Venezuela, and Brazil.

9.7.1   Crustaceans

According to FAO (1994), of the total world catch of marine waters for the year 1992, fish constitute 81.3%, molluscs 10.4%, crustaceans 5.9%, diadromous species 1.7%, and other marine organisms 0.7%. Among the commercially important crustacean fisheries, prawns rank first both in quantity and value. There is a satiable demand for prawns and prawn products throughout the world. They are the most dominant sector of crustaceans and a major resource in many shallow seas of the Tropics. Several species are taken in large quantities in shallow areas off Taiwan, Thailand, the Malacca Straits, Indonesia, Malaysia, India, West Africa, Brazil, Florida, the Caribbean Islands, and many other tropical continental shelf waters. An estimate based on a survey to improve fisheries management shows that 50,000 tons per annum for the Malacca Straits alone seems possible (Johnson, 1976). The nominal worldwide crustacean landings are shown in Table 9.4.

Table 9.4   Nominal Worldwide Landings of Crustaceans (metric tonnes)

Sources	1977	1978	1979	1980	1981	1982	Group Total
Prawns	44,350	99,854	56,181	67,015	71,978	67,930	407,308
Shrimp	83,102	26,205	24,996	167,522	226,233	371,377	899,435
Lobsters	3,998	3,299	4,011	3,115	2,819	3,631	20,873
Crabs	57,840	59,168	33,534	64,695	71,371	63,182	285095
GRAND TOTAL	189,290	188,526	118,722	320,347	372481	560,120	1,677,406

9.7.1.1   Prawns

Crustaceans comprise the second most important item in both magnitude and value of the sea fishery in the Tropics and subtropics. The prawns, shrimps, lobsters and crabs are exploited for both their value and delicacy. All marine prawns are of the family Penaeidae, and commercially the most important species are: *Penaeus indicus*, found in most parts of the Indian Ocean, especially in the inshore waters of India, Bangladesh, Malaysia, Thailand, Indonesia, the Philippines, north-west Australia and east Africa. The species lives in sand-mixed mud at a depth of between 2 and 60 metres. It is a popular species with importers, especially in Japan, USA, and Western Europe. The species is also cultured in the rice fields of Kerala.

Penaeus monodon (jumbo tiger prawn), another large species, prefers a muddy bottom from shallow to deep waters down to 110 metres; (see also Chapter Fourteen on aquaculture potentials) commercially caught along the Arabian Gulf, Pakistan, the west and south coasts of India, Burma, the Philippines, Indonesia, Taiwan, Thailand, Vietnam, and Hong Kong.

Metapenaeus affinis and *Parapenaeopsis* are other important species. The latter two are compared with *Penaeus indicus* in Fig. 9.11. Among all penaeid species, *Penaeus carinatus* is the largest, with lengths up to 30 centimetres, and other species are about 20 centimetres long. Except for the *Parapenaeopsis,* all are essentially migratory. During the post-larval phase of their life history, they usually migrate to estuaries and back waters where they spend much of the time in a nutrient-rich environment and grow until they mature. They are usually benthic feeders, especially on polychaetes, amphipods and copepods. On reaching sexual maturity, they resort once again to the marine environment to spawn and breed. The eggs are usually shed into the sea. Prawns are usually considered to be one

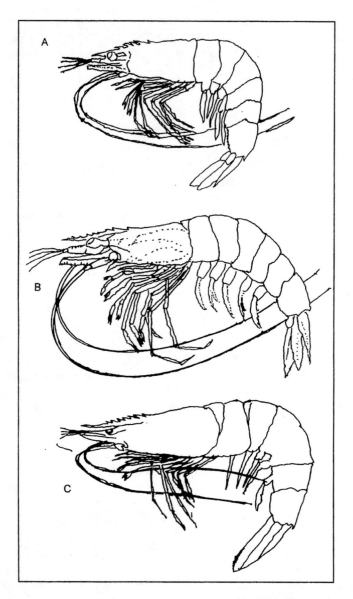

Figure 9.11   The three economically most important species of Prawns;
(A) Penaeus indicus; (B) Metapeneaus affinis, and (C)
Penaeus notialis.

of the most highly-priced commodities and are exported to earn foreign exchange. On the other hand, the penaeid shrimps, *Acetes,* are relatively smaller in size, about 2.5 centimetres in length, with slender elongated bodies, long limbs and long spiny rostrum; they are seasonally caught in very large numbers.

9.7.1.2   Crabs

The commercially important species of marine crabs caught in most tropical coastal waters are *Scylla serrata, Neptunus pelagicas* and *Portunus pelagicus.* The former mostly occupies brackish water, while the latter two can be caught in shallow neritic waters. These are relatively larger in size; the carapace width of adults often exceeds 150 millimetres. They moult periodically and their moulting cycle generally includes premoult, moult, postmoult, and intermoult stages. The moulting process is greatly facilitated by the availability of food, oxygen, hormones, and the size of the crabs. The crabs cease to feed for a short while before moulting. There are many other species which are taken from the coastal waters and soft muddy mangrove areas (see Plate 4.5).

9.7.1.3   Lobsters

Lobsters are highly valued seafood because of their special delicacy and are consumed both locally and abroad. These are the dominant and important export items as 'frozen tails' in the tropical and subtropical parts of the world. The average annual worldwide landings of lobsters from major fishing areas, for statistical purposes, amount to some 0.40174 million tonnes (FAO, 1975– 1982 revision). Lobsters are fairly widespread in the warm tropical waters of the Atlantic coast of Brazil, the Gulf of Mexico, the Caribbean and the Indian Ocean, and occur in large numbers within the relatively shallow coastal waters of depths of 60–180 metres, and beyond the edge of the continental shelves. Although there are several species, the most common species caught for commercial purposes are *Panulirus ployphagus, Palinurus penicillatus, Palinurus longipes* and *Thenus spp.,* in India, Ceylon, Andaman, the Red Sea, Mozambique; *Palinurus laevicauda* on the north-eastern coast of Brazil, *Palinurus regius* along the West African coast, and many other species around Australia, Florida, the Bahamas, the Caribbean, Panama, Mexico,

Guatemala, Indonesia, Hawaii, and Hong Kong. These species usually occupy the engybenthic habitats at depths over 10-30 metres. Deep sea lobster of the genus, *Puerulus,* are relatively less exploited and are found in large numbers on the bottom, below 200 metres. The *phyllosoma* larvae spend more time until several moults are completed. The adults migrate to long distances after their maturity. All these species are important in fisheries (see Table 9.1). Brazil exports the largest number of lobsters, mainly to Japan and USA.

9.7.2   Molluscs

Since the Stone Age, marine molluscs have been highly valued as an important resource. They occur in a wide variety of habitats of littoral and sublittoral zones of the Tropics. Marine species are adapted to inhabit different biotopes with preference for substrates. They are found on rocky shores, on mangrove roots, and on coral reefs, soft muddy and silty bottoms, lagoons, brackish waters; and still others are planktonic. Some are sessile while many have limited movements and the cephalopods are some of the fast swimmers. The total catch of molluscs worldwide was 4,951,718 metric tonnes for the year 1980 (FAO); this included nearly 25% oysters and 17% cephalapods. Edible molluscs are commonly harvested mainly as food and for commerce and industries along the coastal regions. Oysters, clams, mussels, cockles, cephalapods, and gastropods constitute some of the delicacies of the tropical food. A few of these species are also cultured in many tropical countries in south-east Asia and Latin America (see Chapter Fourteen). The bivalves, cephalapods and gastropods are mainly exploited in shallow waters, in intertidal zones, and river estuarine mouths where the salinity may fluctuate with the tidal rhythms and other marine sandy beaches. The worldwide catch of molluscan species is shown in Table 9.5.

Table 9.5 Nominal Worldwide Landings of Molluscs (metric tonnes)

Sources	1977	1978	1979	1980	1981	1982	Group Total
OYSTERS							
Crassostrea	16,799	14,643	11,024	6,597	16,732	26,061	91,856
Scallops	1,697	3,241	2,969	3,258	10,835	6,475	28,475
Clams	582	3,259	3,779	1,571	3,406	3,171	15,770
Cockles	11,145	21,596	14,992	36,121	61,404	86,396	170,130
Cephalapods	78,098	96,864	97,652	153,762	106,478	84,430	540,832
Mussels	145,140	155,721	143,023	134,825	144,276	133,841	856,826
Abalones	9	188	223	107	134	68	729

9.7.2.1   Bivalves

The well recognised bivalves in the Tropics and subtropics are the mussels, oysters, windowpane oysters, scallops, clams, and pearl oysters.   The bivalves of the muddy and sandy beaches include different species such as *Mytilus edulis, Mytilus viridis, Cardium psedolima, Tridacn, Donax cuneatus, Solen,* the clam, *Mertrix metrix,* and several other species.   The shells of most species are largely exported, specially from the coasts of Kenya, New Caledonia, Tahiti, and many Indo-Pacific regions.   The larger size oysters, *Crassostrea paraibanensis, Ostrea rhizophorae, Ostrea cucullata,* are found in large numbers in Brazil, Venezuela, India, and West Africa; they are also a popular food resource in many tropical and subtropical countries (see Chapter Fourteen).

   The pearl oysters of natural oyster beds, *Pinctada vulgaris, P. chemntzi, P. margaritifera, P. anomioides, P. atropurpurea,* and a few others are of great economic importance.   These are found in large oyster beds of the Gulf of Mannar, the Gulf of Kutch, and Palk Bay in the southern coastal waters of the Indian Ocean at depths of between 60 and 72 metres.   They prefer the bottom substratum of muddy sand or rocks or encrustation with dead coral and sand.   The matured oyster grows to a size of about 9×11 centimetres.   The pearls, 'oriental pearls', produced by these species are enormously valued.   But since 1907 the pearl oyster fishery has shown signs of decline.   These were taken traditionally by diving, but, as early as 1958, with the aid of Canadian fishing vessels, the Ceylon Government has tried to catch pearl oysters by dredging, which caused considerable damage to natural oyster beds.   On the other hand, the Tutocorin pearl oyster fishery was stopped in 1926 for reasons of conservation.   Other causes of decline apart from over-fishing, are silting of excess sand into the oyster beds, predatory and disease factors.   One must also add competition with cultured and artificial pearls.   Since 1810 Taumoutu atolls in French Polynesia had exported more than 100,000 tons of the pearl oysters of the species, *Pinctada margaritifera* (Salvat, 1981).   *Placuna plecenta* (scallops), of the Philippines, are also exported in large quantities.

9.7.2.2 Chanks

Chanks are some of the largest among the gastropod molluscs. Man's interest in chanks as food, ornaments, for collection, and commerce dates back to time immemorial. Some species such as *Solonogastrecudia forveata* are found from surface to great depths of 2200–4300 metres (Stramulner, 1985). The species *Xancus pyrum* is one of the most important among chank fisheries. They occur in abundance in coastal habitats around Andaman, Nicobar, and many other islands at 10–25 metres in deep waters. They are gregarious and often prefer the muddy bottom. Dextrally coiled chanks with an average diameter of not less than 8 centimetres are not taken. Ceylon alone exported more than 4 million chanks to India before the Second World War, and since then this also has declined. The 'left-handed' or sinistrally coiled chanks of the species, *Turbinella pyrum,* grow to lengths of about 18 centimetres and are very rare – only about 1 in 100,000. They are regarded by Hindus as the most sacred of all shells. A single sinistral shell can command a price of several thousand dollars. It is believed to bring good fortune to the one who owns it. Other gastropods in demand are the big queen conch, *Strombus gigas, Cowry, Cyprraea tigris* and the nautilus, *Nautilus pompilius,* from the Indo-Pacific regions, the Philippines, and the Caribbean Islands.

9.7.2.3 Cephalopods

Among the molluscs, several species of cephalapods – the squid of the genus *Loligo* and the cuttlefish *Sepia* – are found in the neritic waters with sand bottoms of the tropical and subtropical zones. Giant squid are found in large concentrations along the Azores and smaller species in the coastal waters of the Arabian Sea, Ceylon, India, Indonesia, Africa, Brazil, Chile, and Peru. Squid are exported from California. Very lucrative squid fishery has been developed by the Japanese for decades. Cephalapods are predatory and mainly feed on lantern fish and other small species of fish and plankton. Large squid are the most favoured food of sperm whales and their principal feeding grounds are the Azores islands along the mid-oceanic ridge, and the Peruvian-Chilean coastal regions. *Octopus* occurs in abundance in the Indo-Pacific, along the West African, North and South American coasts. They feed by night and are capable of catching prey bigger than themselves. It has been reported that their salivary glands help to digest the food before they take the prey into the gut. They are a less favoured food, eaten only in some localities and are of less commercial value.

9.7.2.4   Tropical Marine Edible Reptiles

There are only a few species of marine turtles in the Tropics. At least five of the seven recognised species usually nest in tropical coasts. The largest of all cheloniidae, and truly pelagic, is the leathery turtle, *Dermochelys coriacea*. The carapace length extends to about 160–180 centimetres. Others are relatively smaller and include: the green turtle, *Chelone mydas,* the hawksbill, *Eretmochelys imbricate,* the loggerhead, *Caretta caretta,* and the flatback, *Chelonia depress.* Though mostly oceanic, they often migrate seasonally to tropical beaches to nest. Except the green turtle, which is herbivorous, most species are benthic feeders on seaweeds, crustaceans and molluscs, and the planktonic jellyfish. They migrate long distances. It has been recorded that turtles tagged and released in Florida were found along the coastal waters of the north-eastern part of Brazil. The species *Cheloneia depress* seems to be widespread and abundant in the Australian continental shelf, the Torres Straits, Papua New Guinea, Indonesia, and Timor (Groombridge, 1985). Others nest in a wide variety of localities, especially islands in the Indio-Pacific regions and the mainland sandy beaches of Burma, Bangladesh, India; around Palk Bay, the Gulf of Mannar, Andamans, Nicobars and the Maldive Islands, Ceylon, Pakistan, the Gulf of Oman, Yeman, Ascension Island, Mozambique, the Seychelles, Madagascar, Costa Rica, Brazil, tropical Mexico, the Caribbean Islands, the Philippines, Florida and Hawaii. They are generally philopatric, i.e. the mature adults return to the same beaches where they were born to lay their eggs. The adults and eggs are vulnerable as the hunters search and dig the sands where they lay eggs. They are hunted for their eggs, meat and shells; the latter is exported as turtle-shell; its shell is used for ornamental purposes; although these species are widely distributed, most of them are threatened worldwide. The marine turtles are nearly on the verge of extinction and are now considered to be an endangered species.

9.7.2 Coral Fish Community

As already pointed out, the fish species living in the coral reef ecosystem display some of the most diverse forms, richness, and beauty in nature. They are well adapted to varying local conditions. The coral reefs provide both shelter and food for a variety of invertebrates and vertebrates. Most fish species are exotic and have well adapted shape, size, dots, stripes and colours to provide camouflage in the coral environment. Colour is an important adaptation. Some of the wide-ranging and distinct species commonly encountered in the Tropics are: the most ferocious moray eels, *Gymnothorax (Murena) meleagris*; the important food fish 'garoupas' of the family serranidae with several species of, which the most common, *Epinephelus tauvina*, have a large mouth to swallow the prey readily, like to hide in coral crevices and grow up to 2 metres in length. The species is very profitable and is commercially exploited in large numbers in the tropical waters of Brazil, Zanzibar, the Red Sea and the Indo-Pacific up to Hawaii. The family pomacentridae or damselfish are some of the common and brightly coloured coral fish which live in groups, solitarily or frequently commensally with sea anemones. *Dascyllus arunas* and *Chromis sp.*, and *Pomacentrus chrysus* are some of the popular representatives. The common cardinal fish, *Apogon leptacanthus;* the flat fish, *Bathus mangus;* tiny gobies, *Elacatinus oceanops,* are usually found in the coral beds of the tropical world, including Florida. The squirrel fish, *Holocentrus rufuns;* puffer fish, *Tetraodon,* the spotted scat, *Scataphags (Chaetodon) argos* are very common in the Indo-Pacific, the Caribbean, Bermuda and Australian coral reefs. There are many species of butterfly fish, *Chaetodontroloplus mesoleucus,* which graze or nibble on algae encrusted corals. The moorish idol, *Zanclus canesens,* is most beautiful to watch during underwater diving; the trigger fish, pipe fish, file fish, sea horses, parrot fish, angle fish, and others are some extraordinarily curious fish in the coral environment (see Plate 5.1). There are more than 2,800 species over a wide geographical range of the Indo-Pacific and Australia, and the Caribbean waters with such richness in diversity. On a global scale, the highest number of reef fish species has been recorded in the Philippines, New Guinea and other Indo-Pacific regions.

9.8 Tropical Sea Fisheries – Some Ancient and Modern Fishing Methods

In pursuit of food, man has used a variety of fishing gear and methods to take food from the seas. Here only basic information on fishing gear and methods relating to the Tropics and subtropics will be briefly discussed.

The tropical seas are characterised by the variety and abundance of fish species. Accordingly, many different types of fishing craft and gear are in common use. Despite the considerably rich marine fishery resources of the Tropics and subtropics, their potential has not been fully assessed. Most tropical countries have very long coastlines (see Chapter Four, Table 4.3) with significantly numerous bays and inlets and quite extensive and productive areas of continental shelves with rich species of fish, both inshore and offshore, but the deep sea fish resource remains virtually unexploited. Despite the significant number of fishing communities living along the tropical coasts around the world with skill and limited knowledge of the profitable fishing grounds and seasonally migratory fish, the greatest handicap lies in the nature of their craft and gear, which have changed very little over centuries. Their boats are artisanal, mostly dugout canoes made from a range of regionally available timber Fig 9.12.

Fishing gear such as trawls, drift long lines and purse-seines are still unfamiliar for many fisherman, and even the nylon nets and drift long lines have just begun to be introduced. The other more serious problem with fisheries in tropical countries is the lack of quick transport and ice-cold storage facilities to preserve the perishable commodities of valuable fish, prawns, lobsters and cephalapods. The commercial fish resources are mostly confined to the continental shelves and the continental slopes. Until the innovation of fishing technology and the development of powered boats, so as to be able to use trolling pelagic and trawling mid-water and deep sea dwelling fish resources, with more sophisticated equipment aboard, the fishermen seldom employed any modern methods; they were only familiar with rather primitive and archaic methods either on small- or large-scale fishing; they hardly ventured into deep sea or undertook demersal fishing beyond their barriers of depth limits. However, these methods were of considerable interest, though less effective and were often non-profitable. Perhaps the French were the first to use

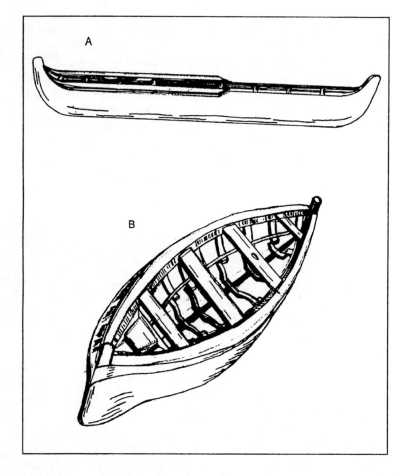

Figure 9.12   (A) Dugout canoe commonly used in inshore waters and in estuaries in Indo-Pacific, along the north-eastern coasts of Brazil, and other Latin American countries, especially by native Indians. (B) Boat built with wooden planks and can be driven by sails, oars or small motors in shallow waters.

steam-powered fishing boats during the eighteenth century whereby the fisherman could spend long hours in the sea, with facilities aboard for freezing the fish catch. With the advent of technological innovations, the Japanese and Russian developed heavily mechanised vessels of most modern 'floating factories' fitted with electronic echo sounders and sonars to find fish on a three-dimensional scale. They are able to exploit the resources of the sea, including those of coastal waters in many regions of Tropics and subtropics, even at the expense of often flouting international laws. In the geographical range of tropical nations and technological and economic terms, in the Atlantic, the fishing fleets are operated mainly by the USA, Venezuela, Guyana, Surinam, Brazil – despite its rugged shelf area –Uruguay, Mexico, Honduras, Panama, Senegal, Guinea, Nigeria and Angola. The coastal area of the Gulf of Guinea and Senegal, up to 6 miles and beyond, is a fishing area exclusively reserved for traditional fishing, using simple craft and tackle for both pelagic and benthic species. Some of the pelagic species include *Sardinella maderensis, Ethmalosa rimbriata, Caranax rhonchus*, and *Pomadas spp.*

The fish resources in the Indo-Pacific regions are now exploited on commercial scales by Taiwan, Thailand, Vietnam, China, Singapore, Australia, Papua New Guinea, Indonesia, the Philippines, Malaysia, Burma, Bangladesh, India, Ceylon, Pakistan, Oman, Madagascar, Saudi Arabia, Somalia, Mozambique and South Africa on commercial scales.

Since recently, though many tropical and subtropical developing countries cling on to their traditional methods, significant impetus is coming into effect, especially in most Latin American and Southeast Asian countries. These countries have often been receiving some foreign aid and UN assistance, in the form of 'technical know-how' and capital investment on trawls and seiners and are able to operate their own fishing fleets with sufficient efficiency. In most tropical countries, due to lack of adequate or less developed facilities in handling, marketing, and storage, besides local consumption, much of the fish products are either dried or salted and dried with significant loss. Only a few countries are now able to introduce canning. Some of the fish are used to produce oil or fertilisers. Despite their importance, the details will not be considered here. Nevertheless, a brief account that may essentially reflect the progress of the fisheries

Figure 9.13   Lobster Pots – usually lowered on the bottom, 6 – 12 at a
                   time, by attaching ropes and the spot is marked by floats of
                   buoyant materials. The pots are used to entrap lobsters,
                   crabs, and a variety of tropical fish.

Figure 9.14   The funnel shaped fence trap. The fence made out of
                   mangrove sticks or bamboo posts. At the narrow one-way
                   end of the enclosure; usually school of fish can be trapped.

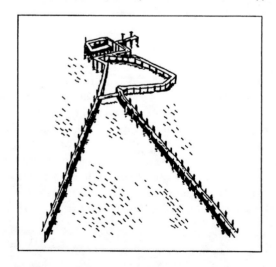

in most coastal tropical and subtropical nations and some of these principal fishing methods designed to catch the kind of fish that live in different depths, habitats, localities and behaviour, which are so significant, may benefit the reader and so will be discussed next.

Briefly, the principal methods used on commercial fisheries in the Tropics and subtropics are:

(1) Trapping.
(2) Long line fishing.
(3) Gill nets and drift nets.
(4) Beach seining and purse-seining.
(5) Dredging or trawling.

9.8.1 Trapping

Though an ancient tradition, by far the simplest method of fishing in the tropical coastal shallow waters of less than 60 metres deep is by trapping the fish, particularly in Southeast Asia, Africa, the Indo-Pacific Archipelagos, South and North Americas. Devices of 'pots' of suitable sizes and shapes for lobsters, crabs, and fish are still in use. These can be divided into two main types:

(1) Cage-like traps with suitable mesh size, made out of canes, flexible bamboo strips, midribs of the palm leaves or plastics or other materials. The pots are designed cleverly with a cone-shaped trapdoor mechanism to prevent the escape of the fish, lobsters, prawns, and other creatures that enter the pots. The pots can be lowered to the bottom or raised to the surface either individually or in pairs at desired areas and depths; the pots are usually left for a period of twenty-four to forty-eight hours. Depending on whether the bottom is smooth, sloped or rocky, a light weight is added to the pot to set it right, and the location is marked by buoyant barrels or polystyrene floats tied to the main rope by which the pots are lowered (see Fig. 9.13). Natural bait often attracts a great many species, especially pieces of octopus meat placed in the centre of the pots before lowering them. The return, or the harvest, is often rewarding.

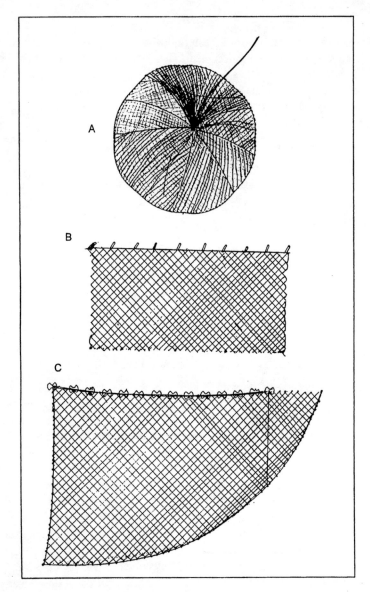

Figure 9.15   Fishing gear: (A) Cast net; (B) Surface gill net; and (C) Drift
net or wall net commonly used in the tropics and subtropics.

(2) This involves piles and weirs. The weir type is an integrated system of a trap and a fence. It requires large areas of shallow coastal waters. A fence-made enclosure of a relatively fixed nature is made out of posts, especially of stakes of mangroves and bamboo; the posts usually stick out above the water. The trap is placed at the far end of the funnel-shaped enclosure, and schools of fish are stopped at a time by the fence and made to converge towards the narrow end of the enclosure where the trap is placed – they can be easily caught in this way, as shown in Fig. 9.14.

9.8.2 Long Line Fishing

Besides cast nets (Fig. 9.15A), perhaps the oldest gear used in tropical countries is hooks and lines. Depending on surface or bottom feeding fish, short lines with suitable size hooks are fastened, about 1 metre apart, to one long line which varies in length. In shallow waters, lines carrying up to 1000-2000, and in other regions 500-1000 hooks may be used. Hooks are baited with small pieces of rather cheaper kind of fish, or nowadays plastic lures are often used – the latter will cut the cost on live bait and will also save valuable time in the sea. The lines are suspended near the surface for the pelagic fish and lowered to the bottom of the seabeds for benthic fish. At a few points the single long line can be tied to finer lines to the surface buoys in order to indicate the fishing place. Usually a small boat with one or two fishermen can operate the whole task of line fishing. The commercial fishing fleets by hooks and line have shown that the catfish of the genus, *Tachysurus,* often dominate the catch; fishermen, by trolling the baited hooks, seek the larger species of other fish, which include different species of sharks, yellowfin tuna *(Thunnus macropterus)*, sier fish *(Scomberomorus commoersoni,, Rasttrelliger kanagurta)*, marlin *(Makaira indica, M. mitsukrii)*, sailfish *(Istiophorus gladius)*, belones *(Belone cancilia, Ablennes hians)*, barracuda *(Sphyraenajello)*, sergeant fish *(Rachicentron canadus)*, queen fish *(Chorinimus lysan)*, and other carangid species – these are caught in waters of depths of up to 100 metres. At times, this type of gear can prove hazardous, particularly with coral beds. Relatively simple and cheaper hand lines are also in use among the rough submerged bottom areas of rocks and

Figure 9.16   Community fishing; operation of beaching seining net.

coral reefs, and this method becomes more profitable, especially when operated on the sight of school or shoals of fish.

9.8.3 Gill Nets and Drift Nets

These are some of the oldest fishing methods in the Tropics and have become refined over the years. Essentially, gill nets are of two types:

(1) Those used by local fishermen in very shallow coastal waters. These nets, with the surface floats, are tied to a head line; the bottom end of the net has no ground rope and so is free. The net can be suspended on the sea surface with the bottom end lowered with small lead bobbins at various intervals. These are temporary and are easily removed; these easily entangle the fish. For a variety of pelagic fish which live closer to the surface, gill nets are used to cause the fish to become entangled by their operculum gill cover (hence the name gill nets). These are commonly made up of fine cotton or of more durable nylon with mesh of different size, with the top rope threaded with smaller polystyrene surface buoys. These are used in many parts of the world, but more so in the Tropics and subtropics, either in shallow waters or in deeper water. To catch fish in the surface layer, the nets are hung vertically with smaller lead sinkers for weight. During hauling, one end of the wing-side rope is fixed while the other end is pulled after lowering the net in a semi-circular manner.

(2) The second type is very similar in principle, but differs in dimensions and is free to be drifted. These are also called wall nets and are used principally for pelagic fish (Figs 9.15, a, b and c).

9.8.4 Beach Seining and Purse-Seining

As noted above, the gill nets entangle the fish, beach seines and purse seines surround the fish, and trawls scoop them up. Beach seining is one of the most important fisheries, which contributes substantially to the economy and total fish landings in tropical countries. Beach seining by necessity is restricted to sheltered bays and to coastal waters that are relatively calm in order to avoid the risk of seasonal

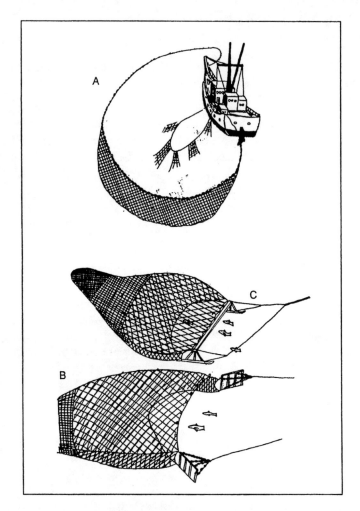

Figure 9.17 (A) Purse seining in operation and is used by large commercial or Government agencies; (B) Purseing involves large nets used for catching pelagic and demersal fishes; and (C) the beam trawl is similar to purse seine net but relatively smaller with the conical end, and the wide open-mouth end can be spread out with the beam and has many devices used for benthic fishes, prawns, cephalopods and other species in the tropical waters.

winds. In tropical and subtropical coasts, including many bays and islands, beach seining is shared by socially bound communities of fishermen. As the whole operation involves a great deal of effort, and many men – between twenty and forty – work jointly at a time. Like purse-seining, several types of beach seining are available. The common net consists of a conical end and a main body with a wing on either side. Each segment of the net has a different mesh size; the smallest size is at the far tapering or 'cod' end and becomes progressively larger and larger, being greatest towards the mouth or free end, where the net is threaded to ropes. Depending on weather conditions, tidal currents, and smoothness of the bottom slope, the length of the net varies, but usually it extends to about 5 kilometres with the hauling rope. Smaller non-mechanised boats are used to lay the net.

This method is used most profitably on sighting of shoals of fish. On sighting, the net is lowered to surround the shoal of fish, and gradually the free end of the net is drawn to the shore towards the fixed end until all the fish are caught (Fig. 9.16).

A variety of school forming species of different size and age, for instance, *Chirocentrus dorab* (wolf herring), *Euthethynnus alleteratus* (mackerel tuna), *Auxis thazard* (frigate mackerel), *Trichiurus haumela* (horse mackerel), *Scomberomorus interruptus* (sier fish), *Engraulis mystax* (anchovy), *Engraluis baelama* (sprats), *Sardinella longiceps* (sardines), *Leiognathus equula* (silver bellies), and many other species are caught.

9.8.5   Purse-Seining

Purse-seines are commercially used for demersal and fish living on or in the bottom over the continental shelves. Purse-seines are relatively larger nets, designed to surround the fish and trawl and scoop them up. It is made out of fine nylon threads with suitable mesh size as recommended by the UN. In tropical countries, not all fishermen abide by this international regulatory requirement of mesh size, but many tend to comply with obligation, and many government agencies impose this as if it were a legal obligation. Catch records for five consecutive years of purse-seining off Bombay (Annigeri, 1982) shows that some species are seasonal, and the species, *Sardenella dayi*, is usually caught in large numbers and supports purse-seining quite well (Fig. 9.17).

9.8.6 Trawling and Dredging

Most richer fishing nations use trawlers, but trawlers are relatively modern to many tropical countries. To offset the operational costs at sea, many improvements have been introduced to increase their efficiency to catch demersal and bottom living fish on the continental shelves. These, with their highly skilled skippers and crew, sail out a long distance from their shores to the potential fish-banks or fishing grounds. Trawling is carried out repeatedly, each operation usually taking 4–5 hours. The trawl or 'bottom trawl' is a cone shaped net with the wide end open and its narrow tip end made with a smaller mesh size. As the net is dragged through huge bodies of water with the lower end of the net sweeping the bottom of the seabed, the pressure of the water contributes to keep the mouth wide open and the fish can be scooped up by the net (Fig. 9.17, B). By this method, a variety of most valued fish of all ages and sizes are caught. These include a variety of bottom living species of flat fish, the larger garoupas, bream, mackerel, catfish, red snappers, rays, sharks, and a number of species of shellfish (especially crustaceans, prawns, lobsters, crabs, cephalopods and other molluscs).

9.9 Present Status of Fishery

Fishery statistics worldwide indicate that fish stocks are on the verge of depletion and fishery prospects are in trouble. The principal causes seem to be over-exploitation of fish stocks and damage to their natural habitats. Despite our belief that the living resources of the sea are infinite and inexhaustible, the harvest of many fish stocks have levelled off, and if the trend of depletion continues it might not be possible to obtain much more protein-rich food from the sea in future. In many parts of the ocean, the most suitable pelagic species such as herring, anchovies (Peruvian fishery, for example), horse mackerel, sardines, tuna, large invertebrates and whales, and the benthic crustaceans, all show signs of depletion. Bottom dwelling or benthic species of fish are also on the decline. Different nations are competing for different fish – the Japanese fishing fleets pursue tuna beyond their boundaries into vast stretches of tropical waters. Now, the commercial fishing of krill in the Antarctic also threatens whale and penguin populations. Large-scale habitat destruction, with the more serious consequence of water pollution of estuaries, river mouths and coastal waters, has an enormous impact on the fish population. Thus, the constraints on relatively cheaper protein available from the sea have been reached. It is therefore necessary, by common consensus, that both over-fishing and environmental protection must be enforced. The highly commercially exploited tuna may need to be protected by some international quotas before they disappear from nearly all oceans. In order to ensure the maximum stability and to maintain a healthy state of reproduction of the fish stock, a rational catch based on scientific data must be agreed to obtain a maximum sustainable yield, i.e. the total amount, in weight, of a given species that can be caught annually without depleting the stock in subsequent years. World total marine catch for the year 1992 was 82,534,300 metric tons and this constitutes 84.1% of the world's marine and inland fish landings put together (FAO, 1994). The marine fish has already reached the best sustainable level, to the extent of seriously damaging the marine ecosystems. According to the latest fishery statistics on catches and landings (FAO, 1994) there has been a steady increase in the fish taken from the marine waters since 1970 and Fig. 9.18 clearly shows the upward trend of catches from 1980 to 1992.

An estimate on maximum sustainable food fish resource has been some 120,000,000 metric tons per year; this together with inland, squid and krill (FAO, 1984). A limit to the size and amount of fish that can be taken should be imposed as mandatory by FAO and other important international agencies. Management should be aimed at controlling the balance between fishing mortality and recruitments; fisheries research throughout the developing tropical and subtropical world must be directed towards this goal and avoid over-exploitation that might result in the extinction of the species.

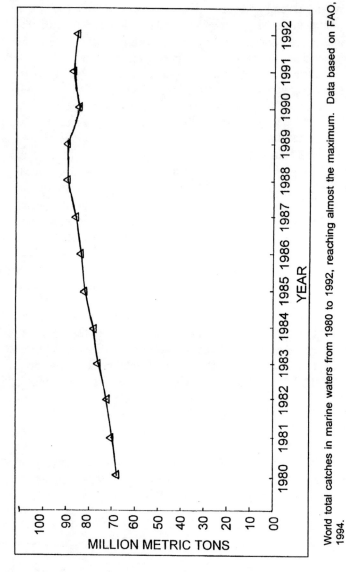

Figure 9.18   World total catches in marine waters from 1980 to 1992, reaching almost the maximum.   Data based on FAO, 1994.

References

Akinrushkin, I. I., *Cephalapods of the seas of the USSR*, Moscow, USSR Ac. of Sci. Press, 1963, pp.1–234

Aleyew, Y. G., Necton, Dr. W. Junk, The Hague, 1977, p.435

Annigeri, G. G., 'On the fishery and biology of Sardinelia dayi a Karawar, India', *J. Mar. Biol. Ass.* India, 24, (1 and 2), 1982, pp.133–140

Bascom, W., 'Technology and the ocean', *Scientific American*, 221, 1969, pp.199–234

Berg, L. S., 'Classification of fishes, both recent and fossil', Trav. Inst. *Zool. Sci.*, USSR, 5 (2), 1940, pp.87–517
Also lithoprint, J. W. Edwards Ann Arbor, Michigan, 1947

Caviedes, C. N., and Fik, J. T., (ed. Glantz. M. H), *The Peru-Chile Eastern Pacific Fisheries and the climatic oscillation in climatic variability, climate changes and fisheries,* Cambridge, Cambridge University Press, 1992, p.355

Chopra, B. N., (ed.), *Handbook of Indian fisheries*, Prepared for the third meeting of the Indo-Pacific Fisheries Council, 1951, p.129

Day, F., 'The fishes of India; being a natural history of the fish known to inhabit the seas and freshwaters of India, Burma and Ceylon', London, Dawson, 1878–1888, vol. I, p.778, and vol. II, atlas containing 198 plates

De Beaufort, L. F., *The fishes of the Indo-Australian Archipelago*, Leiden, E. J. Brill, 1962, vol. II p.481

Dore, I., and Frimot, C., *An illustrated guide to shrimp of the world,* Huntington, New York, Osprey Books, 1987, p.229

Ekman, S., *Zoogeography of the sea*, London, Sidgwick and Jackson, 1953, p.417

FAO, *Year Book of fishery statistics of FAO for 1981*, FAO, Rome, 1981

FAO, *Year book of fisheries statistics catches and landings for 1982*, FAO, Rome. 1984, vol. LIV p.393

FAO, *Year book of fishery statistics, catches and landings*, Rome, 1994, vol. LXXIV, p.677

Fowler, H. W., *Fish of the Red Sea and southern Arabia*, Jerusalem, Weizmann Science Press, 1956, p.240

274

Greenwood P. H., et al., 'Phyletic studies of teleostean fish, with provisional classification of living forms', *Bull. Am. Mus. Nat. Hist.*, 131, 1966, pp.339-456

Groombridge, B., 'Indian sea turtles in world perspective', Proc. Sympos. Endangered marine animals and marine parks, 1, 1985, pp.205-213

Hackel, E., Plankton – studien., Jena, Fisher, 1890, pp.1-105

Holt, S., 'The living resources of the sea', *Scientific American*, 221, 1969, pp.178-194

Issac, J. D., 'The nature of oceanic life', *Scientific American*, 1969, pp.221, 147

Johnson, D. S., 'Prawns of the Malacca Straits and Singapore waters', *J. Mar. Biol. Ass.* India, 18 (1), 1976, pp.1-54

Joseph, K. M., 'A study on the deep seas prawn resources of the south-west coast of India', *Sea Food Export Journal*, 1, (8), 1970, pp.77-83

Keenleyside, M. H. A., 'Some aspects of Schooling behaviour in fish', *Behaviour*, 8, 1955, pp.183-248

Lagier, K. F., Bardach, J. E. and Miller, R. R., *Ichthyology.* Ann Arbor, University of Michigan, 1962, p.545

Longhurst, A. R., and Pauly D., *Ecology of tropical oceans*, New York, Academic Press, INC., 1987, p.407

Lowe-McConnell, R. H., *Ecology of fish in tropical waters*, Edward Arnold, London, 1977, pp.64

McManus, J. W., 'Marine speciation, tectonics, and sea level changes in south-east Asia', Proc. Int. coral reef symposium, 5th, (1984), 4, 1985, pp.133-138

Mayr, E., 'Biological classification: toward a synthesis of opposing methodologies', *Science*, 1981, pp.510-516

Nair, R. V. S., Saoundarajan, and K. Durairj, 'On the occurrence of *Palinurus longipes, Palinurus penicillatus, and Palinurus polyphagus* in Gulf of Mannar with notes on the lobster fishery around Mandapam', *Indian J. Fish.* 20 (2), 1973, pp.333-350

Nelson, J. S., *Fish of the world*, John Wiley and Sons, New York, London, 1976, pp.476

Norman, J. R., *A History of Fish*, London, Ernest Bend Limited, 1975, p.467

Oommmeny, V. P., 'Results of the exploratory fishing in Quilon Bank and Gulf of Mannar', Bull., IFP., 4, 1980, pp.1–49

Pitcher, T. J., (ed.), 'Functions of shoaling behaviour in teleosts', *The behaviour of Teleosts fish,* London, Croom Heim, 1986, pp.294–338

Ryther, J. H., 'Photosynthesis and fish production in the sea', *Science,* 166, no.3901, 1969, pp.72–76

Sale, P. F., 'Ecology of fish on coral reefs', *Oceanogr,* Mar Ann. Rev., 18, 1980, pp.367–421

Salvat, B., 'Preservation of coral reefs: scientific whim or economic necessity, Past, present and future', Proc. 4th Int. Coral Reef Symp., 1, 1981, pp.225–229

Stramulner, F., 'Rare and endangered marine molluscs – A review', Proc. Symp. Endangered marine animals and marine parks 1, 1985, pp.371–382

United Nations, *Demographic Year Book,* UN, New York, 1981

Wenk. Jr, 'The physical resources of the ocean', *Scientific American,* 221, 1969, pp.167–176

Williams, A., *Lobsters of the world,* Huntington, USA, Osprey Books, 1988, pp.186

Young, J. Z., *The Life of Vertebrates,* Oxford Press, 1962, p.820

Chapter Ten
Whales and Whaling

Whales are the largest among the living forms. Although the term 'whale' generally implies something of a large size, for taxonomic purposes, killer whales and narwhals, and a great number of the smaller cetaceans (such as dolphins and porpoises) are also conceived collectively as representatives of whales. Some seventy-six living species (Watson, 1985) have been recognised. They are worldwide in distribution and are found in all oceans and in most of the major seas.

10.1   Origin and Evolutionary Interests

During the Cenozoic, the most spectacular event had been the advent of the diverse and dominant group of animals, the mammals. Perhaps their environment had been more favourable and the mammalian evolution had been relatively rapid. Fossil evidence indicates that generally, in several lineages, there was a trend towards developing larger sizes among mammals. Among the mammals, the carnivores became more successful with well-developed large, especially upper, canine teeth more suitable for tearing and grasping. The tendency for large size canine teeth to tear and grasp seemed to have appeared in three separate times leading to three unrelated groups, the last of which culminated in the well-known Pleistocene *Similodon*. One of these groups of carnivores resorted to an aquatic habitat, giving rise to the modern seals and sea lions. The first and earliest of these lineages seem to be the ancestral group – the enaliarctidae. Following this, in the early Tertiary Period, another branch of the carnivores, the creodonts adapted to the sea and gave rise to the enormous dinosaur-like *Basilosaurus (Zeuglodon)*; these archaic primitive forms of cetaceans became extinct at the end of Eocene (Strahler, 1982), but are from whom the present whales diverged.

10.2 The Ancestry and the Possible Link Between Land Mammals and the Archaic Whales

The recent discovery of early Cenozoic mammalian fossils, including a rear part of the cranium with a piece of lower jaw with three pointed premolars and three detached sharply pointed molars, in Pakistan near the east-west of the Indus River, indicates that these fossil remains were of more primitive transitional forms of whales. It was named as *Pakicetus* because of the country of origin. The site where these relic fossil remains were restored was of great geological significance. During the early Cenozoic period, the then Tethys Sea between the two great continents of the peninsular of India and southern Asia began to be narrowed and eventually disappeared. Under the colliding pressure of the continents, deposits of the shallow marine Tethys seaway of the Asiatic side became squeezed and uplifted into what is now recognised as open folds of 'redbeds', because of the rich iron oxide.

Pakicetus seems the oldest, most primitive, and rather incompletely adapted to aquatic mode of life. *Pakicetus* is even older than the *Basilosurus (Zeuglodon)*. Like the Palaeocene carnivore land mammals and other advanced whales, the *Pakicetus* suggests a possible ancestry of those of the Eocene archaic cetaceans of Archeocets such as *Protocetus* and *Indocetus,* and to the terrestrial carnivore wolf-like Mesomychid Condylartha. However, to draw any conclusions more fossil evidence must be sought from the same strata and of other regions of the same epoch when the Tethys seaway began to be closed.

Despite many uncertainties concerning cetacean origin, it is broadly accepted that in common with many other eutherians, they probably derived from a remotely common ancestral mammalian stock of 'Creodont'. Since the Eocene, about 40–54 million years ago, perhaps, because of their enormous size they became specialised into distinct fish-like mammals, and were beautifully well adapted for a fully aquatic mode of life. Of the three suborders of the cetaceans the archaeoceti, which appeared during the middle of Eocene, had already become extinct during the middle of Miocene – long before man originated. Modern whales are assigned to two suborders: Odontoceti or toothed whales, which probably appeared during late Eocene; Mysticeti or baleen whales, which evolved during Oligocene.

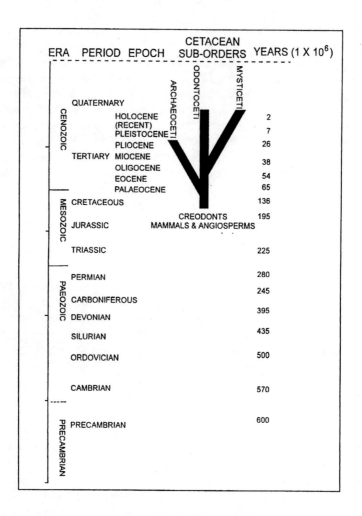

Figure 10.1   Evolution of Cetaceans.

So far, no missing links have been found to bridge the gap between the archaeoceti and the modern whales. The three groups of whales probably evolved independently (Fig. 10.1).

10.3   Characteristics of Living Whales

Whales are exclusively adapted to aquatic life; they are lung breathers and warm blooded; viviparous and suckle their young ones; body varies in length from a few to over 30 metres, fish-like or fusiform; skin is generally smooth and devoid of any hairs, except for the whiskers or few hairs on the chin and supported by a thick layer of blubber; sweat glands are absent; the end of the tail stock is extended into dorsoventrally compressed tail flukes. The head is somewhat triangular and flattened with the skull telescoped; the nostrils or blow holes are either single or double and open from behind the tip of the snout; external ears are replaced by minute apertures just behind the eyes. The eyes are small and dorsolateraly placed. In Odontoceti, teeth are present in either jaw in varying numbers, but absent in Mysticeti, though the embryonic teeth buds are reabsorbed and replaced by the baleen plates of secondary dermal origin from the upper jaw only; the fore limbs are modified into paddle-shaped flippers, the hind limbs are absent, except for the vestiges of the pelvic girides; most whales have a dorsal fin, the ventral aspect of the body is very conspicuous, especially in the Mysticeti whales, by the presence of a variable number of ventral grooves; they are carnivorous or filter-feeders; in males, the penis is retractable, the scrotum does not descend and the penal slit and cloaca are separated, whereas in females the vagina and the cloaca are closely situated in a longitudinal slit, and the teats are retractile. Copulation usually takes place, except in bowheads, in vertical position and the flippers are used for clasping and steering. Most whales have a gestation period of 10–12 months. The blood and muscles have a high concentration of myoglobin; they also seem to have about 11% protein, showing a close relationship with ungulates and carnivores as to their possible descendance. The diploid number of chromosomes are claimed to be 44. Most whales are intelligent with large and highly convoluted brains; their auditory system is well developed and they use echolocation as the principal means of communication; the inferior colliculi are relatively larger than the superior colliculi; olfactory nerves are absent or less developed. The sperm whales can dive to depths of more than 2,200 metres and control their breath for just over one and a half hours. Most whales are epimeletic in behaviour.

10.4   Classification of Cetaceans

Marine mammals include whales, dolphins, porpoises, walruses, seals, sirenians and sea otters, the Greeks being the first to recognise the whales, dolphins and porpoises as mammals. However, it was not until Linneaus, the Swedish naturalist, as early as 1773, that a classification based on true biological features was possible. A comprehensive list of cetacean characteristics will be most helpful and convenient to identify many of the whales which frequent most of the tropical and subtropical waters, particularly during their breeding seasons and when they often get stranded in small groups along many tropical shores.

CLASS: Euthera:

ORDER: Cetacea:

SUBORDERS: 1.   Archaeoceti (= ancient) whales; heterodont, some of their characteristics are closely related to terrestrial mammals.

2.   Odontoceti (= toothed) whales with only a single blow hole; some 5–7 families are recognised with living representatives.

3.   Mysticeti (= baleen) whales; the teeth in the upper jaw are replaced by baleen plates of variable number in the upper jaw only, and with double blowholes.

DISTINGUISHING CHARACTERISTICS
OF THE TWO LIVING SUBORDERS

ODONTOCETI	MYSTICETI
1. Except sperm whales, killer whales and beaked whales, most representatives such as dolphins and porpoises are smaller in size.	Mostly, enormously sized, the smallest being the smaller rorquals.
2. Presence of teeth in either or both jaws – all of the same type or monophydont.	Presence of baleen or plates only in the upper jaw, remarkably suited for filter-feeding.

3. Skull mostly telescoped. The premaxilla and the nasal bones are somewhat asymmetrical.

The skull usually telescoped. The maxilla and premaxilla of the upper jaw are typically arched to shift the nostrils a little backwards.

4. The blowhole is single.

The blowhole is doubled.

5. In some, the melon organ or spermaceti is present.

Spermaceti or melon is absent.

6. Dorsal fin is usually present, except in a few species where it is lacking.

Presence of dorsal fin which is strongly supported by leathery fibrous tissues.

7. The olfactory nerve is absent. or vestigial.

The olfactory nerve is very much reduced.

THE CLASSIFICATION OF TWO LIVING SUBORDERS AND FAMILIES

ORDER: CETACEA

SUBORDER: 1. ARCHAEOCETI (exclusively extinct and only found in fossil forms)

examples:

Pakicetus

Basilosaurus Zegulodon

SUBORDER: 2. ODONTOCETI (toothed whales, homodonts)

Family 1: Physeteridae

Diagnostic characters: gregarious; length varies from 3–20 m; snout projects considerably beyond the lower jaw, a single blowhole; spermaceti organ well developed; 9–30 homodont teeth in the lower jaw, ventral grooves at the throat; found in most of the oceans (Fig. 10.2).

Physeter catadon (Sperm whale) – widely distributed in tropical and temperate waters, but the males often enter cold waters.

Kogia breviceps (Pygmy sperm whale) – worldwide, especially in tropical and subtropical waters.

Kogia simus (Dwarf sperm whale) – a small species found both in tropical and temperate waters.

Family 2: Ziphiidae*

Length 4–13 metres; snout is extended into a prominent* beak; solitary or in smaller groups; caudal flukes without a dividing notch;

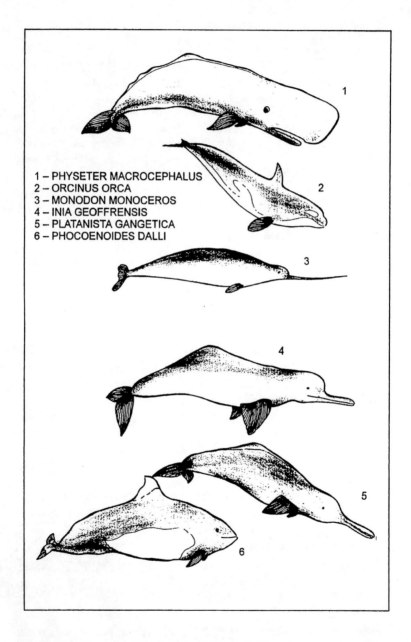

1 – PHYSETER MACROCEPHALUS
2 – ORCINUS ORCA
3 – MONODON MONOCEROS
4 – INIA GEOFFRENSIS
5 – PLATANISTA GANGETICA
6 – PHOCOENOIDES DALLI

Figure 10.2   The Odontocetei or toothed whales.

dorsal fin is reduced; one or two ventral grooves confined to throat; teeth monodont, varies in number from 1-27, present mostly in the lower jaw (but in *Tasmacetus* in both jaws). Found in all oceans and major seas.

Tasmacetus sheperdi
Beradius arnuxii
Beradius bairdii
Hyperoodon planifrons
Hyporoodon (rostratus) ampullatus Ziphinus cavirostris
Mesoplodon (= Dioplodon) densirostris
Mesoplodon (=Dolichodon) layardi
Mesoplodon ginkodens
Mesoplodon grayi
Mesoplodon carlhubbsi
Mesoplodon (= gervaisi) europaeus
Mesoplodon mirus
Mesoplodon stejnegeri
Mesoplodon bowdoini
Mesoplodon bidens
Mesoplodon hectori
Mesoplodon (Indopacetus) pacificus

Family 3: Monodontidae
Comprises the narwhal and the white whales, or beluga – gregarious; moderate length of 3-4.5 metres; blunt snout with the absence of any ventral groove; the dorsal fin is very much reduced or absent; a prominent melon body; two teeth (monodont) in the upper jaw only, and one of which, usually the left one, is extended into a long tusk engraved with spiral grooves, specially in males; these whales are generally confined to the northern hemisphere while a third species[1] is often found in the waters of the Bay of Bengal to the northern part of Australia.

Monodon monoceros (narwhal or unicorn whale) – of economic importance because of its tusk and its oil, which is also taken.
Delphinapterus leucas

---

[1]   *Orcaella brevirostris.*

Family 4: Phocoeneanidae
Although closely resembling the dolphins, they have subtle differences, and the family includes true porpoises. Usually found in smaller groups, often in pairs or singly; smaller in size, 2-9 metres long; the premaxilla is slightly swelled anterior to nasal aperture; dorsal fin generally present, except in one species *(Neophocaena);* on both jaws there are 15-30 teeth which are somewhat spade like; common in most seas, harbours and estuaries of fairly warm waters.

Phocoena (vomerina) phocoena
Phocoena sinus Phocoena spinipinnis
Phocoena dioptrica Phocoena daffi
Neophoecena phocoenoides
Neomeris phocoenoides

Family 5: Deplhinidae
Comprises true dolphins, including the killer and the pygmy killer whales. Usually gregarious; small to medium size (1.5-10 metres); teeth on both jaws, which are simple, with the number of teeth varying from 65-126; ventral grooves are absent; dorsal fin is recurved (absent in *Lissodelphis);* found in most parts of the world; males are generally larger than females and are some of the fastest swimmers.

Orcinus (Grampus/rectipinna) orca (Killer)
Pseudorca crassidens
Fereesa attenuate
Globicephala melaena
Globicephala macrorhynchus – common in tropical and subtropical
                                 waters
Peponocephala eletra
Lagenorhynchus obliquidens
Lagenorhynchus cruciger
Lagenorhynchus australis
Lagenorhynchus obscurus
Lagenorhynchus aclkus
Lagenorhynchus albirostris
Lagenorhynchus hosei
Cephalorhynchus hectori
Cephalorhynchus eutropic

Cephalorhynchus commersoni
Cephalorhynchus heavisidei
Lissodelphis borealis
Lissodelphis peronii
Grampus griseus
Tursiops truncatus
Steno (rostratus) bredanensis
Delphinus delphis (common dolphins)
Stenella coeruleoalba (striped dolphin)
Stenella longirostris (spinner dolphin)
Stenella attenuate
Stenella graffmani
Stenella plagiodon
Sotalia fluviatalis
Sotalia (brasiliensis) guianesis
Sousa teuszii (species endemic to the West African coast and found
        in shallow regions of the coastal environment,
        lagoons and estuaries)
Sousa chinesis
Pontoporia blainvillei
Lipotes vexillifer
Inia geoffrensis
Platanista gangetica
Platanista (minor) indi

SUBORDER: 3. MYSTICETI
Probably descended from toothed archaeoceti-like ancestors. Consists all of the largest known whales; the teeth are replaced by baleen plates of varying numbers (100–450), excluding the smaller ones; the baleen plates are usually suspended from the upper jaws only; the lower jaws are devoid of any teeth or plates. They are also called baleen or whalebone whales (Fig. 10.3).

Family 1: Balaenidae
Usually small and sturdy bodies; ventral grooves below the throat are generally absent; large mouth with numerous baleen plates from the arched upper jaws. They are found in smaller groups or singly and commonly confined to the northern hemisphere, especially in subtropical and temperate zones. Right (Greenland, black right) whales.

1 – ESCHRICHTIUS ROBUSTUS
2 – BALAENOPTERA MUSCULUS
3 – MEGAPTERA NOVAEANGLIAE
4 – BALAENOPTERA ACUTOROSTRATA
5 – BALAENA MYSTICETUS

Figure 10.3   Mysticeti or Baleen whales.

Balaena mysticetus (right or bow head whale)
Eubalaena glacialis (black right whale)
Caperea marginate (pygmy right whale)

Family 2: Escrichtidae
Grey whales of up to 15 metres in length; the dorsal fin is absent; few ventral grooves; dark grey in colour, often with light white patches; baleen plates are rather shorter; jaws are nearly straight or unarched; they migrate in groups mostly in the Pacific.
   Eschrichtius robustus (Grey whale) – hunted to near extinction

Family 3: Balaenopteridae
Commonly called rorqual (= grooved); huge bodies, between 7 and 30 metres long; a distinct dorsal fin; ventral grooves deep and numerous; thick blubber; baleen plates are numerous; the lower jaw is slightly curved; a pointed snout arising from a dorsoventrally compressed head. Found in all seas. Unusually they come to tropical waters to breed, in particular the minke whales.

Balaenoptera musculus (blue whale) – the largest whale found
                                        in all oceans
Balaenoptera physalus (fin whale)
Balaenoptera borealis (sei whale)
Balaenoptera acutorostrata bonaerensis (minke whale) – all oceans,
                but they regularly visit warm tropical waters to breed.
Balaenoptera novaeangliae (humpback whale)
Balaenoptera (edeni) brydei

10.5   Whaling and Whale Products

No reliable data was available on the total whales taken prior to the turn of the century. However, as early as 1900, nearly 30,000, especially blue, fin, sei, humpback, and sperm whales were taken annually in the Antarctic waters alone. An estimate based on FAO (1959) data suggests that of the total world fisheries, which included fish, crustaceans, bivalve and other molluscs, whales, seals, walrus, and other marine mammals, the whale constituted only about 6%. As the uncontrolled exploitation continued and the whale stocks were reduced, the idea of conservation of whales was seriously considered by IWC. Since the discovery of the fossil fuels as mineral oils at Pennsylvania in 1859, the demand for whale oil has been on the

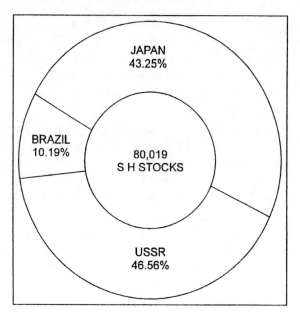

Figure 10.4   A comparison of antarctic minke whales stocks exploited by three whaling nations, expressed as a percentage of the total number caught during 1972/3 and 1982/3 seasons.

Figure 10.5   Foetuses of minke whale encountered in the catch during 1982/83 season by months and numbers.

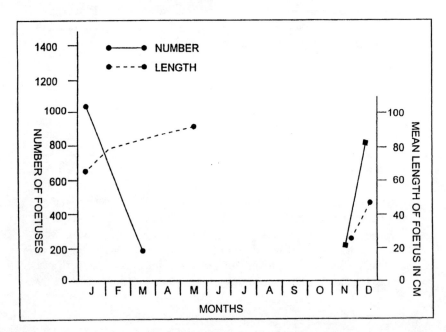

289

decline in the international markets, and the trend for whaling has been reduced quite considerably and only a few nations now carry out whaling under the quota system issued by IWC. Yet, there is an enormous pressure brought upon these few whaling nations by the rest of the world and, at a recent meeting held in Sweden during June 1986 (and fortunately for the whales) an agreement has been reached that a moratorium be imposed on all commercial whaling. The details of the species caught commercially along the tropical north-eastern part of Brazil for nearly a century is shown in Table 10.1. The total catch for the three whaling nations are compared in Fig. 10.4.

Table 10.1   Catches of whales off coast of Brazil:
Costinha/Carbo Frio during 1910–1984

YEAR	Physeter catadon	Balaenoptera musculus	Balaenoptera edeni	Balaenoptera borealis	Balaenoptera physalus	Balaenoptera acutorostrata	Megaptera novaeangliae	TOTAL
	(Sperm)	(Blue)	(Bryde's)	(Sei)	(Fin)	(Minke)	(Humpback)	
(1)	(2)	(3)	(4)	(5)	(6)	(7)	(8)	(9)
1910	= =	=	=	=	=	=	=	
1911							102	102
19,12							342	342
1913							352	352
1914							317	317
1915 }	*	*	*	*	*	*	*	
1923 }								
1924							62	62
1925							42	42
1926							32	32
1927							47	47
1928							40	40
1929 }								
1930 }								
1946 }	=	=	=	=	=	=	=	
1947				⇓ 14			11	25
1948		01		10			21	32
1949				18		01	15	34
1950				98			24	122
1951				151			28	179
1952	01			153			09	163
1953	01			161			08	170
1954	01			183			18	202
1955	01			198			06	205
1956	03			196	01		14	214
1957	02			115				117
1958	04			118	01		05	128
1959	11			294		02	08	315
1960	01/28			500/250	/21		10/03	511/302
1961	05 97			504 453	10		11 02	520/562

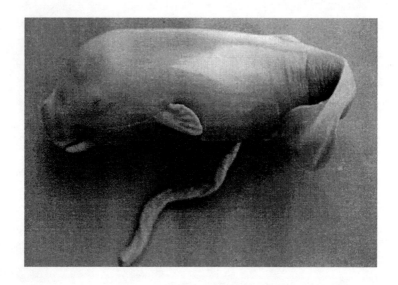

Plate 10.1 Foetus of a sperm whale, Physeter Catadon, male; taken from the mother caught off coast of the north-eastern part of Brazil; measuring 1m long and weighing 18kg. The head is enlarged with well defined single blow hole; eyes smaller, and fully developed mouth. The tail fluke with distinct notch, the fore limbs modified into flippers, while the hind limbs are absent, but the dorsal fin well formed. The penis is protruding and the umbilical cord is relatively short.

Year								
1962	04/81	/01		272/338	/49		08/03	284/472
1963	07 35			253 93	04	02	10 02	272 134
1964	04			256		44		304
1965	13			149		68		230
1966	24			72		352		448
1967	20		06	149		488		563
1968	39		06	158		456		559
1969	75		06	156		617		754
1970	76		03	123		701		803
1971	55		02	118		900		975
1972	66			105	01	702		774
1973	75		01	106		650		732
1974	29		01	102		765		797
1975	54			103		1,039		1,096
1976	09			103		776		788
1977	25			105		1,000		1,030
1978	24					690		714
1979	27					739		766
1980	30					902		932
1981						749		749
1982						854		854
1983						625		625
1984						600		600
Total	=686/241	01/01	25	⇑ 3,715	03/84	13,722	1,542/10	19,922/1,470
				228/1,134				21,392

}	First catch had no record
} *	Whaling was suspended and station closed
⇑ ⇓	Bryde's and Sei were identified mostly as Sei

(After Singarajah, K. V., 1985. Proc. Symp. Endangered Marine Animals and Marine Parks, 1, pp. 131–148)

On the other hand, most catches of the Japanese include many foetuses (Plate 10. 1 and Fig. 10.5).

Despite whaling being a small industry, the economic prosperity has been considerable. In modern industry, practically every part of the whale is utilised profitably. Baleen whales are mostly used as edible meat and the sperm whales are largely used for extraction of oil and for other industrial products. Some of the commercially most important whale products are as follows:

(1) Meat – palatable for human consumption, the greatest proportion goes as food, and the meat contains about 25% of proteins.

(2) Baleen whale oil – as edible fat by hardening; superior quality of edible oil is extracted from blubber and bones of the baleen whales.

Plate 10.2 A & B   (A) The 50m long catcher boat *Cababranco* used in Brazilian
waters, and (B) the harpoon gun.

(3) Though sperm whale meat is used as food, the oil, however, is not suitable as edible fat. Because it is somewhat chemically different from the baleen whale oil and of a very waxy nature, it is converted into a number of commercially important products. The oil is used in soap, illumination, tanning and steel industries, lubrication of sophisticated paint oils, machines such as wrist watches, plane and submarine engines; glycerol, skin cream, cosmetics especially lipsticks, detergents and boot polish, varnish and linoleum; the teeth are used as ornaments.

(4) Spermaceti oil: spermaceti – erroneously thought to be 'sperms carried in the head' and hence the name the 'sperm whale' – is a white waxy substance with a low density which solidifies rapidly at low temperatures, less than 32°C. It is now believed that spermaceti helps to regulate buoyancy by cooling during deep dives. A drop of 3°C has been estimated to give neutral buoyancy.

(5) The baleen plates or whale bones are sold as ornaments or used in synthetic materials which are flexible and resilient especially to produce corsets, umbrellas, fishing rods, tennis rackets and other objects.

(6) Ambergris: a rare and much valued amber colour waxy product, believed to originate in the lower part of the intestine and retained mainly in the colon of the rectum of sperm whales. It is a highly volatile crystalline, sweet-smelling secretion. Recent studies show that it is formed by breakdown of ingested 'cartilagenous beaks' of the giant squids by the bacterium, *Spirillum physeteri*. Ambergris is widely used in cosmetic industries. The purified ambergris is used as a fixative at the perfumery for high-quality scents.

(7) Most by-products are valuable. Some of them include vitamin A from the liver; drugs such as insulin from the pancreas; ACTH from the pituitary glands and other endocrine derivatives. The products, together with some oil, are used in a number of treatments such as asthma, rheumatism, and arthritis.

(8) The low-quality meat and other secondary row materials are converted as animal food and fertilisers.

Plate 10.3   (A) The front view of the whale factory at Cambohina. Underneath the bridge is a large tank where the carcasses of several whales are brought; they are first washed before any treatment. (B) The tug boat which fetches the whales from the catcher boat to the factory. Note one minke whale hanging (far right).

(9)   The white whales frequent the river mouths and are caught by nets. They provide oil and their skin is used as valuable leather.

The essential features of the catcher boat *Carbo Branco* used in the Brazilian waters and the 90-millimetre harpoon gun are shown in Plate 10.2, A and B; and the tugboat which fetches the whales to the factory and the front view of the factory are shown in Plate 10.3, A and B.

10.6   Physiology and Behaviour

The whale's brain, like its body, is perhaps the largest amongst all living animals either on terrestrial or in aquatic environments. The brain-to-body-weight ratio is very varied among whales. The brain net weight also differs in the two living groups. The largest brain of sperm whales weighs some 9 kg while that of the largest Mysticeti – the blue whales – weighs 7 kg and that of smaller minke whales weighs about 4.5 kg. The brain in a fresh condition is very soft and needs hardening before study. The large cerebral hemispheres are very highly convoluted. The cerebellum is also relatively large and folded. The most striking features of the cetacean brain are the absence of the vestigial form of the olfactory nerve which is poorly developed in Mysticeti; unusually larger inferior colliculi and olivatory nuclei, and the amygdaloid body. The unique specialisation of the spermaceti and melon bodies is intimately associated with the auditory system and the complex mechanism of echolocation and, to a certain extent, to regulate the buoyancy mechanism. Whales have an extraordinary capacity to learn and to remember events, as seen in dolphins, and their social behaviour is also unique among mammals.

10.6.1   Food and Feeding

The cetaceans, especially the Balaenopteridae, are huge and usually feed principally on euphausids or krill, calanoids, copepods, and smaller fish, while the toothed sperm whales prefer squid. The great blue whales appear to consume about four tons of zooplankton per day in the Antarctic, while the minke whales feed principally on *Euphausia superba,* particularly between 4–6 a.m. and 4–6 p.m. in the Antarctic consuming about 150–200 kg at a time. This may well coincide with the diurnal vertical migrations of the plankton. For this

Plate 10.4   The Antarcic Krill, Euphausia sp on which the whales feed.

reason, very large concentrations of baleen whales are found in the Antarctic where they feed during summer but they migrate thousands of miles into the Tropics to breed during Antarctic winter. The Antarctic krill, particularly *Euphausia surperba* and other related species, have a moderate lifespan of 3–4 years, growing up to 60–130 millimetres in length and weighing about 2–3 grams (Plate 10.4). It lives in large shoals, usually near the surface down to 100 metres. Plans to use this resource as human food are now developed in Japan and the USSR.

The euphosids are usually active, reach maturity after two years and undertake very extensive vertical migration. During the Antarctic winter, the krill are known to go deeper to 400–800 metres, and during the Antarctic summer they rise to the surface and remain concentrated within the upper 30–40 metres depth. On the other hand, the toothed whales are known to feed on a variety of species of cephalopods, especially squid, cuttlefish and octopus, but their much favoured food items are the oceanic squid such as *Architeuthis sp.*, *Lepidoteuthis grimaldii*, *Cuicioteuthis ungiculatus*, *Onychoteuthis bankasii*, *Tetronychoteuthis dussumieri*, *Moroteuthis robusta*, and other related species. A mature sperm whale can feed on 3–4 tonnes of squid per day.

10.6.2　　Epimeletic Behaviour

Whales are gregarious; this is particularly so in Odontoceti rather than in Mysticeti. School formations are said to be common during migration when feeding in the Arctic or Antarctic and breeding in the lower latitudes. Sperm whales often form a harem with a dominant bull as the chief and they seem to feed on the giant squid along the mid-oceanic ridges. Most whales show epimeletic or 'care-giving' behaviour.

Cetaceans being aquatic and carnivorous, their eyes are also modified to suit the environment for their survival. The eyes are small relative to their huge bodies and are placed laterally. They have sufficient blind spots both in front and behind the body and, consequently, have lost their binocular vision. The eyeball is usually about 10 centimetres in diameter, spherical, and is well protected by the somewhat flattened cornea. The sclera is also unusually thickened, perhaps, to resist compression under greater pressure during diving and swimming. The lens is rounded and well adapted for vision in water; the image is just formed on the retina. The eye muscles play an important part in adjusting the curvature of the lens to correctly accommodate the vision. Their visual fields extend as much as 120–160° on each side. On the other hand, the Platanistic river dolphins of the Ganges, Indus, Yangtze, Amazon and Rio de La Paz have vestigial eyes. The *Platanista gangestica* has significantly reduced eyes or is devoid of lens. Dolphins have remarkably good visual acuity. The visual perception of most whales seems to be influenced by their feeding habits. Dolphins can hunt fast swimming fish; the pilot whales are primarily nocturnal; sperm whales are deep divers, the last two feed on slow-moving squid and cuttlefish, which have smaller eyes but good powers of vision.

10.6.3   Echolocation

Like the flying mammals, whales have developed an extraordinary capacity to send out and to receive sound waves through the dense sea water medium (the speed of sound in sea water with a salinity of 35‰ is about 1,057 metres/second). The echolocation is carried out by frequency modulation. They emit sound pulses of certain frequencies to detect the direction and finer details of the objects in the environment. They are said to generate low frequencies of the range 0.25–1 KHz for orientation and 2–220 KHz to scan out the details. Although these are more common with the Odontoceti whales, the Mysticeti also produce echolocations, but somewhat less efficiently.

10.6.4   Migration

Obviously, the strongest driving factors underlying the migration of whales are their feeding and breeding seasonalities. The larger species were known to breed in the Tropics and subtropics and feed in the Antarctic. The Atlantic bottle nosed whale, *Hyperoodon ampullatus,* the killer whale, *Orcinus orca,* and the narwhal, *Monodon monoceros,* often migrate in search of the squid, *Gonotius fabdcii,* or a concentration of herrings in the North Atlantic or packed ice in the Arctic respectively. The sperm whale, *Pyseter catadon,* particularly the males, move towards the polar regions while the females, often accompanied by the calves, prefer to stay away from the cold. They are also abundantly found in the warmer waters of the Indian, Pacific, and Atlantic Oceans, particularly along the mid-oceanic ridge system to hunt on the giant squids, *Architeuthis.* They generally form male schools segregated especially in tropical waters. A male sperm whale moves on an average of about 1,400 kilometres per day. The annual migrations of baleen whales from the low latitude in the winter to the higher latitudes in the summer are not more sharply defined than those of the larger Odontoceti. The blue, fin, sei, and minke whales concentrate in larger numbers in the Antarctic to feed on the high production of the plankton, especially the krill. The Bryde's whales, *Balaenoptera edeni,* usually feed on the low productive area near Patagonia. The humpback whales, *Megaptera novaeangliae,* appear to prefer the warm water in the Tropics for breeding. They seem to favour a temperature range of about 24–25°C in the northern Pacific. In the southern hemisphere, the females with calves are the last to arrive to feed on the Antarctic krill, and the first to leave to the Tropics to breed again at the end of the feeding season. On the other hand, pregnant females are the first to arrive and the last to leave, probably storing up maximum possible food reserves prior to parturition and lactation. Other factors which influence migration seem to be day length (Dawbin, 1956; 1966; Singarajah, 1984). The right whales seem to keep the coastal landmarks for orientation during migration. The grey whales appear to use the olfactory sense to search the shallow semi-enclosed lagoons or brackish-water enclosures during breeding. For the rorquals, usually the Antarctic and upwelling areas are the feeding grounds; sea water temperature is most important for breeding and they characteristically migrate to warm tropical waters with regularity. Warm water also appears to be important for the breeding of the sperm whales. The pelagic Odontoceti such as *Stenella sp.* regularly forages down to about 700 metres within the 'acoustic scattering layer' on or below the thermocline (see Chapter Twelve) in the tropical waters. The sperm whale has become adapted to a tropical habitat involving even deeper regions of the oceans and is one of the few marine mammals (except perhaps some ziphiids such as *Hyperoodon)* which can readily exploit the cephalopods of the lower regions of the continental slope. The unusually greater abundance of fin whales in the packed ice waters seems to be correlated with greater concentrations of immature euphausids, *Euphausia superba.* Sperm whales may feed more efficiently at night since squid usually rise to the surface layers.

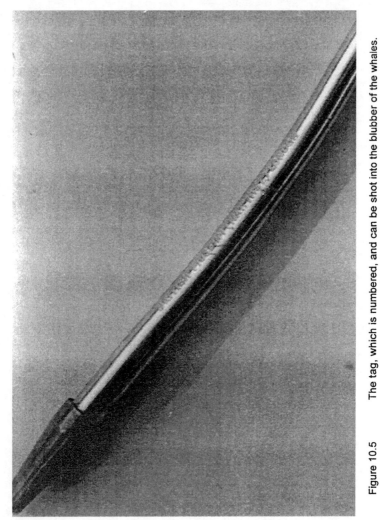

Figure 10.5   The tag, which is numbered, and can be shot into the blubber of the whales.

10.6.5 Stranding of Whales

The periodic stranding of whales, dead or alive, the living ones often with their 'suicidal tendency', has been reported frequently all over the world. The unreported cases are rather more numerous than the reported ones. The stranded whales are often used as food and fuel by the coastal populations. Along many tropical coasts of North and South America, especially Florida and the north-eastern coast of Brazil, South Africa, the Arabian Sea coast, India, Ceylon, the Maldive Islands, Burma, Malaysia, Indonesia, the Philippines, Vietnam, and Australia a number of species of both baleen and toothed whales get stranded. The stranding of the pilot whales particularly, *Globicephala melaena*, either in schools or smaller pots or in pairs, is a common scene along many tropical coasts. Other species include: *Balaenoptera musculus, Balaenoptera sp., Balaenoptera acutotorostrata, Balaenoptera indica, Balaenoptera physallus, Megaptera nodos, Physter catodon, Physter acrocephalus, Balaena australis, Balana glacialis, Orcinus orca, Koya breviceps, Globicephala macroryncus*, etc. Although not all strandings have been recorded, more baleen whales have been stranded than toothed whales, except the pilot whales which often migrate in schools, in which case the whole school occasionally gets stranded. Conservation organisers can often fail in their attempts to divert or drag them from the beaches or shallow waters to the deep waters as they return sooner to the same beaches. The causes of this mystery, which occurs so naturally, are not precisely understood. Various theories have been put forward to explain the phenomenon of the stranding of whales but most are speculative and some are plausible, for instance:

(1) Their epimelitic behaviour, as noted above.

(2) A close social organisation and cohesive gregariousness.

(3) Panic response when they lose the sense of environmental parameters, i.e. depth of water, coastal configuration of the gentle slope or steeper shelves and turbid-muddy waters.

(4) Disorientation due to loss of direction and inability to detect their position by use of echo sounding.

(5) Since the mass strandings of whales often occur following electrical storms during days of the full moon, it is thought

302

that the submerged magnetic lines which the whales use the as reference for movement, may have become distorted, and consequently the whales lend themselves to this predicament of stranding.

Marking whales with small stainless steel tags (Plate 10.5) in the blubber has proved successful to some extent in tracking the migration paths, but more work is required in order to draw any conclusions.

References

Allen, G. M., Models for echolocation, (eds R. G. Busnel and J. F. Fish) Animal sonar system, New York and London, Plenum Press, published in co-operation with NATO Scientific Affairs Division, 1980, pp.625–671

Berzin, A. A., 'The sperm whale', in J. V. Jabalokov (ed.) *Israel programme for scientific translations*, pp.1–394 (transl. from Russian)

Bullock, T. H., et al., 'Electrophysiological studies of central auditory mechanism in Cetaceans', Z. Vergi, *Physiol.*, 59, 1968, pp.117–156

Busheuv, S. G., 1984, 'Feeding of minke whales Balaenoptera acutorostrata in the antarctic', IWS/SC36/Mi36, pp.1–19

Cousteau, J. Y., *The whales, Mighty monarchs of the sea*, Cassel, London, 1972, p.303

Gaskin, D. E., *The ecology of whales and dolphins*, Heinemann, London, 1982, pp.459

Gingedch, P. D., Wells, N. H., Russels, D. E., and Ibrahim, S. M., 'Origin of whales in epicontinental remnant seas: New evidence from the early Eosine of Pakistan', *Science*, 220, 1983, pp.403–406

Horwood, J., *Biology and exploitation of the minke whale*, Florida, CRC Press, Inc., 1990, pp.183

Locidey, R. M., *Whales, Dolphins and Porpoises*, London, David and Charles, 1979, pp.200

Lythgoe, J. N., *The Ecology of vision*, Oxford, Clarendon Press, 1979, pp.369

303

MacKay, R. S., 'A theory of the spermaceti in sperm whale sound production', in Busnel. R. G. and Fish, J. S. (eds.), *Animal Sonar System*, New York and London, Plenum Press, published in co-operation with NATO Scientific Affairs Division, 1980, pp.937–940

Mackintosh, N. A., 'The stocks of whales!', *Fishing News*, London, 1965, p.360

McCormic, J. G., Wever, E. G., Plain, J. and Ridgway, S. H., 'Sound conduction in the dolphin ear' *J. Acoust. Soc. Medca*, 48, 1970, pp.1418–1428

Minasian, S. M., Bacomb, K. C. and Foster, L., *The World's Whales*, Washington DC, Smithonian Books, 1985, p.224

Mitchell, E. D., 'The Mio-Pliocene pinniped Allodesmus', in *Geol. Sci.*, 61, Univ. Calif. Publ., 1966, pp.1–105

Nonis, K. S., and Evans, W. E., 'Directionality of echolocation clicks in the rough tooth porpoise, Steno bredanensis (Lesson)', in Tavolga, W. N. (ed.), *Marine Bioacoustics*, 2, Oxford, Pergamon Press, 1967, pp.305–316

Ohsumi, S. 'Feeding habits of the minke whale in the Antarctic', Rep. Int. Whale Commn., 29, 1979, pp.473–476

Rice, D. H., A list of the marine mammals of the world, NOAA Technical Report, NMFS SSRF, 1977, p.711

Singarajah, K. V., 'Observations and occurrence on the behaviour of minke whales off the coast of Brazil', Sci. Rep. Whale Res. Inst., 35, 1984, pp.17–38

Singarajah, K. V., 'Current status of minke whales and conflicts of interests. Revista bras', *Zool.*, 5(l), an invited paper at the XI Congresso Brasileiro de Zoologia, 1984, pp.101-108

Singarajah, K. V., 'A review of Brazilian whaling: aspects of biology, exploitation, and utilization' Proc. Symp. Endangered Marine Animals and Marine Parks, 1, 1985, pp.131–148

Singarajah, K. V., 'Data analysis of the breeding stock of minke whales and the quata system for Brazil', Rep. Int. Whal. Common., 35, 1985, 531 (Sc/36/Mi4, pp.1–6, + 11 tables)

Slijper, E. J., *Whales*, Ithaca, New York, Cornell University Press, 1979, p.511

304

Strahler, A. N., *Science and earth history. The evolution/creation controversy*, New York, 1987, pp.552

Watson, L., *Whales of the World*, London, Hutchinson, 1981, p.302

Young, J. Z., *The Life of Vertebrates*, Oxford University Press, 1981, p.645

Chapter Eleven
Benthos Tropical and Subtropical

11.1 General Characteristics

In a broader sense, as seen in the classification of the marine environment, the benthonic realm includes all the sea-floors from the highest level of tide mark down to the deepest part of the oceanic trenches. Like plankton and necton, marine benthos are an assemblage of organisms that may belong to different systematic groups which usually dwell or spend most of their lifetime on or in close contact with the sea floor, from the intertidal zone to the bottom of the deepest trenches of the oceans, either as sessile, burrowing, or creeping but with very limited movements. The quantitative distribution of benthic organisms is largely determined by the distance to the proximity of the coast and the depth to which the animals or plants are endemic. In other words, the abundance of benthos generally decreases with increasing distance of the coast and depth; this is mainly because the sedentary, especially the deep sea, benthos derive much of their food from the rains of detritus and bacterial components of the sea. A few mobile benthic creatures are carnivores or scavengers and they feed on dead organisms that sink from the surface or mid-water layers. Tropical and subtropical marine regions have a fairly constant temperature and generally support richer and more diverse benthic organisms than do regions of higher latitudes. A number of representatives of protozoa, porifera, coelenterates, polychaete worms, echiuroidea, molluscs, gastropods, bivalves, crustaceans, barnacles and engebenthic amphipods, isopods; the prawns, shrimps and lobsters, especially on the top parts of the continental slope; echinoderms, tunicates, the nectobenthic ray fish and flat fish, including the seaweeds that grow from the intertidal zone down to the edge of eulittoral zone on the sea bottom, and many others are included in benthos. As early as 1970, the existence of some benthic populations of giant clams and tube-worms was discovered along the active zone of the sea floor spreading near the Galapagos. The theory is that the water that enters into the vent becomes heated and escapes with enough hydrogen sulphide. The hydrogen sulphide is utilised by a group of bacteria to produce organic matter from CO_2 without the need for light, which is required for normal photosynthesis, and are able to support some of the benthic population. However, the general trend is that the benthic biomass decreases from the continental edge downwards to abyssal plains. Below this level the sea floor biomass becomes significantly impoverished. The biomass varies in different regions of the Tropics and subtropics and best estimates are 300–1,000 g/m^2 net weight (see below for hadal benthic organism).

11.2   Diversity of Benthic Community

Although the entire sea floor is covered with the sedimental deposits, the organic matter readily available to the benthic community decreases with depth. Accordingly, the distribution of species depends on the type of sediment, the depth of the sea floor and the extreme conditions of the regions. In the Tropics and subtropics, the species diversity within a community is usually higher, i.e. a natural community has many species, with various feeding levels, but fewer individuals of that species. This point has been discussed in previous chapters and will not be considered further. Despite much diversity of bottom topography, nature of sediments and habitats, nearly 99% of the zoobenthic organisms live in the shallower waters. Most of the sessile and sluggish benthic animals feed mainly by filter-feeding mechanism on phyto-and zooplankton, detritus, and bacteria. The benthic organisms that live on the surface of the sea floor are called epibenthic and those living within the sediments are called sedimentary organisms.

11.3   Major Groups of Benthos

Benthic marine animals were grouped into two types:

(1)   The epifauna living upon or associated with substrates of rocks, stones, shells, piling and vegetation.

(2)   The infauna, comprising all animals of the sandy or muddy layers of the sea bottom (Gunnar, 1957).

The infauna is fully developed in the tropical and subtropical zones where environmental conditions are most favourable. The number of epifauna is also greater in the tropical and subtropical regions. Depending largely on the body profile, size, varying degree of capacity for mobility, habits, habitats, and ecological adaptations to an aquatic mode of life, four main categories of benthos can be currently distinguished:

(1) Benthic proper or macrobenthos, the larger forms which exceed 1 mm in size and are sedentary or fixed also edreobenthic.

(2) The engebenthic or herpetobenthic, which usually crawl,creep or dart on the bottom.

(3) The meiobenthos: the small organisms which fall between 0.1-1 mm in size.

(4) The microbenthos, smaller than 0.1 mm in size, mainly including bacteria.

The benthos may be further distinguished as: nectobenthic and planktobenthic organisms, depending on the mode of their existence in nectonic or planktonic forms as a part of their life cycles. In the Tropics where the water is warm and the temperature relatively constant, and the changes in concentration of oxygen favour the coastal and intertidal pools for richer and more diversified benthic life. As defined earlier, the benthos, rich in number and diversity, occupy the littoral zone where the physico-chemical conditions generally favour the growth of the benthic organisms. The other important factors which determine the distribution of benthos are latitudes, substrates, salinity, temperature, pH, predation and competition. The sedentary benthic animals are such as sponges, hydrozoans, bryozoans, polychaete worms, barnacles, muscles, oysters; the engebenthic organisms with little or limited movements are a variety of crustaceans, especially ostracods, harpactic copepods, mysids and decapods, gastropod molluscs, amphipods, isopods, echinoderms, pycnogonids, ascidians, crinoids, serpulids and nematodes and foramenifera (Fig. 11.1).

Although some benthic animals occur in the eulittoral and sub-littoral zones of the continental shelves, others occur at great depths. However, some of the polychaete worms and britle stars, for instance,

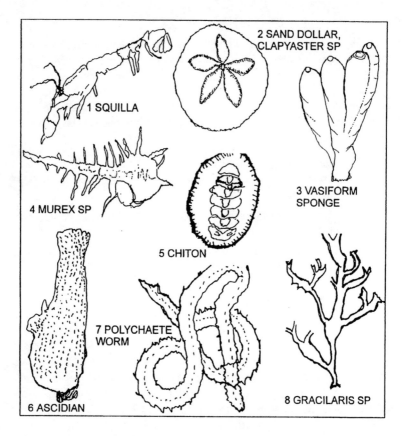

Figure 11.1   Some of the benthic organisms: 1. Shallow benthic stomatapod crustacean which burrows in mud or beneath the stone. 2. Shallow water sandy bottom dwelling sea urchin (sand dollar) can be found in thousands under low tide along most tropical shores. 3. Vasiform sponge with oscular and is often found in coral reef areas. 4. Gastrapod mollusc, Murex, found in tropical sea bottom. It has a tough and spiny shell. 5. Bilaterally symmetrical mollusc, Chiton, with eight calacrious shell plates; commonly encountered on tidal rocks of shady sides, but can crawl during dawn and nights to feed on algae. 6. A simple truncate ascidian, fixed to the sea floor substrate, found in tropical seas. 7. Bottom living polychaete worm often found buried in the mud, and possesses paired segmental appendages; the trochophore larvae are planktonic. 8. Shallow benthic alga with highly branched thallus consisting large quantity of gelatinous material providing agar-agar. It is an economically important alga growing in abundance in the tropics.

are able to tolerate a wide range of depths, from littoral to depths below 5,000 metres (Ekman, 1987) and hence they are called eurybathic.

11.4   Important Dietary Benthos

Benthic invertebrates, principally molluscs, especially oysters, mussels, clams and scallops, crustaceans, echinoderms, tepangs, and many others are often included as dietary items. Because of their popularity and relative ease of cultivation, a number of species are now being cultivated in large scale in Brazil, Korea, Taiwan, Japan, India, France, England, the USA, and many other countries (see Chapter Fifteen on culture prospects).

11.5   Vertical Distribution of Benthos

Although benthic organisms are widely distributed over the bottom of the sea, the distribution is somewhat uneven due to availability of nutritional materials. Vertical distribution of benthos decreases with depth in the tropical waters, although it increases along the coastal regions of Western Australia, Malay Archipelago, Ceylon, the coast of India, Madagascar, Gulf of Aden, West African coast, the Gulf of Mexico, and the north-eastern coast of Brazil.

The microbenthos mainly include bacteria, protozoans, ciliates, and a variety of diatoms and flagellates. The benthos mostly feed on the rains of organic debris of terrestrial origin, small plankton, and a variety of other detritus materials.

Within the limit of the eulittoral zone, some macroalgae are frequently found. The principal phyta that benthos consist of are the anchored macroalgae usually along the intertidal zones of coastal waters and within depths of 20-40 metres; a few may reach a depth down to 60 metres. Some of the macroalgae are of commercial importance while others are food for demersal fish, and suitable substrates for spawning in which many larvae live. Some seaweeds are of industrial application. For instance, the agar and agaroid are extracted in many tropical countries from *Grasileria;* the red algae, *Gelidium* and others. The kelp *Laminaria* is also used as food in many countries, particularly Japan and China.

310

11.6   Hadal Benthic Fauna

The word hadal originates from the Greek (Hadal=hades) – a term for the 'subterranean kingdom of the dead'. Much of the knowledge of benthos from the greatest depths resulted from a number of expeditions. Since 1948, the USSR-sponsored *Vityax* and the Danish-sponsored *Galathea* expeditions collected extensive faunal samples from hadal zones, exceeding 6,000–11,000 metres (see also hadobenthic zone in Chapter Two).

Despite the theory of an azoic zone proposed by Edward Fobes (1839) that below the depth of 650 metres no life continues to exist, first evidence against this came when the underwater cable between Sardinia and Africa was lifted from a depth of 2,160 metres for repair as early as 1860. It was covered with a variety of benthic organisms which had settled on it. The existence of life in deeper waters of all oceans was further demonstrated by a rich collection of deep sea benthic fauna collected by the *Challenger* expedition. Recently, the evidence of life at depths of about 10,900 metres was noted by direct observation during the descent of the bathyscape *Trieste* into Mariana Trench (Piccard and Dietz, 1960).

References

Belyaev, G. M., 'Bottom fauna of the ultra-abyssal depths of the world ocean', (in Russian), Akad. Nauk. SSSR., 591 (9), 1966, pp.1–248

Deitz, R. S., 'The sea's deep scattering layers', *Scientific American*, 207 (2), 1962, pp.44–64

Reichle, D. E. (ed.), *Marine Ecological Processes*, New York, Berlin, Springer-Verlag, 1984, p.546

Thorson, G., 'Bottom communities (sublittoral or shallow shelf)', pp.461–534, in (ed. Hedgpeth, J. W.), *Treatise on marine ecolgly and paleantology*, The Geological Society of America, New York, 1957, vol. I p.1296

Vinogradova, N. G., 'Vertical zonation in the distribution of the deep sea benthic fauna in the ocean', *Deep Sea Res.*, 8, 1962, pp.245–250

Chapter Twelve
Primary Production

12.1   Feat of Biological System

Plants alone are capable of extracting their simple basic organic food from inorganic materials. Their ability to transduce the solar energy into potential chemical energy is perhaps one of the most valuable feats that the biological system has evolved on this earth. Consequently, all living organisms, both plants and animals, including human beings, ultimately depend on the autotrophic plants for the production of organic food materials. The synthesis of the organic food materials of carbohydrates, proteins and fats by the chlorophyll bearing plants such as pelagic and benthic unicellular algae and the shallow water attached seaweeds – using solar energy, water, nutrients, especially CO_2, nitrates, phosphates, silicates, vitamins and trace elements at the initial or primary level in the food chain – is widely regarded as the 'primary production'; in other words, the first metabolic link between living cells or tissue and the dissolved nutrients in the sea water and the available light energy. The rate of this primary production, however, varies considerably in different domains of the sea from season to season. Marine primary production is predominantly produced by phytoplankton.

However, the total amount of production of the organic materials by phytoplanktonic organisms in unit time per unit area of sea surface or unit volume of sea water is often measured, and is usually expressed as both mgC or gC/m^2 or mgC or gC/m^3 per hour or per day (see below). Although the total surface of the sea covers some $361 \times 10^6 \, km^2$, the total annual net marine productivity has been estimated in the order of some 26×10^9 tons of carbon per year. This constitutes about 30% of the total annual global production; more than 60% of this production has been attributed to the Tropics, and the rate of production is continuous, though the estimates are approximations. There is still a serious shortage of data; these are based on piecemeal collected from isolated areas rather than as a whole, which evidence significant regional differences (see Table 12.1). Expressed in calorific value, this amounts to some 2.4336×10^{14} kcal of energy. This value is computed on the assumption that 1 gram of weight of organic production, whether plankton or fish, is equivalent to 1 kcal of energy. Knowing that 1 g C corresponds to 9.36 kcal, then the total amount of production can be converted into kcal.

12.2   Principal Source of Energy

The sun, which is about 100 times larger than the earth in diameter, has an internal source of energy. The sun's surface temperature is about 4,000–6,000°C. This causes the radiation of an enormous quantity of energy in all directions into the surrounding space. The average distance between the sun and the earth is 149.6×10^6 kilometres which is also called the astronomic unit. The largest source of energy available for the phytoplankton for biological productivity is transmitted from the sun in the form of electromagnetic waves which include X-rays and gamma rays travelling at the speed of light of 3×10^6 metres/sec. The solar constant or quantity of heat energy received by the earth's surface from the sun is something in the order of 1.94 calories per square centimetre per minute (or 42.9×10^6 kj/1 metre $2y^{-1}$ rays falling perpendicularly to the surface of the earth). Of the enormous output of solar energy, only a small amount of about 30% or 3.75×10^{20} kcal reaches the surface of the earth and of which only about 1.7% goes into the marine productivity; the rest is either reflected back by clouds or absorbed or scattered in the upper strata of the atmosphere. Sunlight contains photons of many different energy levels. The light energy that reaches the surface is composed of photons of the 'middle-energy-range' wavelengths, i.e. within the range of ultraviolet to infra-red of wavelengths of less than a metre (Fig. 12.1). At the sea surface level about 75% of the total energy of the sunlight is found within a narrow part of the spectrum (see below). Light energy can be quantified in photons as follows: the energy E of each photon equals Plank's constant times the radiation frequency.

$$E = hf = \frac{hc}{\lambda}$$

h = Plank's constant[1]
f = frequency
c = velocity of light
λ = wavelength

As will be seen below, in photosynthesis, chlorophyll pigments use the energy of sunlight to change CO_2 to carbohydrate and it is believed that about 9 photons are needed to change one molecule of CO_2 to carbohydrate and O_2. The global proportions of the various primary energy sources at the present are: coal 27.2%, natural gas 21%, oil 39.6%, nuclear 5.8%, and hydroelectric 6.2%.

However, although the amount of energy reaching the surface varies with latitudes and with seasons, a greater proportion of the light falls on the sea surface, and the regions near the equator absorbs more heat than they lose. Therefore, in the Tropics and subtropics, light can be utilised by the phytoplankton all the year round and at greater depths.

Usually, the rate of organic production is measured in terms of energy. Therefore, for most purposes, the conventional units of calories will suffice. One calorie of energy is the amount of heat energy required by one gram of water to raise the temperature by $1\,^{\circ}C$ (Celsius degree, from $14.5\,^{\circ}C$ to $15.5\,^{\circ}C$); and the higher units of kilocalories (kcal = 1,000 calories) are often useful in organic food production as they involve the energy values: for example carbohydrates = 4.2 $kcal^{-g}$; proteins = 4.2 $kcal^{-g}$; fats = 9.4 $kcal^{-g}$; and alcohol = 7.8 $kcal^{-g}$. The energy of heat can also be measured in joules (j). Quantitatively, 4.18 joules is equivalent to 1 calorie; and 4.18×10^3 j = 1 kcal; thus, heat refers to energy that is transferred from one body to another because of a difference in temperature. In SI units (Systeme International), for heat is the joule, but both calories and joules are interchangeably used.

However, because of the turbulent and rippled nature of the sea surface, a considerable amount of light is reflected and the remaining light incident on the surface is either refracted or absorbed, depending much on the time and position of the sun. As noted earlier, the light

---

[1]   The value of Plank's constant is found to be 6.626×10^{-34} Js or Avagadro number (1.58×10^{-37}) photons. The wavelength of the emitted radiation is inversely proportional to its frequency, or $\lambda = c/v$.

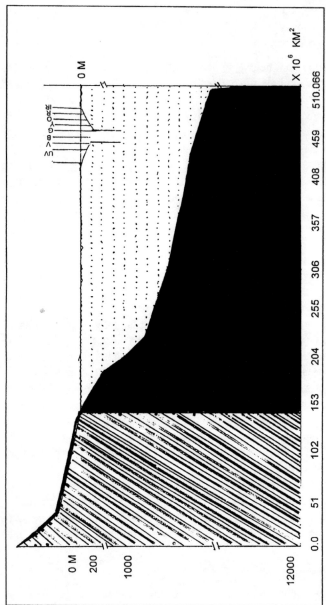

Figure 12.1   The light energy that reaches the sea surface is composed of photons of the 'middle-energy-range' wavelengths, i.e., ultraviolet to infrared. But, with increasing depth, the absolute light intensity decreases exponentially and becomes zero below the depth of 1000m.

that penetrates below the surface is not uniform, but the intensity rapidly attenuates logarithmically with increasing depth to about 1% of its surface intensity. Below this 1% light level, the energy, is insufficient to stimulate any appreciable photosynthetic activity. The depth at which light penetrates the critical limit is dependent much on the spatially and temporally attributable characteristics such as suspended organic and inorganic materials, transparency of water and the mixing processes of the euphotic layer. Because of these inherent diurnal variables, the spectral components of the light are sequestered and the phytoplankton of the upper layer will be able to absorb much of the white (violet – red) light of the most intense region of the solar radiation or, visible part (380nm–720nm) of the spectrum while allowing only the blue-green part of the spectrum to reach the phytoplanktonic organisms which remain suspended in the deeper waters of the euphotic layer. Green light perhaps is the most effectively used part of the spectrum as it predominates in the lower part of the euphotic layer as seen in the absorption spectrum below:

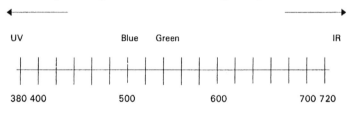

In the Tropics, although light energy is abundant and the photic level will extend to depths over 120 metres, a number of limiting factors come into play. Except closer to coastal and limited upwelling areas, the surface water is subject to intense heat and this reduces the salinity, which makes the surface water become less dense and relatively 'sterile' unless an effective mixing mechanism is brought about by vertical or horizontal currents to replenish the essential nutrients for increased productivity. Nutrients are clearly significant even if CO_2 and light are available at their maximum quantities to enhance the production. Along much of the equatorial belt, however, higher productivity has been reported. This is largely due to the equatorial undercurrents, divergences and convergences which often become pronounced since the effect of the Coriolis force is minimum along the equatorial latitudes. Cushing (1971) observed high productivity in some of the most important areas of the tropical and subtropical regions of the Indian Ocean off the coast of Somalia and in the southern Arabian Sea, as well as the region between Burma and the Andaman islands and found the productivity to be relatively rich, particularly during the north-east monsoon periods, usually during April–September.

12.3   Energy Input and Planktonic Ecosystem

As mentioned previously (Chapter Eight), the principal constituents which contribute to the primary production of the sea are the phytoplankton. These are predominantly represented by diatoms, dianoflagellates, coccolithophores, and silicoflagellates. In nature, the plants alone contain the essential ingredients of the photosynthetic pigments which are capable of intercepting and capturing the light. The planktonic autotrophs possess the unique, mainly green, pigments of chlorophyll a, b, and c, together with other less common pigments of xanthophyll, carotenoids and phycobin which can convert the light energy into chemical energy. For this reason, the phytoplankton are designated as primary producers as they receive the initial input of energy which can either be readily used for growth and to reproduce themselves at a rapid rate or stored as a potential source of energy for the consumers of secondary, tertiary, or higher trophic level feeders of the planktonic ecosystem.

12.4   Photochemical Conversion of Light Energy
         to Chemical Energy

The carbon uptake, fixation, and the complex photosynthetic pathways cannot be considered in detail here. Briefly, CO_2 diffuses into the liquid phase of the pigments which are contained within the chloroplasts. When the light energy is captured it is used in the photochemical process of reducing the CO_2 to carbon, as noted earlier. A chain of carboxylation enzymatic reactions accelerates the conversion of carbon dioxide, through various fixation pathways, into organic materials. A considerable amount of the chemical energy thus stored is converted to utilise for growth or to provide energy to a number of metabolic processes through respiration. An excess of organic (gross) production over respiration is what permits growth, and storage of food reserves (carbohydrates, proteins and fats). The photosynthetic assimilatory process can be summarised by the following fundamental equations:

$$\text{Assimilation: } CO_2 + H_2O \xrightarrow[\text{chlorophyll}]{\text{light}} (CH_2O) + O_2 + 112.3 \text{ kcal}$$

$$\text{Respiration: } (CH_2O) + O_2 \longrightarrow CO_2 + H_2O + < 112.3 \text{ kcal}$$

In respiration, the oxidative process may be considered as the reversal of photosynthesis. Since 1 g $^{-at}$ (1 mole or 22.4 litres of CO_2 at NTP) of carbon locks up about 112.3 kcal of energy to synthesise the organic molecular subunit (CH_2O) in the carbon chain to form carbohydrate molecule, the energy increase for 1 mole of glucose (6 moles of CO_2) in respiration can be represented in the following equation:

$$6CO_2 + 6H_2O + 674 \text{ kcal} \longrightarrow C_6H_{12}O_6 + 6O_2$$

$$C_6H_{12}O_6 + 6O_2 \longrightarrow 6CO_2 + 6H_2O + < 674 \text{ kcal}$$

12.5 Efficiency of Energy Utilization

The efficiency of gross primary production is the ratio between the fixed energy by autotrophs during gross production and the incident light energy for a given unit of area and time, i.e.

$$\frac{kcal \ / \ m^2 \ / \ year^{-1}}{kcal \ incident \ light \ energy}$$

Although the efficiency of each photon of light energy absorbed by the pigment molecule is almost 100% during gross primary production, some of the energy will be dissipated within the chloroplasts in conversion and therefore a little less than the assimilated energy is releasable during respiration.

12.6 Gross and Net Production

The total amount of organic materials synthesised by phytoplankton at a given place per unit time is the gross production. Since respiration is a constant function of all living organisms, a part of the total organic materials is used up for maintenance of their own life through the oxidative breakdown of phytoplankton respiration. The remaining

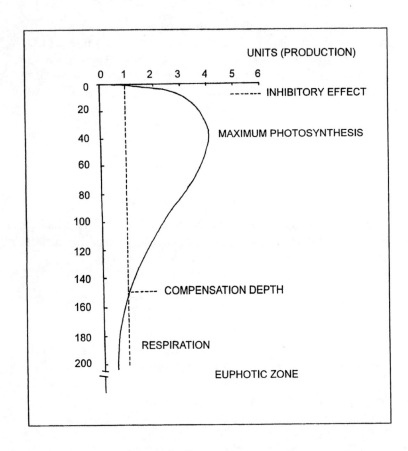

Figure 12.2   Compensation depth in the tropics. The oxygen production
                 by photosynthesis is equal to that consumed by
                 phytoplankton. Depth in m.

part, after respiration, available for growth of the total production corresponds to the net production, i.e. the total energy fixed per unit time during photosynthesis minus energy lost during respiration or the metabolic process of phytoplankton. The biomass or standing crop, the total (wet or dry) weight of all organic materials in the living organisms within a given area or volume of an ecosystem, increases as a result of net production. Planktonic, coral reefs and estuarine ecosystems usually have high productivity. However, the gross production, as already pointed out earlier, is regulated by depth with decreasing intensity of light energy. On the other hand, very high intensity of light, which exceeds the saturation level, near the surface, particularly in the Tropics and subtropics, will inhibit productivity and often prove deleterious. The maximum net production, therefore, is usually confined to a depth where the light intensity is optimal. Productivity decreases gradually until it reaches a point where the production exactly equals the respiration. This depth at which the two independent processes of production and respiration balance, in terms of energy, is generally understood as the compensation depth (Fig. 12.2).

The compensation depth for most phytoplankton species in the Tropics and subtropics often exceeds 120 metres and remain so throughout the year. Below the compensation depth the net production falls short of organic substrate materials required for respiration and the level beyond which the phytoplankton is no longer able to synthesise any organic production and support (both dark and photic) respiration and hence this level is termed the critical depth. On the other hand, there is the thermocline layer in which the vertical temperature gradient occurs maximally during seasonal changes which fluctuate significantly (see below).

12.7   Principal Factors Controlling the Primary Productivity

The two major key factors which determine the magnitude of annual rate of primary production in all seas are: the availability of light and the replenishment of nutrients by circulation in the productive regions. The other basic factors which influence the rate of productivity are: the physical state of the phytoplankton themselves and the temperature.

12.7.1   Light

As already noted, the ability to absorb radiant energy by photosynthetic organisms is facilitated by photosensitive pigments within the chloroplasts. The attenuation of light with depth is the prime factor which affects the productivity of the sea. This may be enhanced by rough weather, light absorbing agents such as algal suspension, detritus and effluents which change the spectral components and their distribution in the subsurface water. Near the surface, in the Tropics and subtropics, excess light usually inhibits productivity or even may cause a deleterious effect and, therefore, the maximum production takes place several metres below the surface. Nearly 50% of the light, including ultraviolet and infra-red, is absorbed near the surface within about 1 metre depth. The rate of decrease of light with depth, the extinction coefficient (= K), is dependent on the turbidity of the water. Therefore, the higher the extinction coefficient the faster the absorption of light (see Appendix III). The importance of light in organic production in the Tropics and subtropics was evident during the *Galathea* Expedition. While a relatively high rate of organic production was found at 'practically' all coastal stations in the Tropics and subtropics, an exception was noticed along the west coast of Africa, particularly near the mouth of the Congo River, where the silt brought down with the river water masses caused considerably low transparency and very poor light conditions for the phytoplankton and, consequently, a very low primary production rate (Steemann Nielsen, et al., 1957). Generally, the productivity in coastal waters is more than the oceanic parts, and the highest production is normally prevalent where there is no marked permanent stratification of waters in the Tropics and subtropics.

12.7.2   Nutrients

In most open oceans of the Tropics and subtropics, the concentration of nutrients is relatively low. As a result, the constraint on the supply of nutrients imposes limits on the rate of primary production, unless the rate of regeneration of nutrients is maintained by vertical mixing within the euphotic zone. The ratio of the nutritional state of proteins, carbohydrates and lipids is 4:2:1, which, in turn, is based on the C:N ratio and the P: chlorophyll concentration ratio. Although the three-neritic, upwelling, and open sea – habitats differ considerably in their abundance of nutrients, areas of upwelling, i.e. the regions where the cold nutrient-rich waters from the deeper parts of the sea rise and mix with relatively nutrient-depleted surface waters, thus are the areas generally rich in nutrients and hence rank highest in primary production and correspondingly support the thriving fishing industry (see Chapter Nine). On average, about 150 gC m^{-2} y^{-1} is fixed during primary production by phytoplankton. To continue the production normally, the carbon, nitrogen and phosphorus must exist in the ratio of 100:15:1 per year near the surface layer (Russell-Hunter, 1970). However, with increasing depth, the ratio of carbon may exceed over ten times its surface concentration and thus the nutrients of N and P will be in short supply to utilise fully the carbon, and hence will limit the organic production of the phytoplankton unless some vertical mixing ensures an even redistribution of the nutrients. Depending on seasons and regions, deep water circulation and vertical mixing lead to rapid replenishment of nutrients in the euphotic layer which stimulates and enhances the phytoplankton blooms.

In tropical oceanic waters, particularly in the tropical eastern Pacific, an inverse relationship between the depth of thermocline and the size of zooplankton standing crop has been noticed (Brandhost, 1958). The magnitude of primary production, as pointed out earlier, is governed by the turnover of nutrient salts which, in turn, depends largely on the depth of the thermocline. However, a strong thermocline seems to exist throughout the year in the tropical oceanic waters, but the depth at which it occurs varies quite significantly depending on the seasons and regions. Correspondingly, the rate of primary production also will vary a great deal. The daily rate is about 50 mgCm^2 at a thermocline depth of 100 metres close to anticyclonic eddies, whereas it increases to 500g mgCm^{-2} in other areas and a still higher rate at upwelling places (Steemann Nielsen et al., 1957).

12.7.3   The Physical State of Phytoplankton

Efficiency of the membrane, concentration of chlorophyll, buoyancy to keep the phytoplankton in suspension within the photic zone and quantitative and qualitative characteristics of phytoplankton can be greatly influenced by the local variables. Barnes (1957) noted a synchronisation of a spring outburst of diatoms and the release of a large number of nauplii larvae of *Balanus balnoides* in temperate waters. He further suggested that the presence of a 'substance' resultant of the metabolic activity of the diatoms may act as a stimulus for the release of nauplii.

12.7.4   Temperature

Next to light and nutrients, the physical state of primary producers, temperature is an important factor. However, the influence of temperature on the size of productivity is indirect and the magnitude of organic production is relatively more in some productive habitats than in others. For example, in the two habitats of open ocean and coastal waters of the Tropics and subtropics, the principal factor limiting the magnitude of planktonic production is replenishment of nutrients, especially phosphate and nitrates, in the productive euphotic layer. The replenishment of nutrients in the open ocean takes place primarily by water circulation, whereas in the shallow coastal waters it is mainly by the bacterial decomposition of sediments from the bottom surface. The regenerative process of nutrients through bacterial enzymatic activities is much dependent on temperature. Thus, the effect of temperature on the recycling of nutrients is significantly greater in the coastal than in the oceanic habitat; in the former the decomposition at a higher temperature enhances the rate of both respiration and photosynthesis during phytoplankton production.

12.7.5   Species Diversity and Productivity

An interrelationship in tropical seas exists between the standing crops of phytoplankton, zooplankton and species diversity, the latter recognisable by composition; i.e. where primary production is high the numerical abundance of zooplankton increases. In the Tropics, particularly closer to the equator, because of the greater range of closely related equitable conditions suitable for marine life, communities have more species. Diversity of species is more evident in local areas because of their own equable environmental parameters. In practice, it is not easy to precisely delimit the community of organisms spatially because of their mobility. In tropical waters, except in the upwelling areas where nutrients by mixing process is fairly stable, the productivity is continuous, but the productivity as a whole is relatively less than in temperate waters. Also, the productivity and the zooplankton crops are higher in neritic than oceanic waters. The phosphate and nitrate concentration in the euphotic layer is also appreciably less, but it increases with depth, being rich in depths beyond 150 metres. On the other hand, there is an amazingly high species diversity of zooplankton; though qualitatively more in species, quantitatively the individual species are less abundant. A rich variety of zooplankton communities consisting of several species of copepods, crustacean larvae, fish larvae, chaetognaths, and ctenophores, usually dominate in productive areas where phytoplankton may bloom from time to time. Several interesting works have been done in temperate waters on the relationship between phytoplankton maximum and the abundance of zooplankton, particularly with the breeding population of copepods (see Clarke et al., 1937: Ussing, 1938; Marshall and Orr, 1952). The copepods are one of the most important groups of zooplankton, in terms of their diversity of species, numerical abundance, worldwide distribution, and they also contribute substantially to the economy of the sea. As stressed in Chapter Eight, they are the principal consumers of the phytoplankton and a vital link between the primary producers and the higher level trophic carnivorous organisms such as many larger zooplanktons, fish, squids and whales.

It is therefore relatively easier to understand the diatom increase and the zooplankton abundance, if one can consider, in particular, the population of potentially breeding copepods. It is now fairly well established that the actual breeding of copepods varies from species to species, and in time and regions. In tropical waters there are a greater number of broods than in temperate waters. The intensity of breeding may also vary in some species more than in others, with clear peaks which reflect the relative abundance of adult copepods and their larvae. Some species, for example, one of the common tropical calonoid copepods, *Paracalanus parvus,* breeds throughout the year while many others only periodically. Because the developmental stages from eggs through nauplii, and a series of copepodites (III, IV, V) into adults are involved, the larval generations vary a good deal both spatially and temporally (Fig. 12.3).

Functionally, too, because the copepodites are smaller in size and more active, their rate of metabolism is greater and they need more food. Although food may not act as a direct stimulus, there is a coincidence between phytoplankton blooms and the breeding time of adult copepods. They consume large amounts of diatoms prior to

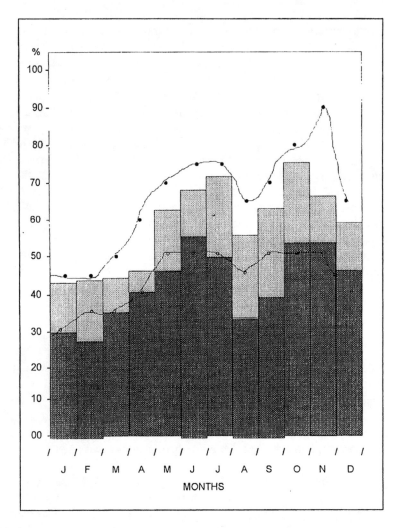

Figure 12.3   The spatial and temporal distribution of generations and the breeding intensity in some of the copepods of the genus, *Paracalanus*, in the tropics. (Phytoplankton, expressed as a percentage of a total of 100,000; and nauplii and copepodites, and adult copepods, expressed as a percentage of a total of 1000, and fish larvae, expressed as a percentage of 100). Based on monthly samples taken over a period of 6 years off the north-eastern coast of Brazil.

laying eggs. Curiously, the spawning of the fish also coincides when abundant copepodites are available in peak periods. Since larval copepods form the most important source of food for many of the fish larvae, the feeding by fish larvae is bound to reduce the copepod populations quite considerably. The trend seems that following the phytoplankton bloom, the copepods breed, and this is followed by the arrival of fish larvae. Aside grazing (see below), other factors may influence which include diurnal, vertical and seasonal, migrations of zooplankton to productive areas; avoidance of any unfavourable conditions; and mortality; thus, the variation in abundance and distribution can be partly explained. However, the productivity is greater in the Indian Ocean where seasonal monsoons enhance the distribution of the nutrients in the euphotic layer. The productivity of the coral reef ecosystem ranks second in the tropical seas, particularly in the tropical Pacific. As noted earlier, the high productivity in the coral ecosystem is determined by replenishment of nutrients locally as a result of a symbiotic association which helps the regenerative process of nutrients, thus enhancing the productivity. Comparatively, with the exception of a few upwelling areas of the Gulf of California and the Gulf of New Guinea, the tropical Atlantic has least production, fewer populations of planktonic larvae and less distribution of corals.

12.7.6   Grazing

From a subjective point of view, real climatic and regional differences within the realms of the Tropics and subtropics occur as can be seen along the coastal waters of the Indo-Pacific, and both sides of the Atlantic. In spite of these, all tropical marine environments exhibit a minimum degree of seasonality relative to temperate waters. Occasional blooms of phytoplankton do occur and in most of these areas where blooms are prevalent phytoplankton stocks are subject to heavy grazing by herbivorous zooplankton crop, and thus they reduce the primary productivity. In essence, the so called 'patchiness' of the phytoplankton crop is a result of the grazing of zooplankton crops. Consequently, an imbalance in the phyto- and zooplankton ratio exists under the prevailing conditions of the localities.

12.8 Energy Transfer at Different Tropic Levels of the Food Chain Within the Marine Ecosystem

Energy transfer means the ratio of food consumed by an organism to its conversion into animal tissue, or energy (Firth, 1969). The intake of energy by the primary producers (autotrophs) is either utilised or lost during their own metabolic growth and reproductive processes, or transferred in a series of successive levels of consumers higher in the food chain or trophic levels within a particular ecosystem. What proportion of the net energy goes into the formation of new tissue of the hetrotrophs will depend much on the particular ecosystem concerned. However, in reality, it is hardly possible for a given kind of organism to feed on only a single species of other organism. Inevitably each kind may feed on many more forms than of a single other kinds of organisms. This series of feeding relationships or trophic levels is generally understood as food chain or food web. Once the energy enters into an organism, it cannot be recycled, but is passed through the organism to the successive trophic levels and ultimately lost into the atmosphere as heat. While the organic compounds of carbohydrates, proteins and fats are broken down into simpler forms and split once again into their initial H_2O, CO_2, N_2 and other trace materials in the environment, and are available to be recycled. It is only a small part of the net production of the autotrophs that will contribute as the major source of energy for the next trophic level of primary consumers (secondary production). Marine phytoplankton generally have less than 0.5% efficiency of primary production and their net primary production is still less efficient. The flow of energy though appears simple initially, but becomes more and more complex at each higher trophic level as the organisms of one level may derive their energy not only from a lower level but also from many other trophic levels of the food chain. The situation is further complicated by the fact that only scanty information is available on the diet of many marine animals. The energy link by micro-organisms, including bacteria, is even less known. The age of the maturity is also important in evaluating energy conversion. The resulting net organic production by autotrophs can either go into the primary consumers or pass through the natural metabolic, growth and death processes. The resulting dead organisms or phytoplankton debris are quickly decomposed by marine bacteria and fungi into recyclable nutrients. A major part of the detritus is also consumed by detrivores.

12.9 Productivity and Organic Cycle

As discussed earlier, the solar energy in the presence of nutrient salts, particularly phosphates, nitrates, silicates, and trace elements such as manganese and iron, together with some organic substances, and CO_2 trigger the process of photosynthesis in phytoplankton. Nitrogen derived from nitrates and phosphorus from phosphates are essential for the synthesis of amino acids and proteins, nucleoproteins and phospholipids respectively; silicon is also important for the construction of the skeletal cell walls by many planktonic organisms, especially diatoms, dianoflagellates, radiolarians and others. Many zooplankton, especially, copepods and a variety of other larvae of crustaceans, cephalapods, fish and fish species, and even whales in the Antarctic, depend on both phyto- and zooplankton as a source of food. Under favourable conditions, through the various processes of excretion, bacterial decomposition of the dead organisms, together with decay of detritus materials, the basic nutrient elements are once again brought to the euphotic layer, and the organic cycle is completed. A schematically simplified representation of the organic cycle is shown in Fig. 12.4.

Of the organic materials eaten by the primary consumers only about 10–20% efficiency goes into assimilation or is turned into their own bodies that may reach the next trophic level. The remaining 80–90% is lost either in maintenance or through respiration and in the faeces and as excretory wastes at each trophic level, i.e.:

Assimilation = Respiration + Net Production

The bacteria undoubtedly play an important role in energy flow as decomposers, using only a relatively small amount of energy compared with the primary producers which spend much of the energy in respiration.

The primary (herbivorous) consumers, zooplankton, which comprise about 80% of copepods and other crustacean larvae, consume a greater proportion of the net primary production, about 40% and 35% of the net production in the open sea and upwelling areas respectively. In the Tropics and subtropics, particularly, there

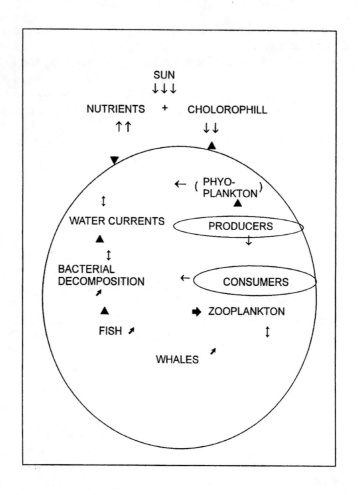

Figure 12.4   The simplified organic cycle.

is little influence of seasonality. The phytoplankton are consumed first by zooplankton which, in turn, are eaten by carnivorous chaetognaths and particularly pelagic fish such as mackerel, herring and tuna. Some pelagic fish, especially Mugilidae, sardines and anchovies are herbivores and they derive their energy directly from the phytoplankton, mostly diatoms and flagellates; and only occasionally some crustacean larvae may be found in their guts. The energy flow and a basic natural food cycle are shown in (Fig. 12.5).

However, it is possible to predict the productivity of phytoplankton under a wide range of conditions fairly accurately when the basic parameters are known. Ryther and Yentsch (1957) have shown a relationship between solar radiation ($kcal/cm^2/d$), the rate of relative photosynthesis (R), extinction coefficient (K), grams of chlorophyll (Ch) per m^3 of the sea water and the constant (3.7) which represents g carbon fixed by 1 g of chlorophyll per hour, i.e.:

$$P(gC\,/\,m\,/\,day) = \frac{R \times Ch \times 3.7}{K}$$

Table 12.1 compares the primary production of the main habitats of the marine ecosystem.

Table 12.1   Comparative estimates of primary production of marine ecosystem

Habitats	Area $km^2 \times 10^6$	%	Primary Production Gross/Net Dry-wt $Tons \times 10^9/Y^{-1}$
Neritic province	27.0750	7.50	4.12
Upwelling areas	0.4332	0.12	0.07
Estuaries and Coral reef areas	2.0938	0.58	0.32
Open sea	331.3980	91.80	50.44
Total marine environment	361.0000	100.00	54.95
			$10 - 3\ gC\ m^{-2}\ day^{-1}$
Sargasso Sea			$100 - 200$
Equatorial Atlantic			$60 - 800$

* Estimates based on data from a number of sources.

The assimilation efficiency and the food chain of the pelagic community will vary according to local conditions endemic to the marine habitats. As pointed out earlier, the highest production, highest efficiency, highest fish production and lowest trophic level will be related to the upwelling areas; whilst the lowest production, lowest

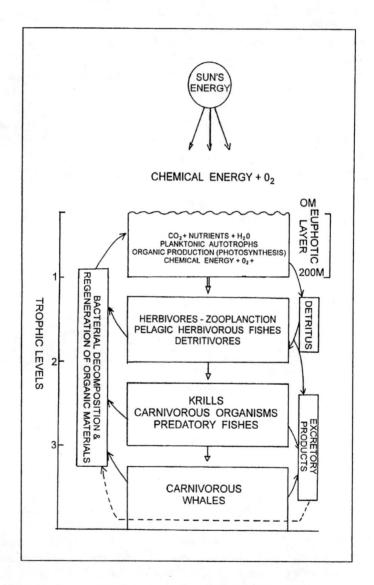

Figure 12.5   The energy flow and the basic natural food cycle.

efficiency, lowest fish production, and highest trophic level will be found in the open sea but the neritic province will fall between these two. The production and efficiency characteristics of these three habitats are compared in Table 12.2.

Table 12.2 Comparison of ratios of productions, efficiency and fish production in the main three marine habitats

	Upwelling Areas	Neritic Province	Open Sea	
Primary Production	6:	2:	1	$(gC/m^2/Y^{-1})$
Assimilatory Efficiency	2:	1.5:	1	(%)
Fish Production	7200:	680	0.005	$(mgC/m^2/Y^{-1})$
Trophic level Nos:	2:	4:	6	

12.10   Thermocline and Productivity

Thermocline is a transitional layer of water which relates to abrupt changes in temperature, and separates the warmer surface waters from the colder waters beneath.   In the Tropics and subtropics, where the seasonal variations are so unpredictable at times, the permanent thermoclines are rather sparse, but when present the nutrient replenishment is slowed down.   In higher latitudes, thermocline is almost absent and there is hardly any difference in temperature between the surface and deeper layers.   However, the thermal property of the water layers varies as depth increases in the Tropics depending on latitude, wind patterns and seasons.   Vertically, the upper layer is isothermal, but well stirred while the lower layer close to the bottom remains stratified and with very slow movement.   The thermocline usually shows vertical gradients of maximum temperature with an average value of 5°C for every 100 metres (Defant, 1961). Thus the thermocline imposes restriction on upward or downward movements of water bodies and acts as a sort of barrier between the surface and bottom layers.   However, any mixing process of the water mass is related to vertical motions.   The vertical motions are so important because they cause in many areas, upwellings and sinking processes.   Therefore, the production is continuous and extends to greater depths.   In addition, the temperature is relatively higher which accelerates the growth rate of many marine organisms and often shortens their life cycles considerably and, consequently, the total annual production is very much enhanced.   Where upwellings are effected by distinct tropical monsoon seasons, divergence usually

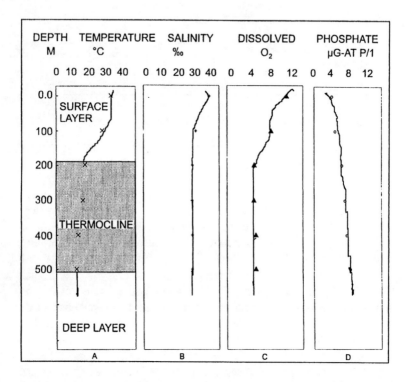

Figure 12.6   (A) Three layered water mass of tropics and subtropics. The top layer is isothermal and is known as THERMOCLINE, where the temperature drops rapidly with an average value of about 5°C per 100m; (B) Salinity variation; (C) Oxygen concentration also varies; and (D) phosphate increases from surface to bottom depending on annual monsoon cycle.

follows convergence. The seasonally variable coastal upwellings and their relation to the thermocline profile and productivity, and the variations in salinity, oxygen and phosphate are depicted in Fig. 12.6. The upwelling areas of the oceans and seas consists only about 0.1% and they contribute roughly about 50% of the world fish catch (Ryther, 1969). In Tropics, the effects of upwelling on demersal fish species have often been stressed (Bunse, 1959). Data collected over several years in upwelling areas off Cochin and Laccadive sea, the latter a zone of consistent monsoons during April–September, usually showed a sudden drop in surface temperature from 30°C–29°C, to the lowest average of 25°C, just before the onset of south-west monsoon and the decrease in temperature is not caused by turbulence due to increased wave action, but is due to upwelling, where the upwelling water comes from near the upper intermediate oxygen minimum layer at depths between 60 metres and 75 metres. The oxygen content is further reduced by increased oxygen consumption on the shelf and this depletion will have some influence on the demersal fish species to migrate away during oxygen depletion and return during the upwelling. These combined factors might add substantially to the high diversity of demersal fish species in tropical upwellings (see Chapter Nine).

References

Banse, K., 'On upwelling and bottom-trawling off the south-west coast of India', *J. Mar. Biol. Ass.* India., 1, (1), 1959, pp.33–49

Barnes, H., 'Process of restoration and synchoronization in marine ecology. The spring diatom increase and the spawning of the common barnacle', *Balanus balanoides* (L.). *Biol. Ann* 33, 1957, pp.67–85

Brandhost, W., 'Thermocline topography, zooplankton standing crop and mechanisms of fertilization in the Eastern Tropical Pacific', *J. Con. int. Explor. Mer.*, 1958, pp.1–16

Clarke. G. L. and Zinn, D. J., Seasonal production of zooplankton off Woods Hole with special reference to *Calanus finmarchius* Biol. Bull. Woods Hole, 73, 1937, pp.464–487

Coombs, J., Hall, D. O., Long, S. P. and Scurlock, J. M. O., *Primary Production Techniques*, Oxford and New York, Pergamon Press, 1985, p.298

334

Cushing, D. H., 'Upwelling and the production of fish', Adv. Mar. Biol., 9, 1971, pp.255-334

Defant, A., Physical Oceanography, New York, Pergamon Press, 1961, vol. 1, p.729

Fairbridge, R. W., The Encyclopaedia of Oceanography, New York, Reinhold Publishing Corporation, 1966, vol. 1, p.1021

Falkowqski, P. G. (ed.), Primary Productivity, New York and London, Plenum Press, 1980, p.531

Firth E. F. (ed.), The Encyclopedia of Marine Resources, New York, Van Nostrand Reinhold Company, 1969, p.740

Foyer, C. H., Photosynthesis, New York, John Wiley and Sons, Inc., 1984, p.219

Jorgensen, C. B., Biology of Suspension Feeding, New York, Pergamon Press, 1966, p.357

LaFond, E. C., 'Marine meteriology and its relation to organic production in Southeast Asian waters', J. Mar. Biol. Ass. India, IV, (1), 1962, pp.1-6

Marshall, S. M. and Orr, A. P., 'On the biology of Calanus finmarchius. VII: Factors affecting egg production', J. Mar. Biol. Ass. UK, 30, 1952, pp.527-547

May, R. M., 'The evolution of ecological systems', Scientific American, September, 1978, pp.160-175

Miller, K., The photosynthetic membrane, Scientific American, October, 1979, pp.100-113

Owen, R. W. and Zeitzsch, B., 'Phytoplankton production, seasonal changes in oceanic Eastern Pacific Tropical', Mar. Biol., 7, 1970, pp.32-36

Parker, S. P. (ed.), Encyclopaedia of Oceans and Atmospheric Sciences. New York, MacGraw-Hill, 1980, p.580

Prescott, G. W., The Algae: A Review, London, Butler and Tamer Ltd., 1969, p.436

Provasoli, L., 'Organic regulation of phytoplankton fertility', In the Sea (Ed. M. N. Hill), New York and London, Interscience Publishers, John Wiley and Sons, 1963, vol. II, pp.165-219

Rabinowitch, E, and Govindajee, Photosynthesis, New York and London, John Wiley and Sons, 1969, p.273

Russel-Hunter, W. D., *Aquatic Productivity*, New York, MacMillan Publishing Co, 1970, p.274

Ryther, J. H., and Yentsch, C., 'The estimation of the phytoplankton production in the ocean from chlorophyll and data', Limnol, *Oceanograph*, 2, 1957, pp.281–286

Ryther, J. H., 'Geographic variations in productivity', *In the Sea*, (ed. M. M. Hill), New York and London, Wiley-Interscience Publishers, 1963, vol. II, pp.347–380

Ryther, J. H., 'Photosynthesis and fish production in the sea', *Science*, 166, 1969, pp.72–76

Steele, J. H., (ed.), *Spatial Pattern in Plankton Communities*, New York and London, Plenum Press, 1977, p.470

Steemann Neilsen, E. and Aabye Jensen., 'Primary oceanic production: the autotrophic production of organic matter in the oceans', Galathea Rep., 1, 1957, pp.49–136

Steemann Neilsen, E., 'Primary production in tropical marine areas', *J. Mar. Biol. Ass.* India, 1 (1), 1959, pp.7–12

Steemann Nielsen, E., 'Fertility of the oceans: Productivity, definitions and measurements', *In the Sea,* (ed. Hill, M. N.), Interscience Publishers, New York and London. 1963, vol. II, pp.129–164

Swell, R. B. S., Geographic and oceanographic research in Indian waters, V. Temperature and salinity of surface waters of Bay of Bengal and Andaman Sea, with reference to the Laccadive Sea. Mem. Asiat. Soc. Bengal, 9, 1929, pp.1–23

Ussing, H. H., 'Biology of some important plankton animals in the fjords of East Greenland', Medd. Om. Greenland, 100, 1938, pp.1–108

Valiela, I., *Marine Ecological Process*, New York, Springer-Verlag, 1984, p.547

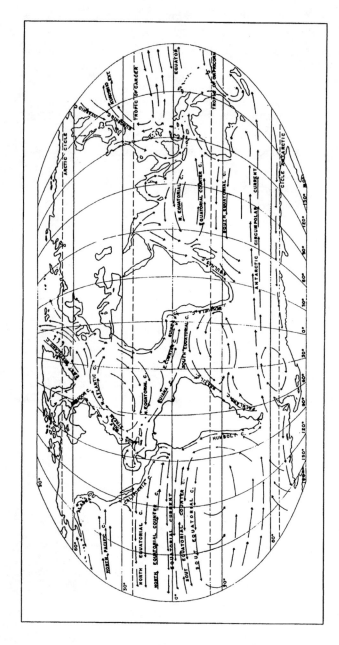

Figure 13.1 General features of circulation: note the North Equatorial Current, South Equatorial Current and the Equatorial Counter Current between them; Gulf Stream, California current, and the Brazil Current.

Chapter Thirteen
Tropical and Subtropical Ocean Currents

13.1　Introduction

Ocean currents have been known for many years, but significant advances have been made, particularly in the Indo-Pacific equatorial regions, only recently. Although the world's oceans can be regarded as a single body of water, the different forces that operate within and without the water keep it in a constant motion. These forces are generated partly by wind and partly by convection and density difference between water masses. The forces that cause the oceans to be restless are governed by a number of factors and the currents of one region may have their origin in some remote areas. Due to this complex and dynamic nature of ocean circulation, it is hard to treat the current system in the tropical and subtropical zones in isolation as a single occurrence without the overriding important features of the general system of circulation. However, greater emphasis will be placed on the current systems that dominate the tropical and subtropical zones. The distribution of currents in the seas (Fig. 13.1) is important not only because it facilitates the determining major areas of convergence and divergence and physico-chemical properties but also because of the consequence of the impact on fertility and the fisheries of the region.

13.2　Complex System of Currents

In order to facilitate the comprehension of the complex phenomenon, the current system can be divided into: warm horizontal currents, and cold vertical currents (Fig. 13.2). The surface currents fall into the first category and are determined largely by the external forces and characterised by direction. Whereas the vertical currents are

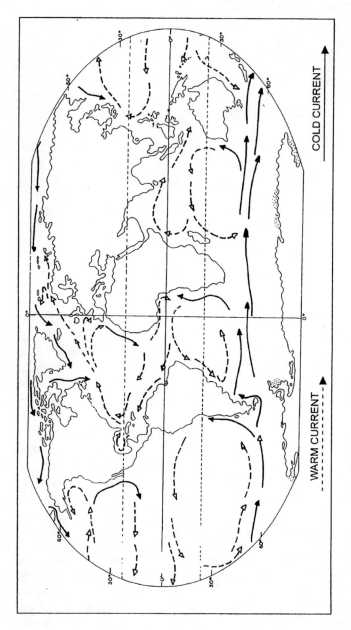

Figure 13.2   Tropical and subtropical Warm and Cold currents.

generated by internal forces largely created by differences in densities, temperature, pressure, salinity, the tidal influence in the coastal regions and bottom topography and other related factors at different levels or strata of water.

13.3   Major Causes of Currents

As has been seen, the surface currents are initiated by a combination of factors such as prevailing winds, temperature variations, pressure differences, salinity fluctuations, tidal flows, bottom topography and boundaries. Wind is the major force to set surface water masses in motion. Once initiated, the currents are influenced by Coriolis force which is caused by the combined effect of the shape and the spinning of the earth on its own axis, to be deflected to the right in the northern hemisphere and to the left in the southern hemisphere. In tropical and subtropical regions, the surface water is subject to intense solar radiation. The warm water loses some heat to the atmosphere and becomes cooler and is driven by frictional force exerted by the prevailing winds. The frictional force is stronger in water than in air and hence the water generates a speed which is only a fraction of that of the wind. Most stronger currents are generally confined to the top layer of about 100–300 metres and the current velocity decreases with depth. A greater part of the heat absorbed in the tropical and subtropical zones is lost in the temperate and frigid zones. The current becomes more effective when the wind blows steadily and transforms the wind energy to water, thus the wind plays a greater part in both the velocity and direction of the current. In the low, tropical and subtropical, latitudes of the northern hemisphere, the direction of the current is determined by the cyclonic and anticyclonic wind system. The westerly flowing north equatorial and south equatorial currents, between $10°$ and $20°$ on either side of the equator are deflected polewards by the Coriolis force to turn to the right in the northern hemisphere and to the left in the southern hemisphere. The course of the wind-driven currents usually turns about $45°$ to the right of the direction of the wind in the northern hemisphere and to the left in the southern hemisphere. The deflection increases with depth due to a decrease in the velocity and eventually the wind-generated surface current may turn to the opposite direction – the 'Ekman spiral', named after Ekman (1902) – and at this depth the functional influence ceases.

13.4 General Features of Surface Circulation in the Tropics and Subtropics

Perhaps the most outstanding feature of the equatorial region is the presence of the two broad north equatorial and south equatorial currents which flow from east to west and are driven by the obliquely blowing south-west and north-west trade winds and these currents. Between them is the equatorial counter current which flows in the opposite direction, towards the east, and is driven by relatively weaker winds but well defined all along the equatorial belt of the Atlantic, Indian, and Pacific oceans and confined entirely to the northern hemisphere about 5° north of the equator. Recently, a large subsurface Cromwell current of 300 kilometres wide and 200 metres deep has been discovered to be running under the equatorial currents eastwards at a depth of over 100 metres and the continuity of this current in the Indian Ocean has been disturbed by the seasonality of the monsoons. The equatorial currents, carrying the warmer waters, are deflected polewards greatly under the influence of the Coriolis force and partly by the continental masses. The deflected currents first trend gradually clockwise towards the north along the coasts of the continents in the northern hemisphere and then move in a circle, or gyre, with the centre in the subtropic. Similarly, in the southern hemisphere, the deflected current curves counter-clockwise turning first towards south, as the slowly moving 'west wind drift' and then to the east and north, thus forming the gyre with its centre near 20° latitude. The Atlantic south equatorial current flows westwards near the equator towards South America. On reaching the Brazilian coast, it divides into two: most of the warm waters merge with the north equatorial current and move northwards towards the Caribbean, passing the Florida Straits to be continued as the Gulf Stream; while the remainder turns southwards forming the weak Brazil current. The Brazil current, carrying warm waters with a relatively high salinity and stretching out to about 100 kilometres wide seawards and about 100 metres deep, meets the cold Falklands current, and they together now turn eastwards as the south Atlantic current.

13.4.1   Currents Related to Tropical and Subtropical Fish Resources

The physical processes and the coastal or offshore upwellings are greatly influenced by seasons. The transportation of water masses from the deeper layer to the euphotic layer brings about the mixing of nutrient salts, dissolved organic matter, and enhances the production-consumption turnover of phytoplankton and zooplankton. This is an important way of providing energy to the surface communities. During the expedition of *Azcher NIRO* (Gololobov et al., 1971) in the Arabian Sea, it was suggested that there exist four different water masses characterised by differences in temperature, salinity, and oxygen content. During monsoons, in summer the subsurface water mass spreads into the shallows along the Indo-Pakistan coasts, and during winter a large proportion of this water mass goes out from the shallows. The depletion of oxygen (0.5–2.5 ml/l) limits the distribution of commercial fish in depths more than 150 metres. Accordingly, the bottom living fish change their habits or habitats during the seasons of the year. The distribution of ichthyoplankton is also influenced by changes of current patterns, temperature and wind. The classical upwelling features are discussed below.

13.4.2   Agulhas Current

Agulhas current – a warm surface current off the east coast of the southern tip of South Africa in the Indian Oceanic part. It comprises mainly the westward flowing south equatorial current, a body of water mass flowing at a velocity of about 2–4 knots and with a temperature of around 22°C–26°C with a salinity of 34.50‰–35.6‰. The Mozambique current between the African coast and Madagascar also contributes to the Agulhas current (see below). The greater volume of the water of the Agulhas current that flows towards Africa turns sharply to the south and then eastwards to join the flow from Africa to Australia. The Agulhas bank is very productive and the fish resource exploited by South Africa consists of the pilchard, *Sardinops ocellata,* the anchovy, *Engraulis capensis,* and the jack mackerel, *Trachurus trachurus.*

13.4.3   Tropical West African Currents

Recent data collected along West Africa, from Mauratinia to Angola, indicate that masses of equatorial oceanic waters undergo significant variations with respect to their dynamic and internal structure; these characteristic variations are more pronounced along the coastal zones. Some of the variations are directly linked with winds (Ekman drift), but at the Gulf of Guinea the magnitude of the same wind effects is relatively reduced.   However, some of the variations seem to be caused by the presence of seasonal oscillations confined within the equatorial zone.   Occasionally, the rare phenomena, similar to 'El Nino' that occurs off the coast of Peru, has been reported in the Gulf of Guinea (UNESCO, 1981).

13.4.4   Benguela Current

The Benguela current flows relatively slowly, northwards in the South Atlantic Ocean off the west coast of Africa as far as 15° before merging with the eastward flowing Atlantic south equatorial current. The direction of the current varies with the seasons and the velocity is about 1.5 knots with a temperature of about 16°C and relatively low salinity.   It is associated with upwelling and is one of the most productive (3.8 gC/m^2/day) areas in the Atlantic Ocean.   Because of the high concentrations of plankton, excellent fishing grounds exist.

13.4.5   Peru-Chile Current

The Peru current is the eastern part of the anticyclonic gyre of the surface current. It has its origin in the subtropical part of the south Pacific Ocean. The water mass divides into two and a portion of the current flows towards the equator at about 200 metres below the surface along the Pacific coast as the Peru current covering a large Peru-Chile area, while the other portion flows polewards. The speed of the current varies with the seasons, but it has been estimated to be 0.3 knots on average. The upwellings occur throughout the year close to the coast and within 100 metres depth from the surface. The water is enriched with nutrients of phosphates, nitrates and silicates but there is an oxygen minimum layer close to the coast; this may be the result of denitrifying bacterial action. The productivity is relatively high at the surface; the phytoplankton productivity has been reported to be around 200 gC/m^2/Y^{-1} and hence the greatest commercial fishery of anchoveta, *Engraulis ringens,* is dependent on high concentrations of phytoplankton.

13.5   Vertical Currents

Upwellings are largely confined to the west coasts of the continents within the tropical and subtropical zones. When winds blow towards the equator, the relatively lighter coastal surface water is caused to move away from the coast seawards. The colder and slightly denser subsurface water from the deeper layer (about a few hundred metres) then moves up to the surface to replace the water that has flowed away. This phenomenon of upward movements of colder waters from deeper layers is called upwelling (see also Chapter Twelve). Upwellings can be local and temporary or may cover vast areas on a permanent basis. Upwelling can also result from deep diverging or converging currents, (Fig. 13.3) or when currents encounter land masses.

Upwelling areas are of considerable ecological and economic importance. The upwelling waters from the deeper layers bring rich nutrients up to the sunlit surface layer where they cause increased phytoplankton growth. This, in turn, supports a complex food chain and enriches the extensive production of fisheries. One of the best such fisheries is just off the coast of Peru, as noted above. The thriving anchovy, *Engaraulis ringens,* fishery depends largely on the Peruvian upwelling current which originates in the Antarctic and flows northwards proximal to the coast. This attracts the mass migration of the sea birds, especially pelicans, gannets, and cormorants whose deposits of guano on the islands off Peru are also of economic importance. When upwelling slackens due to weakening of the equatorial winds, the warm and relatively less saline waters from the equatorial counter current enter the upwelling areas during summer (December–April) around Christmas, and hence El Nino (= 'Christ child'); the consequences are more disastrous, particularly to marine organisms, especially ichthyoplankton, and the fisheries may also fail. Combination of various conditions causing upwelling are much dependent on the hemisphere, coastal configuration and orientation and the wind direction. The well-known areas of upwellings are: the Humboldt (after Humboldt, 1814) or Peru current off the coast of

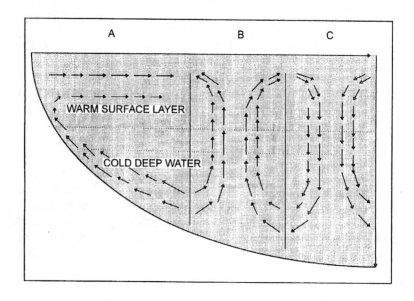

Figure 13.3   (A) Coastal upwelling induced by predominant seaward winds; the warm surface water may extend vertically to depth of 180 – 200m; (B) mixing of waters by divergent currents; and (C) mixing of water by convergent currents.

Peru, the Benguela current off the south-west African coast, the California current off the west coast of North America, the Canaries current off the north African and Spanish coast, and the north-west Australian current; also recently a very active upwelling with a vertical velocity of 10 cm/sec in areas along the Somalia and Arabian coastal belts has been reported (Royal Society, 1969; Zeitzschel, 1973). This current enhances the productivity of the Gulf area and the adjacent regions. Other areas of less intense upwellings are the east coat of India, Thailand and south Vietnam. The Agulhas bank – a shallow plateau off South Africa where water masses from the subarctic and subtropical origin meet and form gyres with effective upwellings of deep waters – favours the growth conditions for hake and other demersal fish.

At the Antarctic convergence, the phytoplankton blooms and supports the abundance of krill population which forms the principal diet of the baleen whales. Warmer currents paralleling the coasts are well exemplified by: the Gulf Stream off Florida which is a warm current about 80 kilometres in width and 640 metres deep; Kuro Shia (= Black) current off Japan, Brazil current off the eastern coast of Brazil; also, as seen above, the Agulhas current, a part of the south equatorial current in the Indian Ocean split by Madagascar into two main streams: one of which flows between the African coast and Madagascar as the Mozambique current which merges again with the other westerly flowing stream on the eastern coast of Madagascar and they together flow as the warm Agulhas current.

The counterparts of the warmer currents are the principal cold currents as represented by: the Kamchatka current, flowing southwards along the coast of the Kamchatka Peninsula and Kurile Islands, the Greenland current moving south along the east coast of Greenland, and the Labrodar current running south from the Bay of Baffin toward Newfoundland, Nova Scotia and New England.

13.6   Deep Sea Currents

The deep sea or vertical currents are primarily caused by density differences. The density of sea water varies with both temperature and salinity, and hence the formation of thermohaline. In the Arctic and Antarctic regions, as the sea surface water freezes it leaves in the cold water below a high contents of salts. This dense water now sinks down slowly and moves in different strata, depending on the temperature, along the ocean floor with very little or no mixing at all. Recently, faster horizontal 'undercurrents' have been discovered at depths below 2,000 metres. The slow sinking process of the denser waters is a common occurrence in the Norwegian Sea, Weddel Sea, Labroder and Irminger Seas and in the Polar Seas.

13.7　Convection Currents

This occurs less frequently in the low tropical latitudes than in the polar seas. Due to an excessively high rate of evaporation the surface waters get cooled and become more dense than the layer of water beneath them. The denser water now sinks slowly to a level where its density corresponds. These rising and sinking currents are called convection currents. The vertical movements can also result in convergence and divergence.

13.8　Tropical Drift and Stream Currents

The drift currents are driven by wind action, and the surface movements are usually of horizontal nature. Depending on the direction of the prevailing wind, the surface water moves in the same direction as the direction of the wind. A good example for the drift is the north equatorial drift which moves towards the Gulf of Mexico through the Caribbean Sea. On the other hand, the Gulf Stream is a pronounced oceanic current caused mainly by the movement of water mass which originates in the Gulf of Mexico, and flows through the Florida Straits and becomes deflected to the right due to rotation of the earth and continues to run towards the north Atlantic. The Kuroshio current, often called the 'black stream' of the Pacific, is the counterpart of the Gulf Stream. Due to various factors such as complex motions of water masses, different salinities and temperature regimes, sinking and upwelling, the speed of the current in different localities differs; these are shown in Table 13.1.

Table 13.1　Current Velocities in Different Global Localities

CURRENTS	MEAN VELOCITIES (m/sec)
Agulhas current	0.60
Brazil currents	0.77
Antarctic drift	0.14

Gulf Stream	2.55
Kuroshio	2.25
N. Equatorial	1.50
S. Equatorial	1.25
E. Counter current	1.25
Cromwell 'undercurrent'	1.25
Deep water currents	0.01

13.9 Effects of Currents

(1) Ocean currents contribute to the redistribution of energy by transporting warmer waters to the higher latitudes and much cooler waters to the lower latitudes.

(2) The ocean current systems help to exchange heat between polar and tropical regions and thus sustain the heat balance. Consequently, the currents have a significant effect on the global climate.

(3) Ocean currents play an important role in the thermal regulation of different parts of the global surface environments.

(4) Ocean currents redistribute food, nutrients, and oxygen and thus support the marine life spatially and temporally.

References

Fairbridge, R. W. (ed.), *The Encyclopedia of Oceanography,* New York, Reinhold Publishing Corporation, 1966, p.1021

Gololobov, Y. K., and Bibik, V. A., 'Vertical structure of the Arabian Sea water and its influence on the distribution of commercial fish', 'The ocean world', Proceedings of Joint Oceanographic Assembly, Abstract, Japan Society for the Promotion of Science, Tokyo, 1971, p.461

Knauss, J. A., 'Equatorial current systems', *In the Sea* (ed. M. N. Hill), New York, London, Interscience Publishers, 1963, pp.235-280

Lehr, P. E, Burnett, R. and Zim, H. S., *Weather,* New York, Golden Press, 1965, p.160

Posner, G. S., *The Peru Current,* Binham Oceanographic collection, Yale University, Bulletin 16 (2), 1957, pp.106-155

'Royal International Indian Ocean expeditions RRS *Discovery* Cruise 3 Report. Oceanographic work in the Western Indian Ocean 1964', London, Royal Society, 1965, p.55

Singarajah, K. V., 'Occurrence and observation of minke whales off the coast of Brazil', Sci. Rep. Whales Res. Inst., 35, 1980, pp.14–38

Stewart, W. W., 'The atmosphere and the ocean. In the Ocean', *Scientific American*, 8–52, W. H. Freeman and Co., San Francisco, 1969

Stommel, H., *Westward intensification of wind driven ocean currents*, Trans. Am. Geophys. Union, 29, 1948, pp.202–206

Sverdrup, H. U, Johnson, M. W. and Flemming R. H., *The Oceans*, Tokyo, Asiatic Edition, 1965, p.1087

Unesco reports in marine science. The coastal ecosystems of West Africa Coastal lagoons, estuaries and mangroves, UNESCO, 1981, p.60

Zeitzschel, B. (ed.), *The biology of the Indian Ocean*, London, Chapman and Hall Ltd., 1973, p.548

Chapter Fourteen

Aquaculture Potentials and Prospects in the Tropics

14.1   Concept

The need for aquaculture in tropical and subtropical countries is obvious. The terms aquaculture, mariculture, and sea farming are often used interchangeably for the cultivation of marine organisms. It is one of the ancient forms of marine science to increase fish yield and fishery products by effectively controlling and improving the methods of input and output of the captive environmental conditions, functions of growth, reproductive cycles, and rate of mortality of the culturable species. Oceans and seas provide about 15% of the world's protein rich resources. Fishery resources from the sea are not inexhaustible, but there are indications that many fishing areas have already been over-fished and seriously depleted.

14.2   Tropical Aquaculture

One of the problems facing the world on a global scale today is the population explosion. The world population for the year 1989 had been estimated to be 5.234×10^9 and it is growing at the rate of about 1.7% a year. This rate is expected to fluctuate between now and the year 2020, and by then the population may be 9.3×10^9. Based on the past trends, other demographic experts predict that the global population by the year 2050 will ultimately stabilise at 14.5×10^9. This implies that the available resources on the biosphere will have to be shared by all populations of the world. In order to avert widespread paucity of food, especially in the Third World or the poorer tropical and subtropical countries, an increased food production is the obvious answer; aquaculture must play an important role. Much of the aquaculture production at the moment comes from south and

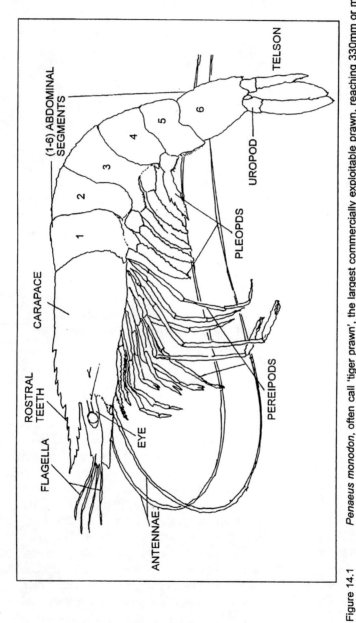

Figure 14.1

Penaeus monodon, often call 'tiger prawn', the largest commercially exploitable prawn, reaching 330mm or more in length. It is a major Indo-Pacific species caught in offshore and inshore waters as well as from tidal ponds. It is also one of the major aquaculture prawn species in Southeast Asia.

south-east Asia, Africa, Latin America, North America and Europe; overall the developing countries in the Tropics dominate in aquaculture enterprise. In sharp contrast, the developing countries in the Tropics produce nearly 74% of the aquaculture products, while the more developed countries produce only 24%. Increased interest in aquaculture on a larger scale is widespread throughout the world. The potential in aquaculture industries for the future lies in tropical and subtropical countries such as Brazil, Venezuela, Bangladesh, Taiwan, Indonesia, the Philippines, Hawaii and West Africa, and many other Indo-Pacific islands where there are many coastal areas which have favourable conditions for culture of marine species. Aquaculture is one of the methods of increasing food supply to our future needs. Culture is the rearing under controlled optimal environmental conditions of marine organisms such as bivalves, particularly oysters, mussels, scallops, whelks; crustaceans such as prawns, especially, *Penaeus monodon* (Fig. 14.1), *Penaeus indicus, Penaeus notialis and Metapenaeus affinis* is extremely profitable; also shrimps, crabs and lobsters.

Some fish species, especially of the family Mugilidae (mullet) and Chanidae (milk fish), Fig. 14.2, are economically viable. Cultivation of some of these species had been an ancient practice especially in China, Taiwan and India, dating back to 475 BC. Gradually this practice has been extended to other Asiatic regions of Indonesia, Thailand, Japan, Malaysia, the Philippines and Korea, and Hawaii; now it is commonly practised in a number of other countries such as the Soviet Union, the United States, France, the United Kingdom, Norway, Greenland, West Germany and Italy; Egypt, Brazil and Mexico (FAO, 1983). Recently, attempts have been made to utilise shallow bays, inlets, and estuarine mouths to cultivate marine items by transplanting. A measure of success has been shown by the Canadians in transporting salmon roe from the Pacific to Atlantic coasts. Striped bass have also been transplanted from the United States to the Black Sea. The total world production from aquaculture was estimated to be some 2.6×10^6 tons in 1976. Aquaculture offers a challenge as a food-producing enterprise. Much progress has been made during the past few decades in the aquaculture industry. Yet much of the potential remain untapped. It is relatively easy to feed the oysters and mullets; the method for unialgal culture is discussed in Appendix V.

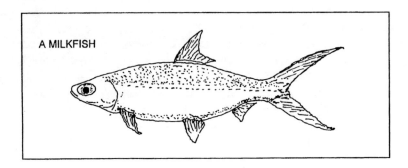

Figure 14.2   (A) The milkfish, *Chanos chanos*, and (B) the common mullet, *Mugil cephalus*, though world wide in distribution, found in abundance throughout the tropical and subtropical regions; both are commercially valuable and cultivated in many parts of Southeast Asia and Latinamerica.

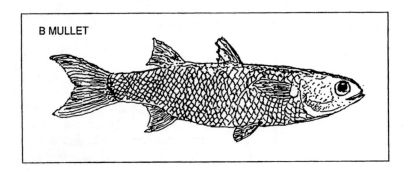

14.3   Seaweeds Potential

Seaweeds, or 'macroalgae', provide an additional but essential culture potential. Production of marine algae has been increasing in China, Japan, Korea, the Philippines, Norway, Indonesia, Chile, and many other countries (FAO, 1994). Seaweeds are a valuable source of carbohydrate, protein, fats, minerals and vitamins. In the Tropics, many coastal nations and islands have used seaweeds as food, medicine and fertilisers for centuries. China and Korea use seaweeds as food and medicine; the Indian subcontinent as both medicine and in farming, as soil neutralisers and fertiliser and, occasionally, to extract iodine; Japan, besides its home consumption, exports a considerable quantity of dried seaweeds; Brazil and USA use seaweeds for both food, especially in agar, alginates, and carragheen for ice cream industries, and as a preservative, an emulsifier, in other pharmaceutical products and as fertiliser on a commercial scale. Nevertheless, seaweed has been a relatively neglected area and less trapped resource. Some of the most valuable and commonly found seaweeds in the Tropics and subtropics are of the genus:

Gracilaria
Dityotadichatoma
Dictylomertensii
Dictyopter delicatum

14.4   Site Selection

One of the most crucial decisions involved in aquaculture is the selection of the right site. This site should be naturally endowed with many favourable parameters to productivity. Aquaculture needs the selection of site to be more profitable. It is important to remember that in trying to create a captive environment or enclosures for culture, the ecosystem, i.e. the complex community of organisms and its environment functioning as a natural ecological unit, is more likely to be altered. However, the most important parameters to be considered are:

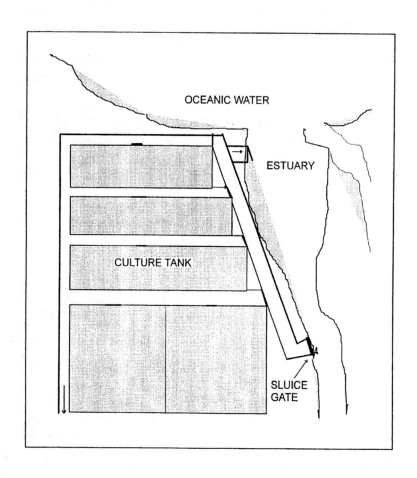

Figure 14.3   A prototype of culture site and culture tanks with sluice gates.

(1) Availability of good quality water continuously.

(2) Constant supply of nutrients.

(3) Shelter from the adverse stormy weathers and tidal fluctuations so that water can be diverted directly from the estuaries into a series of ponds constructed with the mud banks.

(4) Optimal temperature.

(5) Prevention of pollution from risks of domestic sewage and from any industrial discharge.

(6) Food supply.

14.5 Hatchery Techniques

Depending on the species and the culturing environment, the culture techniques vary considerably. One of the most widely practised methods is the hatchery type. This method generally involves construction of a series of large tanks 2×2 hectares with slew gates facing the main canal (Fig. 14.3). The water is diverted from the main canal of marine or estuarine waters during high tides, or, in some areas, large quantities of water can be pumped directly from the sea or estuaries. The latter method is more useful because the water can also be filtered relatively easily, especially for the cultivation of sessile and engybenthic organisms such as prawns and shrimps and sessile oysters and seaweeds. The culture tanks should be initially cleaned of predators so that different species can be segregated as the situation demands.

14.6 Hatchery Techniques for Profitable Species

Owing to the great importance of the hatchery techniques in bivalve, crustacean, and teleost cultivation, the design and operation procedure to produce 'seed oysters' or spat, crustacean and fish larvae on a large scale, are necessary. The details are beyond the scope and cannot be considered in this chapter, but a brief summary may be given. Some of the privately owned hatcheries are capable of producing as much as 100 millions of spats per annum to meet the demands by both local oyster cultivators and the export markets overseas. Consequently, the methods by which many routine procedures that could be made simple

Plate 14.1 A & B   The culture of mussels. Method of obtaining seeds in the
wild and growing species in the natural estuarine
environment. Exposed under low tide.

and inexpensive and the general layout of the sea water storage tanks, together with devices to pump as much as several thousand litres of the sea water per day, will be very useful. Since the good quality of sea water considerably influences the good growth, healthy reproductive potential and survival of larvae, those aspects such as the need for keeping the canals and pipes for water circulation as short as possible will be essential. On the other hand, large volumes of water are necessary to ensure effective water circulation and replenishment of oxygen, elimination of wastes and to maintain a favourably optimal temperature all the year round for the more agile teleosts such as mullets and milk fish, and the engebenthic crustaceans of prawns and shrimps.

14.7   Selection of Species

The choice of the species should be related to the selection of a particular culture site because it is important to ensure the survival and the success of the productivity of the culture industry. It is also important to have a comprehensive knowledge of the physiology, reproductive pattern and general life cycle of the species in question for culture, together with the environmental conditions.

14.8   Seed Stock

Seed stocks are the potential source of profitable productivity. A steady source of stock should be available all the time. Some may be rigidly seasonal. Seeds of some species such as oysters and mussels may be easily obtained when they spawn naturally in the wild, or by artificially inducing in the laboratory. Crustacean post-larvae and fish fry may need to be transported from distant nursery stations (see Plates 14.1, 14.2).

14.9   Feeding the Stock

Feeding the stock is one of the main and most expensive tasks in the culture industry. The nutritional preferences of different species should be recognised. Therefore, the right choice of food must be ensured. When stocks are confined to culture tanks, the density of food, particularly algal material, introduced must be carefully monitored. Depending on the feeding, some species can mature

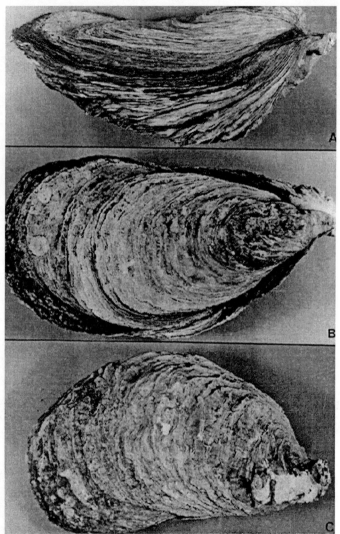

Plate 14.2   A large oyster, *Crassostrea paraibanensis*, with great culture
potential, and can be cultured relatively easily. An adult on
an average measures 206 length, 96 width 72 height (mm)
and can weigh just over 1050 grams. (A) Lateral view; (B)
with the right valve up; and (C) with the left valve up. These
oysters prefer soft sand mixed muddy bottom well below the
tide mark of estuaries. Native of north-eastern coasts of
Brazil.

earlier than others. Because of the extensive unialgal culture and the massive production larvae, especially of the brine shrimp, *Artemia salina*, in the laboratories as food for culturing organisms such as filters-feeders, crustaceans and fish larvae in hatcheries, the operational cost for the production of food is often very high; 50–60% of the running cost. Budgetary provision must be made for quantitative and qualitative variations of dietary items (see Table 14.1).

Table 14.1   Common Food Items and Analysis of Gut Contents

PLANKTONIC FOOD	COMMON CULTURE SPECIES OF		
	Bivalves Oysters	Fish Oysters	Mullets, Diaptemus rhomberus, Chanos chanos, Spyraens sp.
Flagellates	+	+	+
Diatoms:			
Coscionodiscus centralis	+	+	+
Chaetoceros affinis	+	+	–
Campilodiscus sp.	+	+	+
Ditylum brightwelli	+	+	+
Gyrosigma sp.	+	+	+
Licomophora juergensi	+	–	+
Navicula litoricola	+	+	+
Nitzchia bilobata	+	+	+
Stroptotheca tamesis	+	+	+
Synedra sp.			
Talassiosira subtilis	+	+	+
Dianoflagellates:			
Peridinium depressum	+	+	+
Oscillatoria erythrea	+	+	+
Zooplankton			
Ciliaztes	+	+	+
Tintinids	+	+	+
Rotifers	+	+	+
Naupli of cirripede	+	+	+
Copepodites	+	+	+
Detritus materials	+	+	+

+ = Present and – = Absent

Some of the common culturable species are listed in Table 14.2

Table 14.2   Some Common Species of Aquaculture in Marine and Brackish
Waters
of Tropical and Subtropical Regions

Group Phylum Class	SCIENTIFIC AND COMMON NAMES		FOOD DIET SPECIES	LOCATIONS
SHELLFISH				
Mollusca				
Bivalvia				
Crassostrea paraibanensis	Oysters		(1–7)	Brazil
Crassostrea rhizophorae	Oysters		(1,2,3,4,7)	Brazil, Venzuela
Crassostrea cuculata	Oysters		(' +)	Indian-subcontinent
Mytilus samaragdinus	Mussels		(1,2,3,5)	Philippines, Thailand
Perna indica	Brown Mussel			India, Indo-Pacific
Perna perna	Brown Mussel	(' +)		Malay-Tai Archipelagos
Andra granosa	Cockle	(' +)		Malaysia, Taiwan
Arthropoda				
Crustacea				
Penaeus indicus	Prawn	(' +)		Malaysia, Indo-Ceylon
Penaeus monodon	Giant-shrimp	(' +)		Indonesia, Philippines, Taiwan Indo-Pacific
Penaeus japonicus	Giant-shrimp	(' +)		Indo-Pacific
Metapenaeus moyebi	Giant-shrimp	(' +)		Brazil
Penaeus aztecus	Camarao	(' +)		Brazil
Penaeus duorarum				
Chordata				
Osteichthyes				Brazil, south-east Asia
Mugil cephalus	Grey mullets	(1,2,3,4+++)		Brazil, south-east Asia
Mugil waigensis,	Mullets	(1,2,3,++)		Brazil, south-east Asia
Mugil oeur	Mullets	(1,2,3,++)		Brazil, south-east Asia
Mugil brasilensis	Mullets	(1,2,3,++)		Brazil, south-east Asia
Chanos chanos	Milk-fish	(1,2,3,4,9++)		China, Philippines, Indonesia, Taiwan, India, Hawaii, etc.
Diapterus rhombeus		(' +)		Brazil, south-east Asia
Elops surus		(' +)		Brazil, south-east Asia

Seaweeds	Global locality
Hypnea musciformis	Afro-Asia, south-east Asia, Brazil, India
Sargassum sp.	Afro-Asia, south-east Asia, Brazil, India
Sargassum vulgarae	Afro-Asia, south-east Asia, Brazil, India
Gracilaria verrucosa	Afro-Asia, south-east Asia, Brazil, India
Gracilaria cylindica	Afro-Asia, south-east Asia, Brazil, India

Dietary selections - Phytoplankton : *1. Dunaliella primolects, 2. Isochrysis galabana, 3. Monochrisis luthei, 4. Pheodactylum triconutum, 5. Procentrum micans, 6. Skeletonema costatum, 7. Tetraselmis suecica, 8. Ditylum brightwelli, Cosinodiscus, Nitzia, 9. Pleurosigma,* etc. +Mixed food: zooplankton larvae of: *Artemia salina, Cirripede nauplii, Tintinids, Rotifers* a variety of crustacean , especially copepods and the unialgal species of phytoplankton listed above and often detritus. Based on laboratory experiments and analysis of gut contents. The seaweeds are cultivated in large tanks; or in inshore waters; and they enhance the yield, particularly crustaceans, when introduced into culture tanks.

14.10   Risks in Culture Industry

In any aquaculture industry, on a small or large-scale, there are many risks involved due to natural calamities over which there can be very little or no control. Careful planning and management are important. Major risks are involved in all stages of site selection, species selection, seed and feed stocks, disease and pollution. Also, storms, droughts, monsoon floods, and predators are the other unpredictable risks, especially in the Tropics and subtropics.

(1) *Pollution*: apart from the physical site, the water quality is the most basic requirement for the successful industry. Wherever one may plan to establish culture enterprise, the water is subject to many different types of pollution such as from the local sewage, industrial wastes, pesticides, insecticides and oil. The stocks can be directly affected. The growth and reproductive processes can be inhibited. Mortality may be on a massive scale.

(2) *Disease*: because of the very nature of culture media, waterborne diseases are fatal to most aquatic organisms. Indiscriminate, without a sort of quarantine, transfer of potential breeding stocks can cause serious problems to the culture industry. The seed stocks should be kept in isolated conditions until checked that they are free of any unusual behaviour or disease. If detected, the problem can be dealt

with immediately before it can infect the rest of the natural stocks. High density of the stocks can be easily avoided in the beginning. Occasionally, depending on seasonal and geographical regions, the red or brown tides may affect the stocks, especially the filter-feeders and the herbivorous fish, *Mugil cephalus, M. brasiliensis, M. waigensis, M. oeur* and the related species and the toothless, herbivorous *Chanos chanos,* may cause mass mortality. Most of these fish can withstand the tropical shallow warm (26–32°C) estuarine or lagoon waters.

(3) *Difficulties*: in most of the tropical countries, although culture prospects are promising, the operational costs are very high, particularly in the initial stages. There is very little incentive for private owners, and government support is needed frequently.

In the initial stages of establishing an aquaculture industry, careful planning on a sound scientific and technical basis and many fundamental requirements are involved; for example, the preparation of the culture tanks, introduction of suitable substrates, construction of water canals and sluice gates, pumping of water, basic buildings and laboratories, equipment, suitable vehicles etc. Administratively provision must also be made for such expediency as farm-owning or land-lease, use of water from a coastal or estuarine source, or sheltered bays; labour, export and import licence etc. In most tropical and subtropical countries government imposed restrictions and other bureaucratic hurdles exist. Because of the enormous potential in aquaculture, cooperation on a national and international scale to transfer training, research and technology will be of great benefit to the coastal aquacultural nations of the Tropics and subtropics.

References

Bliss, D. E. (ed.), The biology of crustacea, vol X, Economic aspects: Fisheries and culture, New York and London, Academic Press Orlando, 1985, pp.1–315

FAO, Year book of fisheries statistics; 1978, report no.46, Rome Unipub, New York, 1979

FAO, '1981 Year book of fishery statistics – catches and landings', Food and Agriculture Organization of the United Nations, Rome, 1983, vol. LII, part 1

Gaitsoff, P. S., The American oyster, Crassostrea Virgenica Gmelin, Fish Bull, Fish Wildl. Serv., 64, 1964, p.480

Korringa, P., *Farming the cupped oysters of the genus Crassostrea*, New York, Elsevier Scientific Publishing Co., 1976, p.224

Liao, I. C., 'Notes on some adult milk-fish from the coast of southern Taiwan', *Acquaculture*, 1 (3), 1971, pp.1–10

Loosanoff, V. L. and Davis, H. C., 'Rearing bivalve molluscs,' *Adv. Mar. Biol.*, 1, 1963, pp.2–136

Mathieson, A. C., 'Sea weed aquaculture', *Marine Fisheries Review*, 37, 1975, pp.2–14

Pillay, T. V. R., 'Progress of aquaculture', Ocean Year Book 1, (eds. E. M. Borgese and N. Ginberg), Chicago, Chicago University Press, 1979, pp.84–101

Pope, J., 'Stock assessment in multi-species fisheries, with special reference to trawl fisheries in the Gulf of Thailand', South China Sea Fish Dev. Coord. prog. 19, 1979, p.106

Ryther, J. H. 'Photosynthesis and fish production in the sea', *Science*, 166, 1969, pp.72–76

Singarajah, K. V., 'Culture of estuarine and nectonic fish trophic level feeders', Bolm. Inst. *Oceanogr.*, Sao Paulo, 29 (2), 1978, pp.361–365

Singarajah, K. V., 'Some observations on the spat settlement, growth rate, gonad development and spawning of a large Brazilian oyster', Proceedings of the National Shellfisheries Association, 70, 1980, pp.190–199

Walne, P. R., Culture *of bivalve molluscs*, Fishing News (Book) Ltd., Farnham-Surrey, England. 1979, p.189

Chapter Fifteen
Human Impact on Marine Environment

15.1   Global Range of Pollution

Life is dependent on the environment. The science of the sea has expanded immensely. Like the earth, the atmosphere and the hydrosphere are constantly changing. Pollution of the sea has become a problem on a global scale. There is a constant interaction between the biotic components and abiotic factors. Therefore, man always had a varied response to challenges to his environment. Because of his inquisitive nature, he desires to conquer and to control the environment with respect to natural calamities such as earthquakes, typhoons and droughts; fight against shortage of food and shelter. In these insatiable desires, man tries to harness the energies, explores the deepest parts of the sea, and exploits the resources of the sea, and at the same time pollutes the sea and the air, despite the technological advancement which has led him to develop unprecedented mastery for altering the environment. With increased population growth in the Tropics and subtropics and with the advent of the larger industries along the maritime nations, the magnitude of pollution by waste disposal, bacterial detritus, pesticides, oil leakage, heavy metals, radioactive materials etc., is ever increasing.

15.2   Massive Problem of Oil Pollution in Tropical and Subtropical Countries

All organisms must live in harmony with habitats of an environment to which they are most suitably adapted and where their ancestors existed for generations. One of the striking examples which causes concern to the marine environment is pollution by adding extraneous materials and disrupting the natural ecosystems. It has been reported that the aquatic environment such as lakes, rivers, estuaries and seas have been polluted nearly six times more today than they were about sixty years ago. A minute concentration of 1 part per million (ppm) of pollutant seems insufficiently small and almost negligible, but it has now been reported that DDT in the sedimentary deposits of estuaries and coastal waters in such concentration as low as 290 ppm can biologically result in concentration up to 2,000 ppm in the body of the sea gulls and penguins feeding on fish, which in turn graze on the primary and secondary trophic level feeders in the bottom. Recent experimental as well as field observations have shown that DDT has a profound, deleterious effect on the enzyme system. Higher concentrations of DDT levels lead to liver cancer in a number of animals, including mammals. The deleterious effect of DDT is not confined to aquatic animals alone. The rate of photosynthesis by the primary producers such as marine phytoplankton has been reduced quite considerably (Wurter, 1968). A large body of evidence now exists to show that harmful effects of DDT at a very low environmental level of 50 parts per trillion can easily kill newly hatched crab larvae. Adult females of the winter flounder, *Pseudopleuronectes americanus,* had concentrated DDT in their ovaries as spawning approached and mortality of post-yolk reached 100%, which has significant impact on the population abundance (Sindermann, 1982). Most tropical and subtropical countries have massive problems and although short term benefits are to increase the yield, the toxic substances eventually find their way to the sea. Many coastal water herbivorous fish are directly affected. Like DDT, mercury also has an adverse effect on photosynthesis. In Brazil, for example, large quantities of mercury are being used at the gold straining industries and the mercury finds its way to the sea through a number of rivers and enough mercury has accumulated in the coastal waters with enormous effect on the marine life and marine ecosystem. Oil pollution is a major problem in the Gulf states, especially Iraq, Saudi Arabia, and Iran. Pollution due to industrial waters on a large scale is common on the east and west coasts of India, Indonesia, Vietnam, Taiwan, Australia, Venezuela, and even a most technologically advanced country like the USA.

15.3 Ecological Disasters by Oil Spills

The major chemicals, particularly from the coastal states, enter the seas via rivers and estuaries. Many of the organic pollutants such as proteins, fats, soaps, carbohydrates, oils, tars, dyes, and many synthetic detergents discharged into the sea accidentally or intentionally will have far reaching consequences with irreparable loss. The detection of toxic chemicals such as DDT (dichlorodiphenyl-trichloro-ethane) and PCBs (polychlorinated biphenyls) in penguins, whales and other sea mammals has spurred scientific concern (see below). Effects on spawning, hatching and larval survival are particularly critical. When the oil tankers *Amoco Cadiz* off the coast of Brittany in 1957, *Torry Canyon* in March 1965 off the coast of Cornwall, and *Tampico Maru* in 1964 off the coast of Mexico were wrecked respectively, the effects were enormous. Nearly 60% of the world's oil production comes from the offshore wells. The disaster of offshore well accidents near Santa Barbara, California in early 1969 resulted in an oil leakage of 10,000 barrels and had an enormous impact. The Japanese supertanker *Showa Maru*, carrying 237,000 tons of crude oil from the Persian Gulf, ran aground at the southern tip of the Malacca Strait just outside Singapore harbour in 1975, spilling the bulk of the oil. More recently, the *Exxon Valdez* ran aground on 24 March 1989, about 35 kilometres south of Valdez, Alaska. It spilled nearly 11×10^6 gallons of crude oil into Prince William Sound, creating the largest spill and havoc in North American history. The Iranian oil tanker *Hart II* in January 1990 off the coast of Morocco and the oils spills at the Ilha Mare off Salvador have caused considerable damage by spilling millions of tons of oil, spreading to nearly 300 square miles on the sea surface and the consequent hazards are incalculable. The spillage of oil of 70×10^6 barrels in the Arctic Circle by Russians about fifty years ago still has an adverse effect with tremendous consequences and the American expertise is required to dispose of it. In spite of the regulations and international agreements, oil tankers and cargo vessels empty their ship's bilges containing oily materials into the sea. The effects of oil pollution on marine population of fish, shellfish, sea mammals and sea birds are so damaging that a recovery from the toxicity may not be possible for many decades to come. In most tropical and subtropical inshore waters, oil pollution lasts several years and results in long-term disaster to many forms of marine life through the odorous toxic crude oil which profoundly affects the sensory system in particular of even the migratory fish. Apart from the immediate mortality of many millions of life, the population, particularly marine birds and marine mammals, is considerably reduced. The other major group of animals easily affected are the filter-feeders such as oysters and mussels. Due to tidal movements, the oil pollutants especially the carcinogenic toxins, even enter the sheltered bays and estuaries and affect large populations of oysters rendering them inedible. The whole world now experiences oil spills almost every day, though not on the scale of Iraq which released over 500×10^9 barrels per day into the Arabian Sea during the recent war, but the effects will be seen for many years to come. Spills of any size or magnitude deposits highly toxic hydrocarbons into the coastal waters thus causing long-term harmful effects to marine life throughout the coastal waters; the sea floor on marine life and the tropical and subtropical regions are more vulnerable as they lack much of the modern technology to deal with the non-biodegradable materials. The highest pollution seems to exist in the Indo-Pacific regions. The most severely polluted seas within the Tropics and subtropics are listed in Table 14.1.

Table 14.1   Marine Environmental Pollution, particularly of the Major Tropical and Subtropical Regions

COASTAL COUNTRIES AND SEAS	PRINCIPAL SOURCES OF POLLUTION	EFFECTS ON MARINE ORGANISMS
Saudi Arabia, Red Sea	industrialisation and oil products	detrimental to fish
Iraq, Persian Gulf Arabian Sea	oil spillage from ships, discharges of refineries and wars	Irreversible loss of marine birds and fish resources
India, Indian Ocean	discharge of domestic wastes and heavy metals	health hazards
Bangladesh, Bay of Bengal	discharge of sewage and industrial wastes	health hazards
Sri Lanka, Indian Ocean	oil slicks from vessels and tar leakages	affects fisheries
Malaysia and Singapore, Indian Ocean	oil spills	detrimental to fish
Thailand, Indian Ocean	discharge of raw tar and tapioca	affects fisheries, industries and outbreak of epidemics
Vietnam	discharge of sewage,	health hazards, shell fishery
Indian Ocean	oil spills	is badly affected
China, Indian Ocean	Industrial wastes, heavy metals	toxic effect on marine organisms
Philippines, Indian Ocean	mining wastes, and sewage	bad effects on shell fish and fisheries

Indonesia, Indian Ocean	Indonesia wastes, oil and untreated sewage	health hazards, destruction of marine life
Bougaineville	copper mining	affects fishing industries, especially tuna and sea birds
Pacific		
Marianas Trench	dump of radioactive substances	dangerous effect on all marine organisms
Australia	Faulty sewage treatment	health risks
Pacific Ocean, USA (tropical part), Pacific Ocean	heavy metals and oil spills	health hazard
Mexico, Atlantic Ocean	heavy metals and oil spills	health risks and fisheries affected
Br. Guyana, Atlantic Ocean	discharge of mining wastes	health risks, affects fisheries
Brazil, Atlantic Ocean	heavy metals from gold mines, pesticides, industrial effluents, and sugar cane factories, coffee industries and sewage	health risks, affects fishing Industries
Peru, Pacific Ocean	oil spills	affects fisheries
Mururoa Atoll	French nuclear tests	radioactive, contamination through fish resource
Pacific Ocean		
West Africa, Atlantic Ocean	oil spills, discharge of agricultural industrial wastes	damage to mangroves and fishing industries
South-east Africa	Tar, oil spills, sewage	damage to fish and sea birds

15.4   Air Pollution

Besides nature, man pollutes the atmosphere with extraneous materials which even in trace concentration will have a harmful effect on living organisms.   It is now a common experience to encounter toxic substances in the atmosphere that irritate our eyes, nose, and respiratory passages; and some deleterious effects are irreversible. Natural volcanic eruptions are some of the principal causes of airborne dust (see below) and gaseous pollutants which can cause considerable damage and impair visibility of the environment.   Air pollution can easily modify the marine ecosystem.   Recent reports indicate that the burning of a large number of used tyres in Canada produced toxic fumes which escaped into the atmosphere, while a thick oil formed out of the residue eventually found its way to the sea, especially along the coast of Alaska through a series of streams and rivers. This, besides being a direct threat to the marine organisms, contributes to the destabilising of the marine ecosystem. More recently, natural hazards such as unpredictable droughts, floods, cyclones, and hurricanes as witnessed by England, France and other European countries, including Russia, and more frequently in tropical regions such as China, Florida, the Caribbean, Bangladesh and Indonesia, and earthquakes and volcanic explosions are other examples (see below).

15.5   Natural and Man-Made Disasters

Volcanic eruption and emission of dust from time to time, volcanic gases, dusts, molten and solid substances often escape from the vent of the crust of the earth on continents and beneath ocean bottoms. The changes associated with volcanic eruptions can profoundly affect the climatic conditions and ecology of the regions at varying distances. An abrupt increase in the intensity of volcanic activities can cause the emission of vast amounts of dust into the atmosphere and spread out on a global scale.   Greatest volcanic eruptions occur often in the Tropics.   For example, the most enormous volcanic eruption ever recorded was that of Tambora on the island of Sumbawa, Indonesia, in 1815.   About 37 cubic miles of dust and debris were thrown out. Again in 1883, the eruption of the east Indian volcano Krakatoa, lying in the Straits between Java and Sumatra, erupted with such violence and liberated substantial amounts, nearly 4 cubic miles, of volcanic dust into the atmosphere that the sunsets all over the world were conspicuously red for nearly two years. The fine volcanic dust carried by the prevailing winds around the globe formed a thin layer in the stratosphere (Fig. 15.1) and began to spread out on either side of the equatorial zone and obliterating the solar energy reaching the earth. During the first year, following the eruption, the solar energy was reduced by 20% and then 10% for the next two successive years. It took a few more years for the dust to settle down on the surface of the sea and the continents, and to eventually clear the barrier and the global climate to return to normal. The volcanic dusts are also causes for forming the ice age.

There is evidence that two great volcanic eruptions occurred twice

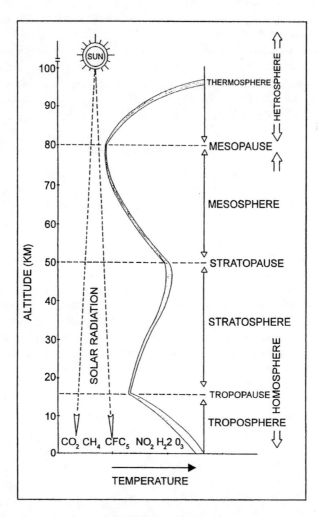

Figure 15.1   Solar radiation through the atmosphere.   During volcanic eruptions the dusts often form a thin layer in the stratosphere.

on the island of Thera, north of Crete, in the past; one during the mighty ice age about 25,000 years ago and the other about 3,400 years ago. Many believe that the latter 'Santorini' eruption of Thera wiped out the Minoan civilisation of Crete. Others, from archaeological and climatological findings, suggest that the enormous quantities of dusts, pumice and other debris carried across the sea were contributory to the disappearance of an ancient agricultural civilisation of the Harapan people of the region of the Thera desert between the north-west of India and west of Pakistan. Apart from the natural air pollution, the accidental man-made disasters are often irrevocable. For example, atom bomb experiments at Bikini Atoll in the Pacific Ocean in 1954 caused the rise of radio-active effects in the sea water and atmosphere to a widespread and unprecedented scale. The deleterious effects may last for many generations to come. During December 1984, in Bohpal City, India, trimethyl acid cyanide explosion caused more than seven thousand deaths and over 25,000 people are still suffering from irreversible disabilities, disturbed vision, lung cancer and a variety of other incurable diseases. Similarly, in USSR the nuclear reactor explosion of Sinobel caused more than 2 thousand deaths and the effects of radiation has extended for more than 25 km^2 and more are still being reported.

15.6   Effect of Global Atmospheric Pollution

As noted earlier, due to increased human activities and the ruthless exploitation of the global natural resources, particularly the fossil fuels such as coal, oil and natural gases, and the destruction of the tropical virgin forests several millions of tonnes of carbon dioxide are constantly released into the atmosphere. From 1960 to 1985 worldwide emission Of CO_2 rose at an average rate of 4% per year (Repetto, 1985). In nature, in addition to CO_2, a few other trace gases exist in the atmosphere. Some of these are beneficial to the living creatures of the earth while a few, derived from human activities, such as chlorofluorocarbons, methane and nitrous oxide, are harmful. However, the atmospheric concentration of these gases has been increasing since the pre-industrial time of about 1,750 years ago (Table 15.1).

   If the continuous increase of these gases is not controlled, the 'insulating' effect of the lower atmosphere will be affected and the global temperature may increase and consequently, the sea level will rise (see below). Thus, the atmospheric interception of infra-red radiation and the warming up of the atmosphere on a global scale is generally conceived as the greenhouse effect. However, the analogy to the ability of greenhouses to retain heat is not completely satisfactory because the glass in a greenhouse acts as a mechanical barrier to the escape of warm air, but the glass plays only a minor role in absorbing infra-red radiation.

Table 14.1   The Most Important Atmospheric Trace Gases that Enhance the Greenhouse Effect

Green House Trace Gases		Pre-industrial era (1750) Concentration (*ppmv/**ppbv)	Current time %	Present rate of increase
01 Carbon dioxide *	(CO_2)	280	351	0.48
02. Methane ***	(CH_4)	800	1,700	1.00
03. Nitrous Oxide **	(N_2O)	280	310	.035
04. Atmospheric vapour	(H_2O)	–	–	–
05. Man-made Chlorofluorocarbon				
	** (CFC11)	00.0*	0.26	5.00
	** (CFC12)	00.0*	0.44	5.00

ppmv = parts per million by volume; and **ppbv= parts per billion by volume 05. Virtually non-existent in the atmosphere before about 1930.

Obviously, the two main causes for increase in concentration of these greenhouse gases, particularly CO_2 are: the indiscriminate destruction or 'deforestation' of the vast areas of tropical forests; for instance, the burning and clearing of rainforests for agriculture and land uses in Brazil, Mexico, Costa Rica, El Salvador, Paraguay, New Guinea, Sumatra, Malaysia including Borneo, Burma, some parts of India, Ceylon, Congo, Zaire, Cameroon, the Ivory Coast, Nigeria, several islands in the Pacific, and the subtropical forests of Madagascar, the Philippines, the Caribbean Islands, etc. Secondly the burning of fossil fuels, especially coal, gas, and oil for energy by both developed and developing countries; the former more than the latter. Next to CO_2 in order of enhanced greenhouse effects is the increase in concentration of the man made chlorofluorocarbons (CFCs) for the use in specific chemical industrial uses such as aerosol, solvents, can-propellants, as blowing agents for foam packing and especially in such products of hair spray, cheese, paint, toothpaste, shaving lotion, deodorant, cologne, insecticide etc. The propellant usually contains a freon fluorocarbon or a volatile hydrocarbon. Much controversy is centred over the safety of the ozone layer of the stratospheric ozone layer. The freons in the form of microparticles of about 0.5 microns either in solid or liquid droplets are partly responsible for the air pollution; the methane (CH_4) due to decomposition of the organic debris of the paddy crops and defecated materials of a variety of ruminants by aerobic and anaerobic bacteria and relatively a small percentage due to extraction and combustion of fossil fuels; the increase in nitrous oxide (N_2O) seems to be associated with a rapid increase in the use of chemical fertilisers, especially in expanding agricultural countries of the Tropics and subtropics. In environmental terms, the increase in CO_2, for example, can be absorbed partly by aquatic phytoplankton and terrestrial plants during photosynthesis and partly by the oceans in the formation of sediments by combining the minerals discharged into the oceans and seas by rivers. It has been predicted (Study Group, MIT, 1970) that by the year 2000 the concentration of CO_2 in the atmosphere would increase and this excess of CO_2, about 18% of the present level, would warm up the globe by 0.5°C. If there were no equilibrating forces, this could be sufficient to change the climate to the extent of melting the ice-caps of Greenland and the Antarctic resulting in the sea level rising by 60-120 metres, submerging most of the coastal regions, particularly a greater portion within the tropical belt. The fuel consumption by the tropical countries amounts to some 26%; but the rest is being consumed by developed industrial nations. Unless some global agreement to regulate the increase of the freons or gases of the greenhouse effect is made, there will be a rise in the global temperature and will even affect the ozone layer.

15.7　Oxygen and the Ozone Layer

As noted earlier, the atmospheric air contains 20.95% of oxygen by volume (see Table 15.2). The O_2 is an important element on this planet to support life. Much oxygen is used in respiration of living things, combustion and disintegration of organic materials. Thus the loss of O_2 during the oxidative processes is restored by gaining the O_2 through photosynthesis by green plants and by a recycling mechanism. Unlike other gases, there is no evidence of depletion of atmospheric O_2 and its concentration has remained the same since 1910. However,

O_2 can easily react with other substances to form simple compounds or oxides, for instance: $C + 2O = CO_2$, $C + O = CO$, $S + 2O = SO_2$, $N + 2O = NO_2$, and water $2H + O = H_2O$, to mention a few.

Contrary to constant concentration of oxygen, the chemical actions of certain pollutants through human activities can have serious and far reaching consequences to the atmospheric conditions. The immediate effect is the depletion of ozone in the stratospheric zone (see Fig. 15.1). As seen in Chapter Twelve, the solar radiation that reaches the surface of the earth contains high energy photons in the ultraviolet range which, unless absorbed before penetration of the earth, could be injurious to life. In fact, most of this energy is absorbed in a series of chemical reactions involving smaller concentrations of ozone in the upper atmosphere. Because O_2 is highly reactive, the molecule can react with an extra atom of oxygen to form ozone or trioxygen, by a series of oxidative processes.

Ozone absorbs UV radiation within the range of wavelengths between 242 nm and 332 nm and the energy gained breaks down the ozone to one oxygen molecule and one oxygen atom usually in the presence of a third energy absorbing particle often designated as 'M'. The O_2 and O recombine and release energy during regeneration. This energy is transferred to the 'M', and when energy is absorbed by 'M' it becomes converted to an excited state. The excited molecule M, in turn, gives up energy as heat (see the equations below). Although the whole chemical process provides no net change, it helps to convert the hazardous UV light to heat, thus cutting off the shorter wavelengths below about 300 nm. This means that only the relatively less harmful UV radiation with long wavelengths reaches the earth. Any pollutants that interfere with the stratospheric ozone layer can cause exposure to the more energetic shorter wavelengths resulting in greater risk to life on earth. The ozone cycle can be represented sequentially as follows:

(1) $O_2 + UV^{radiation} + O = O_3$
$$\lambda \rightarrow < 242nm$$

(2) $O_3 + "M" + h\nu = O_2 + O + $ energy

(3) $O_2 + O + (excited)M$ heat $\uparrow$

Recent studies of the upper atmosphere have shown that there are several depletion path ways of ozone. The discussion will be limited to only a few. One utilises the stratospheric water vapour and another involves nitrogen oxide. The depletion of stratospheric ozone by noncatalytic moisture can be represented by the following equation:

(4) $O + HOH \rightarrow 2H\ O^{[hydroxyl\ radical]}$

(5) $HO + O_3 \rightarrow H\ -O-O^{[peroxide\ radical]}$

Depletion of stratospheric ozone by catalytic NO is:

(6) $NO + O_3 \rightarrow NO_2 + O_2$

(7) $NO_2 + O \rightarrow NO + O_2$

The net result is:

(9) $O_3 + O \rightarrow 2O_2$

Since the function of the nitric oxide is mainly catalytic, even smaller quantities are sufficient to trigger ozone depletion. In 1970, it was realised that hydrogen oxide generated by water vapour emissions from the proposed fleet of high altitude supersonic transport jets SSTS (Plate 15.1) could pose a threat to the ozone layer. It was discovered that oxides of nitrogen in the stratosphere can also destroy ozone catalytically. The possible destruction of the ozone by nitrogen oxide emissions from SSTs played a role in halting US development of the aircraft in the early 1970s, although economic considerations were probably the crucial factor. In 1974, Mario Molina and F. S. Rowland theorised that CFCs used as propellants in aerosol spray cans could reach the stratosphere, release chlorine, and destroy 20% of the ozone (CFCs sometimes known by trade name Freon).

The modern Concorde jet planes flying in the stratosphere, covering a large air space, especially across the Tropics and subtropics, would upset the ozone balance. Nevertheless, some consolation can be derived from the fact that rates of ozone depletion involving NO and H_2O from the jet exhaust effect are relatively less significant. On the other hand, the freons, chlorofluoromethanes, $(CFCl_3\text{–}11)$, used as a propellant in aerosol cans, and $CF_2Cl_2\text{–}12$ (a refrigerant) are much more serious components of depletion of the ozone layer. The reactions involve Cl atoms to catalyse as follows:

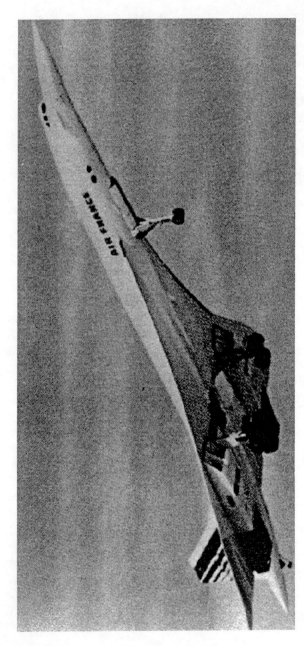

Plate 15.1 The British-French Concords: A Supersonic Jet liner first flew from Heathrow at supersonic speed of about 1900 kmph on 14th September, 1970 and is on regular flights between continents since. The supersonic transport (SST) was expected to fly in 1972, but the US Congress cut off the funds for this project rather prematurely because of some evidence of ozone layer being affected.

(10) $Cl + O_3 \rightarrow ClO + O_2$

(11) $ClO + O \rightarrow Cl + O_2$

(12) $O_3 + O \rightarrow 2O_2$

The main sources of stratospheric atomic chlorine are the freons. These are fairly stable in the lower atmosphere and take a relatively longer time to diffuse into the stratosphere. There when they become exposed to the solar UV radiation and get energised, the $C - Cl$ bonds are broken to release chlorine atoms. Recent evidence indicates that this chlorine pathway has greater influence on depletion of the upper atmospheric natural ozone layer. Overall, these highly reactive chemical components in the stratosphere have been the direct cause and consequence of the 'ozone holes' over Antarctica in the southern hemisphere. Ozone exists in nature and is produced by naturally occurring photochemical reactions and by the passage of electricity during lightning through air in the stratosphere of the atmosphere (see Fig. 15.1). The adverse effects of ozone depletion to life on both land and sea are of serious concern.

15.8   Effects of Oil Pollution on Marine Ecosystems

The effect of oil spills on the marine ecosystem is profound, particularly on plankton, pleuston, neuston, necton, intertidal organisms, corals, sea birds, marine mammals, and other bottom-dwelling benthic organisms. The death toll of sea birds runs into thousands. Even within smaller areas, the oil directly affects the feathers and consequently incapacitates the birds, inhibiting their natural buoyancy and their ability to fly or run and so renders them likely to drown and die. This results in an irreparable loss to a variety of species of tropical birds. The South African penguins are currently in danger of extinction due to oil spills from tankers.

Among mammals, the whales and dolphins and many other species which often resort to coastal waters during breeding are often susceptible to oil pollution. They often get suffocated by oil and other oil products. Their reproductive capacity can be impaired.

Fish, planktonic eggs, and larvae of a variety of species, including invertebrates, which mostly drift on the surface of the sea are seriously affected, although the adult may escape either by swimming away from the polluted area or resist to some extent by the mucus secretion from their gill and body surface. Nevertheless, some species, especially the herbivorous mullets, sardines and anchovies may contaminate through feeding, while the benthivorous feed on bottom sediments, algae and smaller invertebrates such as molluscs and crustaceans. Toxic hydrocarbons have been reported to be retained by a number of species of benthic filter-feeders such as oysters and mussels for more than six months (Blumer, 1971).

Plankton: phytoplankton seems to be affected by depression of photosynthesis and inhibited growth of diatoms by oil pollution (Galtsoff et al., 1935). The larvae and juveniles are usually more sensitive to pollution than are adults.

Neuston and pleuston: live in intimate contact with the interface between air and sea surface and are the victims of oil spills as they come in direct contact with the floating oil. *Sargassam* weeds also become contaminated.

Intertidal organisms: oil spills often directly poison the intertidal organisms, in particular their visual systems are affected, although functional implication is little understood. Although algae of the rocky pools within the intertidal zone, despite some resistance, may become coated with oil and finally perish. Many other intertidal organisms such as barnacles, other crustaceans and molluscs are easily affected.

Benthic organism: the oil spilled out eventually sinks to the bottom and becomes mixed with sediments resulting in extensive damage to benthic invertebrates principally molluscs, crustaceans and echinoderms. The heavily oil-laden bottom sediments, make the benthic population nearly disappear as a result of accumulation of hydrocarbons in their body. The lethal effect of oil pollution on benthic organisms is considerably devastating.

Marshy wet lands: many estuarine shores in the Tropics and subtropics are characterised by swamps of several species of mangrove trees. The oil spills can cause immense damage to these mangroves by penetration of the toxic substance through pneumatophores, finally blocking the conducting channels of pholem and xylem, and the general functioning of these plants and thus result in the destruction of their habitat. A variety of population of epiphytic algae, bivalves, barnacles, fish, including eels and birds are also affected.

Coral reefs: the enormous oil tankers often sail through tropical waters between China, Australia, Korea, Malaysia, Singapore, Indonesia, the Philippines, Burma, India, Ceylon, Pakistan, the Gulf of Aden, and African and South and North American coasts. Most accidental oil spills have caused considerable damage to corals. Although the relatively warm tropical environment is susceptible to bacterial action of the oil, the damage to coral reefs is extensive, especially in Indo-Pacific oceans.

Estuaries: are extremely productive sites for shellfish, crustaceans and fish. Many estuaries form convenient ports on national and international scales, receiving frequently large oil tankers. The estuarine ports are often turbid with suspended sediments, and any oil spill becomes easily absorbed with the detritus of waters, and is then dispersed thus affecting considerable areas.

15.9   Control of Oil Spills and Pollution

It has been estimated that between $1.8-10.0 \times 10^6$ tons of oil leaks out into the marine environment (Porrticelli et al., 1971). This magnitude, however, does not take into account the variable oil spills in many isolated cases and by refineries as a result of either human or technical error.

The control of oil spills in high seas and coastal waters is difficult. The weather, surface currents, waves, and direction of wind can make the clean-up process still more difficult and cumbersome. Several methods exist in handling oil spill, but most of them seem not entirely satisfactory. Dispersants and oil often cause greater damage to marine life. In modern times, the oil spill is controlled by floating very large inflatable tubes which act as barriers, surrounding oil leaked area and consequently enclosing a pool of oil. The contained oil is then pumped into a well equipped clean-up ship. However, this method involves considerable difficulties and risks, particularly in high seas and during rough weathers.

15.10   Greenhouse Effect on Global Climate

As noted above, the trace gases, or as they are collectively called 'greenhouse gases', are able to absorb and emit radioactive waves, and consequently influence the global natural ecosystems. The atmospheric concentration of the gases has been increasing for the past

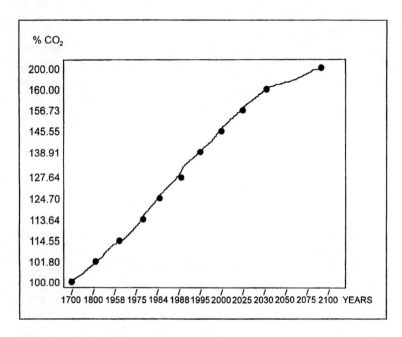

Figure 15.2   Progressive Increment of atmospheric carbon dioxide,
                   expressed as percentage of the concentration in 1970 as
                   100%.

few centuries. Carbon dioxide has increased by about 25% since the pre-industrial era of about 1750, from 280 ppm to about 351 ppmv by 1988. Table 15.3 summarises the steady increment of atmospheric carbon dioxide.

Table 15.3   Greenhouse Trace Gases, particularly CO_2 and the Progressive Increase in the Atmosphere

From Year To Year	Gas	Concentration
1700	CO2	275 ppmv
1800	CO2	280 ppmv
1958	CO2	315 ppmv
1975	CO2	340 ppmv
1984	CO2	343 ppmv
1988	CO2	351 ppmv
1995	CO2	382 ppmv
2000	CO2	400 ppmv
2025	CO2	410 ppmv
2030	CO2	431 ppmv
2050	CO2	20 GtC/year*
2075	CO2	440 ppmv
2100	CO2	550 ppmv

* GtC/year 10^9 tons carbon/year
Present emission CO_2 = 5 GtC/year

The progressive increase of the carbon dioxide over the period of 400 years is shown in Fig. 15.2.

The other radio-active components, including aerosol, of the atmosphere are collectively considered as 'greenhouse' gases. The future changes of global climate will depend on the atmospheric concentration of the green house gases. The estimates based on observations since the beginning of this century, suggest that the global warming of 1.5°C to 5°C will lead to a sea level rise of 20 centimetres to 165–149.11 centimetres, and the major contributing factor to such a rise would be thermal expansion of the sea water. The global average sea level has risen by 12±5 centimetres during the twentieth century (Bolin et al., 1986). Furthermore, the climatic predictions suggest that the highest warming to occur at higher latitudes, due to increased concentration of greenhouse gases relative to a smaller rise in temperature in tropical or lower latitudes, would result in areas more sensitive to precipitation. Even by a conservative estimate, the world will warm by about 2–2.5°C and the ocean will expand and the sea is likely to rise to a level about 14–15 centimetres higher than it was in 1985 (Warrick et al., 1990). Consequently, the spatial characteristics of terrestrial, coastal and marine landscapes, together with the distribution of vegetation and fauna, will be adversely affected. The corals which continued to flourish uninterrupted will have devastating effects. There will be a considerable depletion in the phytoplankton production which will lead to a significant reduction in living resources, especially in fish stock, bivalves, cephalapods and crustaceans. Equally affected will be the benthic and littoral communities. Imagine the horrors of the possible nuclear war which threatens the life not of the marine environment alone but of this planet as a whole, and therefore the entire world should share the responsibility for the long-term environmental problems and be able to scrutinise and survey continuously the profound effects on global ecosystems. Man made pollutants are also on the increase; tropical forests are decreasing and significant changes to the marine systems on a global scale are invading even the deepest sea.

15.11   Effects of Sea Level Rise

Of the measurable consequences of sea level rise, the most obvious one is shoreline retreat which results in permanent flooding or inundation of low-lying areas (Bath et al., 1984) or temporary flooding and erosion; salt intrusion can cause considerable damage to marine animals and marine algae.

References

Bath M. C and Titus, J. (eds.) *Greenhouse effect and sea level rise*, Van Nostrand Reinhold Company, 1984, p.325

Blumer, M., 'Scientific aspects of oil spill problem', *Environmental affairs*, 1, 1971, pp.54–73

Boech, D. F., Hershner, C, and Milgram, J. N., *Oil Spills and the Marine Environment*, Cambridge, Mass., Bellinger Pub. Co., 1974, p.114

Bolin, B., et al., *The greenhouse effect, climatic change and ecosystems*. Chister, New York, John Wiley, and Sons, 1986, p.541

Burns, K. A., and Teal, J. M., 'Hydrocarbons in pelagic Sargassam community', Deep sea Research, 20, 1973, pp.207–211

383

Comel, D.W., 'The Great Barrier Reef. A case history' Biological Conservational, 3, 1971, pp.249–254

Galtsoff, P. S., Prytherch, H. F., Smith, R O. and Koehring, V., 'Effects of crude oil pollution on oysters in Louisiana waters', Washington DC, Bulletin of Bureau of Fisheries, 48, 1935, pp.143–210

Hohn, M., 'The use of diatom populations as a measure of water quality in selected areas of Galveston and Chocolate Bay', University of Texas, Publications of the Institute of Marine Science, 9, 1959, pp.404–453

Holum, J. R., Topics and terms in environmental problems, New York, John Wiley and Sons, 1977, p.729

North, W. J., Neushul, M. Jr, and Clendenning, K., 1965, 'Successive biological changes in marine cove exposed to large spillage of oil', Symp. Commission Internationale Exploration Scientifique Mer Mediterranée, Monaco 1964, pp.335–354

Porticelli, J. D., Keith, V. F., and Storch, R. L., 'Tankers and Ecology', Transactions of the society of Naval Architects and Marine Engineers, 79, 1971, pp.169–221

Repetto, R., (ed.) The global possible resources development, and the new century. New Haven, and London, Yale University Press, 1985, p.538

Sindermann, G., Ocean Health and Georges Bank. In Georges Bank. Past, present and the future of a marine environment, (eds G. C. MacHead and J. H. Prescott), Westview Press, Colorado. 1982, pp.77–92

Smith, J. E., 'Torrey Canyon' pollution and marine life. A report by The Plymouth Laboratory, XIV, Cambridge University Press, 1968, p.196

Warrick, R. A., Barrow, E. M., and Wigley, T. M. L., 'The green house effect and its implications for the European Community', Report 12707 EN, ECSC-EEC-EAEC, Brussels, 1990, p.30

Wurster, C. F. Jr, 'DDT reduces photosynthesis by marine phytoplankton', Science, 158, 1968, pp.1474–1475

Wurster, C. F. and Wingate, D. B., 'DDT residues and declining reproduction in the Bermuda petrel', Science, 159, 1968, pp.979–981

Chapter Sixteen
Functional Implications in the Behaviourof Marine Animals

16.1 Introduction

To understand the basic marine biological problems, physiology is very valuable. Physiology in a broad sense means functions of all living organisms in nature. It deals with not only biology, but its very foundation and unifying principles depend on chemistry, physics, physico-chemical laws, statistics, mathematics and a host of other subjects. It is a more pragmatic subject based on experimental evidence and closely associated with many other biological sciences, requiring quantitative precision and knowledge of continuous life processes in exactness. Many of the marine organisms behave in an adaptive manner. Physiology helps to understand these intricate, complex and diverse problems. Adaptation is the ability of organisms to adjust to environmental conditions through heritable traits, to increase the chances of survival and to reproduce successfully; it is the essential basis of evolution. It is not possible to discuss the subject as a whole here because of the diversity and abundance of available data, and it is beyond the scope of this book, but those essential life phenomena concerning the marine organisms in relation to their environment, which seem of especial interest to the theme of this chapter will be focused on briefly here.

16.2 Physico-Chemical Factors

The biochemical and physiological processes of living organisms are based on physical and chemical properties of highly organised complex molecules and ions that constitute all living systems. Water is involved in all functional stages and its conservation and regulation of chemical composition of the internal environment of the body is of vital importance to life (see below).

16.2.1   Diffusion

Diffusion is a process by which many living organisms exchange gases, some absorb nutrients and many others eliminate excretory products. Thus, diffusion in the body can take place through gaseous and liquids media and across the protoplasm and cellular membranes. It plays an important role in translocation of materials in the living tissues. When two salt solutions of different concentrations are separated by a membrane which is permeable equally to both salt solutions and water, the salt passes from the region of higher concentration to the region of lower concentration until both sides have the same strength. This is simply known as diffusion. However, the rate of diffusion is influenced by a number of factors and these are demonstrated in Fick's Law:

$$R = DA\ (C_1 - C_2)$$

R = the diffusion rate,
D = the diffusion coefficient,
A = the cross sectional area through which the diffusion takes place, and
C_1-C_2 is concentration gradient (see Fig. 16.1)

16.2.2   Osmosis and Osmotic Pressure

Life on this planet is thought to be of marine origin. All organisms consist of a greater percentage of aqueous solutions of different concentrations of solute in their body. Perhaps, except the unicellular organisms, most of the metazoans have their bodies divided into compartments. Protozoans, for example, have intracellular fluids and vacuoles, whereas multicellular organisms have both fluid-filled intracellular and extracellular compartments. The biological membrane (plasmamembrane) which divides cells or tissues has the remarkable property of being able to allow some molecules or ions to pass through, but not others. The French physiologist, Claude Bernard (1855), remarked that the extracellular fluid constitutes an internal environment; and the maintenance of a constant internal environment is important for the cells to function efficiently. The

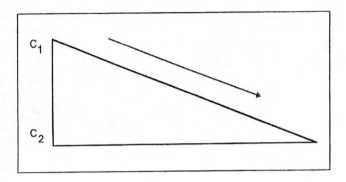

Figure 16.1   Diffusion, the movement of a solvent or solutes through a membrane along the concentration gradient.

Figure 16.2   Osmosis and the osmotic pressure. (A) Initial level (1) and (B) Final Level at the end of the experiment (2). The column (level 2 – level 1) is due to solvent migration and is equal to osmotic pressure.

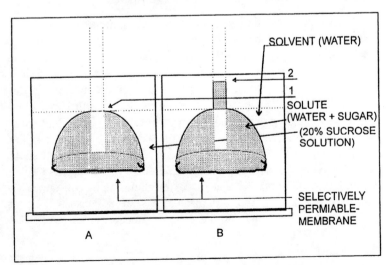

distribution of water among the various fluid compartments of the body is determined mostly by the solute content of these fluids. However, most solutes permeate cell membranes relatively slowly compared with pure water; the latter tends to establish an equilibrium between cells and their environment.

Osmosis is an immensely important physical process. It is the process by which when two solutions of different concentrations are separated by a membrane (semi-permeable or selectively permeable membrane, i.e. permeable to a solvent such as pure water but not to solutes such as sugar or salt), water molecules pass from a region of low to higher concentration (see Fig. 16.1). An osmotic membrane allows only water molecules to diffuse through and any dissolved substances contained in the water are left behind. This movement of water is called osmosis. If, however, pressure is applied to the side of stronger solution, the movement of water into it will be opposed. The pressure required to stop the movement of water from the weaker solution to stronger solution is called osmotic pressure. Thus, the osmoregulation is concerned with the control of osmotic pressure within an organism. One of the methods used in the desalination to convert the sea water to fresh drinkable water is 'reverse osmosis'.

The movement of the solvent molecules into the solute is because the membrane is selectively permeable to the solvent (water) and not to the solute (solution) to which the membrane is impermeable. However, as seen in Fig 16.2 below, the difference in the two columns is due to entry of solvent into concentrated solute. If, however, a pressure necessary to prevent the movement of solvent molecules is applied, diffusion of solvent molecules can be stopped. This excess of pressure necessary to oppose the movement of solvent molecules and keep the solution in equilibrium with the solvent is called osmotic pressure of the solution. It is due to reduction of activity or 'freedom' of the solvent molecules in solution. When a substance is dissolved in a solvent, the activity of the solvent molecules is reduced. In fact, the solvent molecules will diffuse from a region of greater activity to a region of lesser activity; i.e. the activity is greater in more dilute solutions while the activity is lesser in the more concentrated solutions. In any case, for osmosis to take place a semi or more appropriately selectively permeable membrane is necessary. Like other vapour pressures, freezing point depression etc., it involves colligative property of solutions. The osmotic

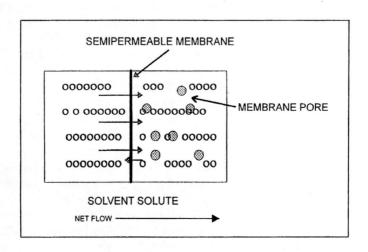

Figure 16.3   Osmotic flow from pure solvent (water) to solute (solution).

pressure of a solution will depend on the number of undissociated particles, molecules, or ions, per unit volume (litre). For example, in the non-ionizing compound like glucose molecules, the osmotic pressure is developed by virtue of the number of glucose molecules present. Conversely, sodium chloride would dissociate into sodium and chloride ions as shown in Table 16.1.

Table 16.1. The Compounds that Nondissociate and Dissociate

Non-ionizing (undissociating) Compounds	Ionizing (dissociating) compounds
Glucose $C_6H_{12}O_6$	Sodium chloride ($Na^+ Cl^-$) Sodium Sulphate ($Na_2SO_4 = Na^+ Na^+ SO^{--}$) Magnesium Sulphate ($Mg^+ + SO^{--}$) Urea CO ($NH_2$)2 deaminated product)

Van 't Hoff showed that osmotic pressure of a dilute solution was equal to:

$$P = \frac{nRT}{V}$$

n = number of ions or molecules (is the molarity of the solution = number of grams moles of solutes per litre)

R = gas constant (0.0826 1 / atm)

T = absolute temperature,

V = volume in litres

Since P is approximately proportional to the concentration of the solute, but independent of identity of the solute, it is a colligative property.

The term tonicity is often used to express the effective osmotic pressure.

(1) Isotonic solution = when two solutions of the same osmotic pressure are separated by a selectively permeable membrane, no resultant flow occurs across the membrane because they are isotonic with each other.

(2) Hypertonic = solutions with higher osmotic pressure.

(3) Hypotonic = solution with low osmotic pressure.

As far as human plasma is concerned, a 0.9% saline solution of (0.16 M Nacl) or 5% glucose will be isotonic with human plasma. The higher is the osmotic pressure, the lower is the concentration of solvent as can be seen in Fig. 16.3.

The advantage of the colligative properties is that it enables to measure the number, n, of moles of solute present in the solution. If we know the mass of the solute, we can calculate the molecular weight or molar mass. W/n where W is in grams and, therefore, W/n is in grams per mole. Freezing point depression provides easiest of the properties to measure experimentally. The molecular weight in grams also expressed as gram mole dissolved in 1 litre gives an osmotic pressure of:

$$22.4 \text{ atmospheres} = 22.4 \times 760 \text{ Hg} \qquad \text{Eq. 1}$$

As noted above, the same solution has a freezing point of 1.85°C. It gives a relatively easy way of calculating the osmotic pressure. For example, the molecular weight of NaCl is 23+35.5=58.5. To prepare a 0.9% saline solution will require 9 g NaCl per litre of water.

From Equation 1 above, the osmotic pressure of this solution exerts:

$$= 2 \times 22.4 \times 9/58.5 \times 760 \text{ mmHg}$$
$$= 5,238 \text{ mm Hg.}$$

Note: since NaCl dissociates into Na^+ and Cl^- ions, each has its own osmotic pressure and hence it is necessary to multiply by 2. On the other hand, the molecular weight of glucose is 180, and it does not dissociate and therefore will exert nearly the same osmotic pressure and they are said to be iosotonic with each other. Ringer solution is also used in many laboratories as a physiological solution.

16.3   Osmoregulation

As seen in earlier chapters, the living organisms had their origin in the shallow salty seas billions of years ago, and the water played an essential role for all physiological processes and in maintaining a stable internal environment in equilibrium with the external environment. Water provides the most ideally suited medium for interactive processes of many chemical substances within the living system, and to establish the ionic gradients and the electric potential difference, particularly in excitable cells such as nerves and muscles. Thus a stable internal osmotic environment is essential for the survival of all organisms. During the course of evolution many animals have developed interesting structural and functional adaptations to cope with even the most subtle or extremes of changes in their environment and the ability to control the osmotic pressure within an organism is generally known as osmoregulation. Thus, osmoregulation necessitates the exchange of water and salts between extracellular environment and the external medium.

Many organisms have developed osmoregulatory organs. The higher vertebrate animals like fish have gills, salt glands, integuments and kidneys, and others have an effective circulatory system, and mainly kidneys and integuments. In relatively simpler organisms, these organs are distinctly absent and the mechanisms to maintain homeostasis (the maintenance of a constant internal environment) or to regulate the water content and solute concentration in their body may vary considerably. The osmotically most active substance or molecules are those of sodium, chloride, potassium, urea and proteins molecules, though the latter are somewhat restricted in their movement across the membranes but contribute substantially to intracellular concentration. Most animals are tolerant to a higher concentration of urea and retain large amounts of urea molecules in the body fluids in order to increase the internal osmotic pressure and to reduce the water loss.

16.3.1   Osmoregulators

Many marine animals whose internal body fluids are different from the external medium in which they live are called osmoregulators, i.e. they are able to maintain their difference of internal body fluids within a narrow range of osmotic pressure. For example, marine teleosts, marine reptiles, and marine mammals (whales) have hypotonic plasma with regard to sea water. On the other hand, marine elasmobranchs (sharks, rays and skates) have plasma which is isotonic with sea water.

16.3.2   Osmoconformers

Most animals maintain their ionic concentrations of intra and extracellular body fluids (Fig. 16.4) by regulating the osmotic pressure. A number of marine animals which maintain their extracellular body fluids similar to that of their environment are

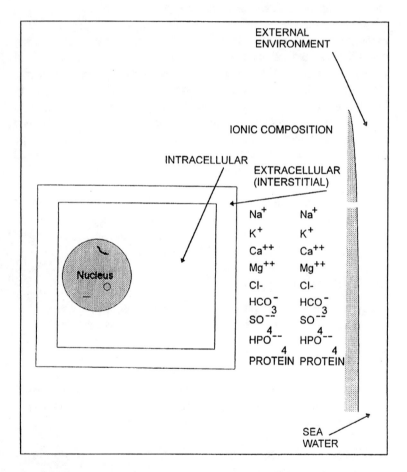

Figure 16.4   Ionic composition of intra- and extracellular fluids and the
                     surrounding external sea water medium.

generally known are osmoconformers, i.e. the concentration of their internal body fluids conforms with the external ambient.

16.3.3   Osmoregulation in Fish

Although the kidneys are the principal organs of excretion in many higher animals, other organs are also important in the regulation of osmotic extraction of water content.   In fish, apart from the respiratory function, the gills play a vital role in controlling the osmotic problems.   Since first reporting the extrusion site of chloride cells in marine teleosts (Keys and Willmer, 1932), a number of workers have contributed to elucidate the functional nature of the chloride cells, or 'ionocytes', the latter rather recent term because of direct evidence that they are, in fact, the sites of transport, specially of Na^+ and Cl^- ions. The chloride ions are pumped out of the cell by active transport and sodium by passive transport mechanisms.

As pointed out at the outset, behaviour is a fascinating and exciting area and has been associated more with adaptive nature for an end purpose than with survival value.   It is almost impossible to distinguish the adaptive behaviour without neural function. It is now fairly well established that many animals, including marine forms, exhibit physiological endogenous rhythms or 'circadian' behaviour. The reactions of an organism to a stimulus from an internal or external environment results in a response or an outward expression which we usually conceive as behaviour.   Diurnal vertical migration in planktonic and many benthonic larvae is a common phenomenon; and light seems to be the triggering and controlling factor.   There is no doubt, as Sherrington (1947) pointed out, that any behaviour involves nervous integrative function, above all the others.

16.4   Riddles of Nature

Behaviour is basically of two types: inherited and learned.   The former is claimed to be the basis of genetical traits or inheritance: the latter on the basis of learned behaviour.   Among vertebrates, only reptiles and birds have developed unusually large telencephalon or corpus striatum.   This part of the brain seems to be controlling the complicated behaviour called stereotyped, or behaviour by instinct, as we can see in the mating rituals of many birds.   This behaviour appears to be choreographed, and neural actions along specific nerve pathways in the corpus stratum are involved. However, all have adaptive advantages to individuals.

Fish, including the fry of herrings, shads, mullets, sphyraenids, pomprets, mackerel, tuna, flying fish, marlins, and several hundreds of others, display remarkable schooling behaviour. Many marine fish, reptiles, and mammals migrate across thousands of miles either in search of food or mates, often in schools or pods. Vision seems to be highly adaptive in schooling behaviour of fish (Shaw, 1962). There are subtle differences and intricacies where function and behaviour interact and overlap. Obviously, the sensory system, especially sight, taste and smell, and the hormonal system are often implicated. For example, the visual system plays an important role in the orientation of fish swimming parallel to one another in such precise proximity and elegance during schooling. The lateral line organ, well innervated with nerve fibres, in fish is thought to be responsive to vibration and water movements. Many larvae become orientated to swim against the smaller currents. For, example, the ephyra larvae of *Aurilia aurita* orientate less certainly than many others. They can swim against slow currents with the body slightly tilted, but when currents become stronger the ephyrae tend to cease pulsating and let themselves be swept into the turbulence. Hardy (1958) recorded them as probably finding prey by water disturbance. Trochophore larvae of *Nephtys sp.*, which usually swim in a spiral manner, can reduce their spiralling and swim effectively against the lowest velocity currents. Polydora larvae can swim more vigorously, and can successfully resist low velocity currents. The pelagic stage of *Autolytus prolifer* with its body exhibiting a sinuous movement; the nauplii (stage VI) of *Elminius modustus* and *Balanus balanoides*, which swim in a circular manner almost ceaselessly, can swim against smaller currents more effectively. Most copepods, *Calanus, Centrophagus, Tisbe, Acartia,* are extremely sensitive to water movements and can swim against water currents, able to detect the velocity of the movement of water. The *Eurydice pulchra* is a strong swimmer and can resist a high velocity of 11.9 cm/second and can cause sufficient turbulence to disturb other planktonic organisms (Singarajah, 1991). Other larval species include zoea of *Portunus, Carcinus, and Cancer pagurus, Mesopodopsis, Neomysis, Crangon crangos, Aegeon fasciatus,* and several others. It is interesting to note that the ability to swim fast and escape evidently depend upon good neuro-muscular coordination

(Singarajah, 1969). The planktonic crustaceans can escape more effectively than can a simpler organisation, such as coelenterates and polychaetes. Furthermore, with some interesting exceptions, the older larvae are more effective than younger ones. Copepods are extraordinarily good at swimming (26 cm/second) and escaping. Among vertebrates, especially marine mammals, whales and dolphins are epimelitic (Singarajah, 1984), with almost human-like strong bonds between mothers and calves. They have developed high sensory abilities to scan the environment by echolocation and to communicate. They have a complex social organisation with dominance of hierarchy (Wursing, 1979). The distinct advantages in this adaptive behaviour probably are: to find food, and the ease with which they can find the mates to reproduce. In circadian rhythm, it seems that daily variation in light and temperature operate as functional factors (Pengelly et al., 1971).

Migration has evolved in many marine organisms. They migrate for many reasons, but the main ones are associated with feeding or reproduction. Perhaps the greatest contrast between eels and salmon are the nature of their migration; the eels descend while the salmon ascend upstream. The anadromos salmons lay their eggs on pebbly substrates in river bottoms, and the fertilised eggs when hatched out grow as smolt in well oxygenated small rivers and descend to the sea until they mature and return to spawn in the same rivers where they were born. There is some evidence that salmon, during their migration upstream, use the sense of smell (Hasler et al., 1955) based on memory of odour.

There is evidence, however, that the direction of transport can alter in those diadromous fish which migrate between sea and freshwater and vice versa; perhaps a functional adaptation to tide over the changes in salinity during their spawning migration. For example, the catadromous eels, living in freshwater, descend to the sea and migrate thousands of miles from freshwater seawards to spawn and die (see Fig. 16.5), i.e. the larvae, or 'elvers' as they are called, which were born in the high seas, once again return to freshwater ambient and grow there until they become fully mature. Conversely, the anadromous fish ascend rivers from the sea for breeding, e.g. salmon and many species of clupeoid family. Among the invertebrates, only a few migrate between fresh and sea water. A good example is the estuarine mitten crab, *Eriocheir sinensis*, which spends much of its

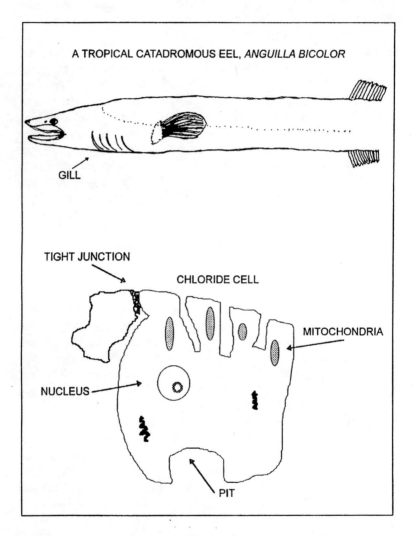

A TROPICAL CATADROMOUS EEL, *ANGUILLA BICOLOR*

GILL

TIGHT JUNCTION

CHLORIDE CELL

MITOCHONDRIA

NUCLEUS

PIT

Figure 16.5   Chloride transporting epithelial cell in teleost gill as an osmoregulatory-organ.

time in freshwater environment and returns periodically to the sea to breed, as the larvae prefer the higher salinity in the sea than are able to tolerate the freshwater.

Many fish produce electrical impulses. For example, the electric rays, torpedo, *Electrophorus, Malapteus,* and some catfish. The nerves of the cephalopods, especially the giant squid, sensory nerves in the *Limulus,* and many others are excellent examples. In fact, Hodgkin (1963) was able to measure the electric potential across the membrane of the giant axon of the squid, Loligo, and explained the characteristics of formation of an impulse.

16.4   Membrane Structure and Properties

The membrane structure is of paramount importance in understanding the various mechanisms which determine the transport of ionic materials into and out of cells. The exquisitely thin, highly organised, homogenous, bimolecular lipoprotein, which is about $75(\pm 20)$Å, thick and selectively permeable, is the functional boundary which separates ionic solutions and severely restricts the interchange of materials. According to the unit membrane theory of Robertson (1958), the living cell membrane is 'to be regarded as a biological constant common to the surfaces all of kinds of cells'. Chemically, the living cell membranes are primarily composed of a mixture of complex substances, lipids and proteins. The nerve membrane consists of alternating layers of lipids and protein molecules stacked at right angles to each other. In the lipid layer, the long, thin, lipid molecules are closely packed with their long axes parallel and orientated to perpendicular to the membrane. The lipid layer is bimolecular where the non-polar ends of lipid molecules are opposed. The polar ends of the lipids carry highly enlarged, round, complex phospholipid molecules. The phospholipid molecules consist of two parts: a long straight chain of carbon and hydrogen atoms, no electrical polarisation (hydrophobic), attached to a more complicated structure containing atoms of carbon, hydrogen, oxygen, nitrogen, and phosphorus. This group, as a whole, is electrically polarised (hydrophilic). The hydrocarbon chains are soluble in organic solvents such as ether and chloroform but are insoluble in water. The more complex end of the molecule containing phosphorus and nitrogen is not soluble in these substances, but it dissolves readily in water. On the external surface

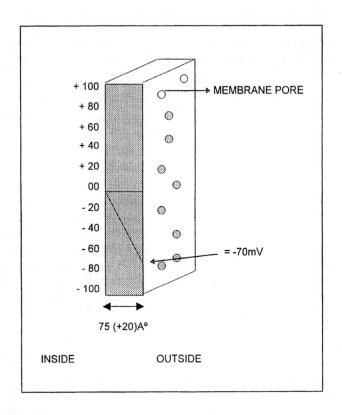

Figure 16.6   The membrane pores and the membrane potential.

of most cells is a basement membrane, a thin homogeneous layer. The phospholipid molecules have a hydrophilic end which is attracted to the water and a hydrophobic end which is not. The protein complex, on the other hand, is a monomolecular layer, lying on either side of the lipid core. The central zone of the lipid layer is about 35Å thick when is in turn bounded externally by another thick, dense line about 20Å thick. The whole structure measures about $75(\pm 20)$Å across and is called unit membrane. The membrane is perforated by water-filled pores about 7Å in diameter and is more permeable to some ions than to others (Fig. 16.6).

Table 16.2 ionic concentrations (in mM) of
an excitable cell

INTERSTITIAL (Exterior)		INTERCELLULAR (Interior)		RATIO (Extra:Intra)	PORE:HYDRATED IONS (*Diameter)
	mM		mM		Å
Na^+	150	Na^+	15	10:1	7:5.12
K^+	5.5	K^+		1:30	7:3.96
Cl^-	125	Cl^-	09	14:1	7:3.86
HCO_3^-	27	HCO_3^-	08	3:1	–
A^-	00	A^-	155	0:155	–
					Water molecule 7:3.00

Based on Solomon, 1960

16.4.1   Donnan Theory

A condition of the Donnan distribution is that the products of the concentrations of ions on each side of the membrane are equal:

$$(K^+)i \times (Cl^-)i = (K^+)o \times (Cl^-)o \; \; or$$
$$(K^+)i / (K^+)o = (Cl^-)o / (Cl^-)i \tag{1}$$

$$Ez\frac{RT}{zF}Ln\frac{(K^+)o}{(K^+)i} = \frac{RT}{zF}Ln\frac{(Cl^-)i}{(Cl^-)o} \tag{2}$$

$$Ez^+ = \frac{RT}{zF}Ln\frac{(K^+)o}{(K^+)i} \tag{3}$$

16.4.2   Equilibrium Potential and Nernst Equation

Equilibrium potential is the voltage drop that develops across the membrane when the influx and efflux of a diffusible ion are equal.

For example, the potassium potential $E_k{}^+$ is determined by the concentration ratio of the K^+ ions inside and outside the cell, $(K^+)o/(K+)i$. $E_k{}^+$ is proportional to the logarithm of the concentration ratio of the two solutions and also varies with the difference in cations and anions mobilities. If mobilities (ionic influx and efflux) are equal, no potential is developed. The relationship can be quantified as in K ions. The quantitative relationship between concentration ratio and equilibrium potential is called Nernst Equation:

$E_k{}^+$ = Equilibrium potential

R = Gas constant = 8.316 joules per degree

T = Absolute temperature = 273+t (20°C room/37°C body temperature)

z = Valency of the element K^+ = univalence

F = Faraday = *96,500 coulombs per Mol. of charge

Ln = Symbol for natural log = 2.3 (from table)

( ) = Concentration i inside and o outside

This can be simplified to:

$E_k{}^+$ –58 log (K+)o at 20°C or –61.5 $\log_{10}$ $(K^+)o/(K^+)i$ at 37°C

Constant field equation (Goldman (1943), Hodgkin and Katz (1949) – GHK):

The magnitude of the membrane potential at any given time depends on the distribution of K^+, Na^+, Cl^- ions and the permeability of the membrane to each of these ions. This relationship is shown in the equation:

$$E_{KNaCl} = \frac{R_t}{F} \, Ln \, \frac{P_k(K^+)o + P_{Na}(Na)o + P_{Cl}(Cl^-)i}{P_{ki}(K^+)i + P_{Na}(Na^+)i + P_{Cl}(Cl^-)o} \qquad (2)$$

where

P_k, P_{Na}, P_{Cl} are the permeability coefficients for the corresponding ion species, for example:

$$P_k:P_{Na}:P_{Cl} = 1:0:04 = 0.45$$

The equilibrium potential due to Cl^- is –70 mV.

The rate of flow of the ions through the membrane is determined by:

(a)   Concentration (differences) gradients.

(b)   Electrical gradients (except forces on charged particles).

(c)   Permeation forces (Stein, 1967):
    (i)   Simple diffusion (passive transport)
    (ii) Active transport of Na^+ and K^+
    (iii) Facilitated permeation, involving carriers or pores.

The outward diffusion of K^+ due to concentration gradient separates negative charges and thus generates an electric field or voltage gradient.

16.5   Thermal Effects on Reproduction and Distribution

One of the important variable and governing parameters of the marine environment is temperature.   In many marine animals, especially invertebrates, the environmental factors such as light, temperature nutrition and seasons play an important role in growth, reproduction and distribution of larvae.   In the tropical waters, most organisms tend to spawn during warmer seasons (Stephensen, 1934).   Gonads' growth seems to depend primarily on nutrients, temperature and photoperiod; the development of gametes to maturity in many invertebrates, particularly bivales, might be a function of the ambient temperature (Gessner, 1970).   The breeding period tends to lie within a narrow range of thermal conditions.   A critical temperature seems to be necessary for a number of invertebrates for the development of eggs. For example, in the scallop, *Acquipecton irradians,* for early clevages of the fertilised eggs a critical temperature lies between 15°C and 20°C (Sastry, 1966).   Several other examples have been reported by a number of workers.   On the other hand, the fish are poikilothermic or thermal conformers.   However, gonads development in fish was found to be increased in the relatively warmer tropical waters.   From a viewpoint of growth and physiological adaptation, the fish are usually larger in the Tropics than in higher latitudes.   For example, the elasmobranch sharks and teleostic tuna, marlin, and many others.   The adaptive function of large size may be correlated with conservation of body heat (Lindsey, 1966).

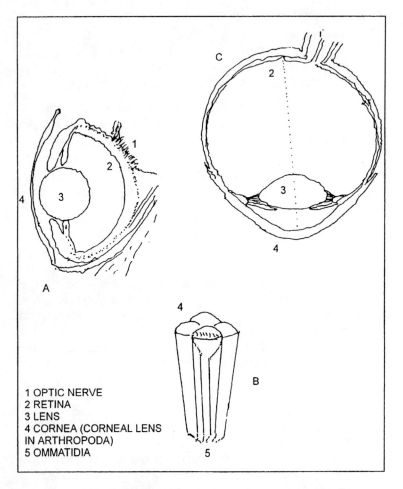

1 OPTIC NERVE
2 RETINA
3 LENS
4 CORNEA (CORNEAL LENS
IN ARTHROPODA)
5 OMMATIDIA

Figure 16.7   Comparison of the three types of eyes in: (A) Cephalopod,
                   (B) Arthropod and (C) Human

Temperature largely determines the distribution of many marine organisms (Gunther, 1957) and their abundance. Zoogeographically, fish are often considered as tropical, temperate and boreal. The distribution of fish larvae is rather restricted to warmer waters than that of the adults; and the limits of distribution closely correspond with a particular isotherm (Brett, 1970). The changes in isotherm inherently accompany the physico-chemical and dynamic properties of the sea water such as monsoonal variations, water masses, surface currents, upwelling, thermocline, salinity, concentration of dissolved oxygen, concentration of phosphates and primary production.

16.6 Buoyancy and Pressure Sensitivity

The problems of buoyancy and pressure sensitivity have been considered briefly in Chapter Seven and will not be considered any further. Sensitivity to hydrostatic pressure of several marine organisms are well documented (Knight-Jones and Qusim, 1955; Knight-Jones and Morgan, 1966; Singarajah, 1966; 1991a; 1991b). In addition to other devices, many fish have developed swim bladders to function as a hydrostatic organ and to control buoyancy. Swim bladder is a derivative of the alimentary canal (oesophagus) and lies dorsal to it and below the vertebral column. The bladder may be single or bilobed. Inside the bladder is a concentration of fine blood vessels forming a gland-like structure called rete mirable which is capable of secreting gases. The bladder is usually filled with gas, especially oxygen and, in some cases, with oil. The swim bladder, when present, is linked with the Weberian apparatus of the inner ear. Any changes in hydrostatic pressure alters the volume of the swim bladder which, in turn, stimulates the sacculus and the fish are able to sense the water depth and maintain themselves there without much muscular effort, thus economising energy. Perhaps, this is an adaptive behaviour during ebb and flow of tidal flow. However, the swim bladders are distinctly absent in elasmobranchs and in some deep sea bony fish.

16.7 Visual Adaptations

For many animals, vision is an important sensory function to link to their photic environment. Unlike terrestrial animals, marine organisms live in an environment where light is at great premium.

Bioluminesene is also common in some deep sea creatures. Except perhaps a few deep sea creatures, most marine animals are able to adapt both structurally and functionally to survive in varied conditions of the seas. The distinct features of the photoreceptors are that they have highly specialised intricately lamellated compact membranes with large surface area. In invertebrates, these membranes become evaginated as microvilli. There are two major types of receptors; the ciliary and the rhabdometric; the former derived from membrane of non-motile cilia while the latter is derived from plasma membrane forming into microvilli. For this and other reasons, photoreceptors are thought to have evolved independently (Eakin, 1963). However many organisms may contain both receptors, for example in *Pecten* (Barber et al., 1967). Because of the greater diversity of invertebrates than vertebrates, the light sensitive membrane bound visual pigments differ in different groups of animals. In vertebrates, the receptor cells called rods and cones are distributed over the retinae. Invertebrate organisms like jellyfish, sponges, corals, annelids, arthropods, molluscs, and echinoderms can perceive direction of light but are not able to form true image. Despite diversity, some have one thing in common, i.e. photon absorbing receptors. The rods and cones of vertebrates are relatively more uniform than the diverse photoreceptors found in invertebrates. On the basis of their ability to form an image, the visual organs are called as simple or complex eyes. The complex eyes are three basic types: cephalapod, compound and vertebrate eyes Fig. 16.7.

The primitive eyes of the horseshoe crab, *Limulus,* show some similarity with the already extinct fossilised trilobites (marine arthropod), which lived in the ancient seas nearly 250 million years ago and appears to be the first animal to have eyes capable of a high degree of resolution. Among other marine organisms, the smallest copepod, *Copelina,* is able to form image, other crustaceans, crabs, lobsters, prawns, mysids, and the molluscs, especially squids and octopus and most vertebrates have well developed eyes.

Radiant energy exists in the form of photons or quanta travelling through time and space. The energy (E) per quantum is given by the well-known formula:

$$E = hv\frac{he}{\lambda}$$

h = Plank's constant
v = frequency
λ = wavelength
c = the speed of light

Its SI unit (Système International d'Unitès) is the (J), although photon energy is often reported in ergs (1 erg = 10^{-7} J or calories = 4.184 J). While the intensity of light is the rate of delivery of photons, the work that a single photon can do is proportional to its wavelength. Light is required only to trigger an impulse. However, the range of human vision is between 380 nm and 760 nm, if the lens of the eye is removed as in corrective surgery for cataracts, the lower limit shifts to 350 nm (Wald, 1945). While in marine fish, the photic conditions vary with habitat. Fluctuation in the intensity of underwater light that results from the action of waves and ripples as lenses can influence the visibility of objects (McFarland and Loew, 1983). The cephalopods have two sets of pigments, one with maximum wavelength absorption 475 nm, while the other with an absorption maximum of 490 nm (Sinclare, 1985). Using a single-beam, wave-length scanning, dichoricmicrospectrophotometer, Singarajah and Harosi (1992) determined on isolated receptors three visual pigments in the demersal sea bass, *Centropristis striata*. They found, among other properties, the wavelength of peak α-band absorbency (λ_{max}) of pigments of the receptor cells. The shortest-wavelength-absorbing type (λ_{max} 463±2 nm) found only in single cones, both members of the double cones contained the longest wavelength-absorbing pigment of the three, with λ_{max} 527±5 nm. Rods were found to be typical rhodopsin, λ_{max} 498±2 nm. Thus, the retina of this predatory fish appears to use three closely spaced visual pigments, with λ_{max} clustering about 500±30 nm (Fig. 16.8).

Among the crustaceans, the lobsters have the largest compound eyes. The insects also use compound eyes and they can see ultraviolet of the spectrum. Electrophysiological studies have shown that the spectral sensitivity of the eye to different wavelengths can be closely related to the absorption (Singarajah, 1987). Among the cephalopods, squid, cuttlefish, octopus, giant squid and nautilus, the largest eye is found in the giant squid, *Architeuthis,* on which the sperm whales feed. Many are vulnerable for a variety of predators. Their speed and well developed visual system perhaps reflect the adaptive values to escape from predators. Despite some optical similarities between the

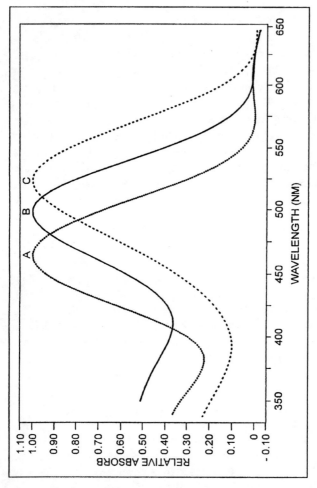

Figure 16.8   The relative absorbance spectra determined from the three visual pigments present in the black sea bass, *Centropristis straita* (after Singarajah and Harosi, 1992).

cephalapods and vertebrates, they evolved quite independently. The photoreceptors in cephalapods are large and long, most fish have flattened cornea, and the lens also has a refractive index of that of sea water 1.43, an adaptation to decrease water resistance. An interesting example can be seen in the family of flat fish, where the young ones are born with eyes on the side of their head, but as larvae grow and seek the bottom habitat; as a result one side of the head grows more rapidly than the other and the eye migrates resulting in both eyes coming to lie on one side. The rods are responsible for scotopic vision at low light intensities, while the cones for photopic vision at intensities above the threshold for the rods. While many species have duplex retinae with rods and cones, some elasmobranchs and deep sea fish have purely rods (Blaxter, 1970) and retinae with pure cones seem to be found in Pacific salmon, *Oncorhynchaus sp.* (Ali, 1959) and in herring, *Clupea harengus* (Blaxter and Jones, 1967). Vision is important as an adaptive mechanism since many species use visual cues for courtship and mating, breeding, spawning, to adapt to the surrounding substrates as seen in flatfish, recognition of food, schooling, escape from predators, and avoidance of nets. Lack of colouration in most fish larvae may be an adaptation for camouflage during their pelagic stage.

References

Ali, M. A., 'The ocular structure, retinomotor and photobehavioural responses of the juvenile Pacific salmon', *Can. J. Zool.* 37, 1959, pp.965–996

'Membrane structure and its Biological application', Annals of the New York Academy of Sciences, vol. CXCV, (ed. D. Green), 1972, p.519

Barber, V. C., Evans, E. M. and Land, M. F., 'The fine structure of eye of the mollusc', *Pecten maximus*, Z. Zelifosch, Microsk. Anat. 76, 1967, pp.295–312

Bently, P. B., *Endocrines and Osmoregulation*, New York, Heidelberg and Berlin, Springer-Verlag, 1971, p.300

Bernhard, C., *Introduction of experimental medicine* (English translation, 1949) New York, 1855

Blaxter, J. H. S. and Jones, M. P., 'The development of the retina and the retinomotor responses in the herring', J. Mar Biol Ass. UK, 47, 1967, pp.677–697

Blaxter, J. H. S., 'Light', in (ed. Kinne, O), *Marine Ecology*, London, New York, Sydney and Toronto, Wiley-Interscience, 1970, pp.213-320

Brett, J. B., 'Temperature, animals and fish', in (ed. Kinne, O), *Marine Ecology*, London, New York, Sydney and Toronto, Wiley Interscience, 1970, pp.554-560

Douglas, R. and Djamgoz, M., *The Visual System of Fish*, London, New York, Chapman and Hall, 1990, p.526

Eakin, R. M., 'Lines of evolution of photoreceptors', in *General Physiology of Cell Specialization* (eds. Mazia, D. and Tyler, A.), New York, McGraw-Hill, 1963, pp.393-425

Gessner, F. 'Temperature', in (ed. Kinne, O.), *Marine Ecology*, London, New York, Sydney and Toronto, Wiley-Interscience, 1970, pp.363-616

Gunther, G., 'Temperature', in (ed. Hedgpeth, W. H.), *Treatise on Marine Ecology and Paleoecology, Ecology*. Mem. Geol. Soc. Am., 67, 1957, vol. I, pp.159-184

Hardy, A. C., *The Open Sea, Its Natural History: The World of Plankton*, London, Collins, 1958, p.335

Hasler. A., and Larsen, J. S., 'The homing salmon', readings from *Scientific American*, W. H., San Francisco, Freeman and Company, 1955, pp.265-268

Hodgkin, A. L., 'Conduction of the nerve impulse', The Sherrington Lecture VII, Liverpool University Press, 1971, p.108

Keys, A. B. and Wilimer, E. N., 'Chloride secreting cells in the gills of fish with special reference to the common eel', *J. Physiol.*, 76, 1932, pp.368-378

Knight-Jones, E. W. and Quasim, S. Z., 'Responses of some marine plankton animals to chances of hydrostatic pressure', *Nature*, London, 173, 1955, pp.941-942

Knight-Jones, E. W. and Morgan, E., 'Responses of marine animals to changes in hydrostatic pressure', *Oceonogr.*, Mar. Biol. 4, 1966, pp.267-299

Krog, A., *Osmotic Regulation in Aquatic Animals*, Cambridge University Press, reprinted by Dover Publications, inc., New York, (1965), 1939, p.242

Lindsey, C. C., 'Body size of poikilotherm vertebrates at different latitudes', *Evolution*, 20, 1966, pp.456-465

McFarland, W. N. and Loew, E. R., 'Wave-produced changes in underwater light and their relations to vision', *Env. Biol. Fish*, 8, 1983, pp.173-184

Pengelley, E. T., and Asmudson, S. J., 'Annual biological clocks', readings from *Scientific American*, San Francisco, W. H. Freeman and Company. 1971, pp.105–112

Ruch, T. C., Patton, D., Woodsbury, J. W., and Towe, A. L., *Neurophysiology*, Philadelphia and London, W. B. Saunders Company, 1968, p.538

Sastry, A. N., 'Temperature effects in reproduction of the bay scallop, *Aequipecten irridans* Lamarck', *Biol. Bull. Mar. Biol. Lab.*, Woods Hole, 130, 1966, pp.373–386

Schmidt, R. F. (ed.) *Fundamentals of neurophysiology*, Berlin, New York, Springer-Verlag, Heidelberg, 1975, p.293

Schmidt-Nielsen, Slat glands, *Scientific American*, 200 (1), 1959, pp.109–116

Sinclare, S., *How Animals See*, Melbourne and Auckland, Heinemann Publishers, 1985, p.46

Singarajah, K. V. and Harosi, F. I., 'Visual cells and pigments in a Demersal Fish, the Black Sea Bass *(Centropristis striata)*', Bulletin; 182, 1992, pp.135–144

Singarajah, K. V., 'Behaviour of Pleurobrachia pileus to changes in hydrostatic pressure and the possible location of baroreceptors', Mar. Behav. Physiol., 19, 1991a, pp.45–59

Singarajah, K. V., 'Responses of zooplankton to changes in hydrostatic pressure', *J. Mar. Biol. Ass.* India, 33 (1 and 2), 1991b, pp.317–334

Singarajah, K. V., 'Escape reactions of zooplankton: the avoidance of a pursuing siphon tube', *J. Exp. Mar. Biol. Ecol.*, 3, 1969, pp.171–178

Singarajah, K. V., 'Spectral sensitivity of motion-sensitive units of the butterfly ventral nerve cord', *J. Insect. Physiol.* 34 (11), 1988, pp.1005–1012

Singarajah, K. V., 'Observations on the occurrence and behaviour of minke whales off the coast of Brazil', Sci. Rep. Whales, Res. Inst., 1984, no.35, pp.17–38.

Solomon, A. K., 'Pores in the cell membrane', *Scientific American*, California, W. H. Freeman and Co, San Francisco, December, 1960, pp.1–9

Stephensen, K., 'The breeding of reef animals', part. 2, in 'Vertebrates other than corals' Rep. Gt. Barrier Reef Comm., 3 (9), 1934, pp.247–272

Wald, G., 'Human vision and the spectrum', *Science*, 101, 1945, pp.653–658

Wursing, B., 'Dolphins', readings from *Scientific American*, San Francisco, Freeman and Company, 1979, pp.79–87

Appendix I
Salinity Determination

Definition

Salinity varies from region to region and from season to season. Salinity is defined as the total salt content, in grams, dissolved in 1 kg of sea water, expressed as g/kg, and symbolised as:

$$S = ‰$$

This definition theoretically takes into account that all iodate and bromides are converted into chlorides (see Chapter Six on Chemistry of sea water).

Reagents

(1) Silver nitrate ($AgNO_3$) solution.

(2) Dissolve 27.25 g of silver nitrate in distilled water and dilute to 1 litre.

(3) Potassium chromate indicator. Dissolve 8 g of reagent grade potassium chromate (K_2CrO_4) in 100 ml of distilled water. Store in a clean dropping bottle.

Materials

(1) Knudsen burette assembly or 50 ml burette to read to the nearest 0.02 ml.

(2) Pipette 5 ml.

(3) Magnetic stirrer.

(4) Filter paper (Waterman no.1).

(5) Erlenmeyer flasks 200 ml capacity.

(6) Funnels.

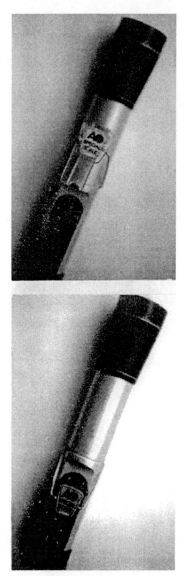

Appendix plate 1.1   Hand held refractometer to record salinities rapidly and of wide range.

Procedure

(1) Take 10 ml of filtered sea water in an Erlenmeyer flask.
(2) Add 4 drops of chromate indicator.
(3) Titrate, using 50 ml burette, until the entire solution becomes red-to-peach colour. Stop the addition of silver nitrate solution until the colour becomes pale yellow. Now add drop by drop silver nitrate solution, and the colour changes to dirty-orange colour which is the end point.

Results

Read and record the burette reading. Take the average of two readings and compare the result with that of the average of salinometer readings.

Measurement of Salinity by Salinometer

A precise and more rapid method for determination of salinity is by the use of a refractometer, particularly where range is likely to be considerable and when several samples need to be analysed at different stations at relatively short time.

A hand-held 'Pat Pend' refractometer is manufactured by the American Corporation, Keene, H. N., USA. (Catalogue No 10419) with salinity range between 0 and 160‰ and refractive index from 1.3330 to 1.3730 (Appendix Plate I.1). This is an extremely handy instrument to record salinities of a wide range, especially when working with turbid waters such as estuaries, bays and coastal waters, where suspended matter is always present and on board ship.

A small beaker containing a filtered sample of sea water or estuarine water can be measured by immersing the narrow end of the refractometer until the glass window gets wet. The overlapping special flap is now placed in contact with the surface of the glass window so that a thin film of water sample is trapped between the glass window and the cover flap. Now the refractometer can be raised in the air and viewed through the broader end and the salinity can be read out directly from the scale. For more precise work, the

414

refractive index[1] also can be noted and by doing a small calculation the salinity can be corrected to an accuracy of 0.001%.

References

Barnes, H. N., in *Apparatus and Methods of Oceonography*, part 1, Chemical, London, George Allen and Unwin Ltd., 1959, pp.178–199

Grasshof, K., 'Chemical methods', in *Research Methods in Marine Biology*, (ed.) Schlieper, C., London, Sidgwick and Jackson, 1972, pp.1–25

Knudsen, M., (ed.) 'Formulae and explanation of the Table', in *Hydrological Tables* (ed.), Copenhagen, Tutein and Kock, 1903, pp.1–22

Martin, D. F., 'Salinity'. in *Marine Chemistry*, vol I, Analytical Methods, New York, Marcel Dekker INC, 1968, pp.81–96

Riley, J. P. and Chester, R., 'Salinity, chlorinity, and the physical properties of sea water', in *Introduction to Marine Chemistry*, London and New York, Academic Press, 1974, pp.10–36

---

[1]   Refractive Index: the ratio of the speed of light in a vacuum to its speed in sea water.

Appendix II
Laboratory Methods for Dissolved Oxygen Content

Besides the importance of understanding the biological activities in the marine environment, the estimation of oxygen is necessary for a number of experimental works both in the open ocean and in the laboratories. For example, when facilities for C^{14} method is not readily available for estimation of the primary production, the only other reliable alternative is the standard Wingler (1888) method of oxygen determination. However, despite several modifications, though a little laborious for routine analysis, the Wingler technique gives sufficient accuracy, more than many other methods. The concentration of oxygen in sea water varies from 0 to a little over 10 ml per litre, depending on whether the water is stagnant or supersaturated with O_2 by photosynthetic activities by phytoplankton.

Reagents

(1) Manganous sulphate – dissolve 365 g of $MnSO_4.4H_2O$ in distilled water and make up to 1 litre in a volumetric flask.

(2) Alkaline iodide solution – dissolve 330 g of NaOH or 462 g of KOH and 100 g of KI in distilled water and mix the two solutions and make up 1 litre.

(3) Sodium thiosulphate – dissolve 2.48 g of $Na_2S_2O_35H_2O$ in recently boiled out distilled water and make up a litre. Add about 3 ml of chloroform as preservative and store in an amber coloured bottle. This gives N/100 (0.01N) sodium thiosulphate solution. This may be standardised against potassium-bi-iodate solution.

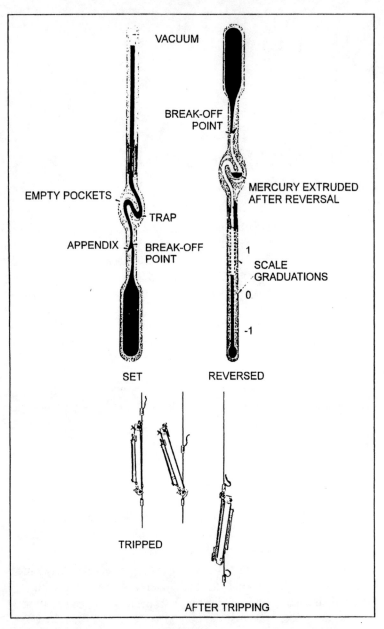

VACUUM

BREAK-OFF POINT

EMPTY POCKETS

TRAP

APPENDIX

BREAK-OFF POINT

MERCURY EXTRUDED AFTER REVERSAL

1

SCALE GRADUATIONS

0

-1

SET

REVERSED

TRIPPED

AFTER TRIPPING

Appendix figure II.1   Nansen bottle with reversing thermometer, showing the three operational positions during collection of sea water sample at desired depths.

(4) Potassium hydrogen bi-iodate solution – dissolve 0.325 g of $KH(IO_3)2$ in distilled water and make up to 1 litre. This gives about 0.01N. The hydrogen bi-iodate should be previously dried at 105°C and cooled before weighing.

(5) Sulphuric acid – concentrated H_2SO_4 (36N/1.84 s.g.).

(6) Starch solution – dissolve 1 g of soluble starch powder in 100 ml of distilled water and boil for a minute and filter if necessary. Add 2 ml of chloroform as preservative. It is advisable to prepare the solution each time fresh.

Materials

(1) 6 amber coloured sampling bottles of 250 ml capacity.

(2) 2×100 ml pipettes; 2 pipettes calibrated to deliver 0.5 ml.

(3) 1 automatic burette and automatic pipette.

(4) 4×1 litre volumetric flasks.

(5) 6 Erlenmeyer 250 ml flasks.

(6) Reagent bottles, with stoppers, with suitable capacity and colour.

(7) Hot plate.

(8) Oven.

(9) Nansen or Van Dorn bottle with reversing thermometer, optional (Appendix Fig. II.1).

Procedure

(1) *Collection and treatment of samples*:
The 250 ml amber coloured sampling bottles are first rinsed with sea water samples and drained before filling. A surface water sample from a bucket or deep water bottle with the help of thin flexible rubber tubing, provided that care is taken not to let any air bubbles in. This can be done by placing the end of the tube so as to stay diagonally in contact with the bottom or side of the bottle and drawing slowly until the sample bottle overflows. Expel a little water by placing the stopper gently.

(2) *Treatment of the sample*:

It is much more convenient to carry to the field manganous sulphate and the alkaline potassium or sodium solution in 1 ml ampules previously filled and sealed, separately. Remove the stopper carefully and first add 1 ml of manganous sulphate by breaking the ampule or by a pipette; this is soon followed by 1 ml of alkaline iodide solution. Replace the stopper, making sure that no air bubbles are left in. Shake well and a mucus precipitation forms. Store the bottle in suitable boxes until brought to the laboratory. During this time a mucus colour precipitation settles down and fills nearly ¾ of the bottle.

Standardization of Sodium Thisulphate

Measure out precisely 97 ml of boiled out water and transfer it to a 250 ml Erlenmeyer flask with a stopper. Add 1 ml of concentrated sulphuric acid and mix all. Now add 1 ml of alkaline iodide reagent and mix again; this is followed by 1 ml of manganous sulphate solution and mix thoroughly. Finally, add 25 ml of standard potassium hydrogen bi-iodate solution, swirl it gently and restopper the flask, and then leave it in the dark for about 10 minutes. Using a micropipette or burette, titrate the liberated iodine with the sodium thiosulphate solution until a pale straw colour develops. Now, add 0.5 ml of starch solution and complete the titration until the blue colour disappears. Note the burette reading, which should not normally exceed 0.1 ml.

Standardization of Normality of Sodium Thiosulphate

The calculation is as follows:

$$N_1 \times V_1 = N_2 \times V_2$$

N_1 = Normality of standard $KI(IO_3)_2$ = 0.01N
V_1 = Volume of standard $KH(IO_3)_2$ = 25 ml
N_2 = Normality of the thiosulphate solution = to be calculated
V_2 = Volume of the thiosulphate solution = Burette reading

Take the average of the two best readings.

Analysis of the Treated Sample

(1) When the samples are brought to laboratory, the mucous precipitate fills nearly three-quarters of the bottle. Introduce 1 ml of concentrated sulphuric acid well below the surface of the sample in the bottle. Restopper the bottle and shake it well; the precipitate should be dissolved completely by mixing. The liberated iodine is now ready to be titrated.

(2) Using the oxygen pipette, withdraw 100 ml of this clear golden brown colour liquid and transfer to a 250 ml Erlenmeyer flask.

(3) Titrate with standardised sodium thiosulphate solution until the pale straw colour appears.

(4) Now add 3 ml of starch solution, a blue colour develops.

(5) Continue the titration until the blue colour just disappears.

(6) Record the volume of the thiosulphate solution used. Repeat the titration and take the average of at least two volumes for the final calculation.

Calculation of Oxygen Content

(1) Volume of treated sample:

A 100 ml aliquot of treated sample was titrated. The volume of the sea water sample really with reagents was 250-2 ml.

i.e. $$\frac{(250-2) \times 100}{250} = 99.2 \, ml \qquad (1)$$

(2) (a) The oxygen content is expressed in ml/litre. At standard temperature and pressure (STP/NTP), one mol of every gas will occupy the same volume of 22.4 litres. This statement holds good for every gas, regardless of its molecular formula (Avagadro's Law). Therefore, oxygen occupies 22.4 litres at STP, i.e. 22,400 /32 = 700 ml.

(b) 1 ml 1/10N (0.01N) thisulphate is equivalent to 0.08 mg oxygen at NTP.

(c) Let volume of thiosulphate used $= V1$.

(d) Let the volume of the sample used $= V2 = 99.2$ ml as in Equation 1, above.

(e) 1 ml N/100 thiosulphate

$$= \frac{0.00008 \times 22,400}{32} \, ml \, O_2$$

$$= \frac{V_1 \times 0.00008 \times 22,400}{V_2 \times 32}$$

(f)

$$= \frac{V_1 \times 0.056 \times 1,000}{V_2} \, ml \, O_2 \, / \, I$$

References

Barnes, H. N., in *Apparatus and Methods of Oceonography*, part 1, Chemical, George Allen and Unwin Ltd., London, 1959, pp.178–199

Fox, H. N. and Wingfield., 'A portable apparatus for determination of oxygen dissolve in small volume of water', *J. Exp. Biol.* 15, 1930, pp.437–444

Martin, D. F., 'Analytical methods: Dissolved oxygen content', *Marine Chemistry*, New York, Marcel Dekker INC, 1968, vol. I, pp.87–96

Riley, J. P. and Chester, R., 'The dissolved gases in the sea water. Part: Gases other than carbon dioxide', *Introduction to Marine Chemistry*, London and New York, Academic Press, 1974, pp.105–151

Strickland, J. D. H. and Parson, A. 'Manual of the sea water analysis', Fish Res. Board of Canada, Bull, 125 (no.125), 1965, p.202

Appendix III
Measurement of Extinction Coefficient

Light is one of the most important parameters of the marine environment. Light is essential for basic organic production in the sea. When light strikes the surface of the sea, depending on the position of the sun, a certain amount of light is reflected back into the atmosphere while the remainder penetrates into the deeper layers of the sea. However, all the light that penetrates below the surface does not reach the bottom but gets diminished or attenuated with increasing depth. The rate of decrease of light intensity with increasing depth is called extinction coefficient. This can be represented by the relationship:

$$\frac{2.3 \log(I\lambda d_1 - I\lambda d_2)}{d_2 - d_1}$$

Where, $I\lambda d_1$ and $I\lambda d_2$ are the intensities of light of a given wavelength (lambda) at different depths d_1 and d_2, and d is the depth in metres at which the Secchi disc is just hardly visible.

The attenuation of the light is caused by a number of factors. The principal factors are: absorption and scattering. Absorption is the transduction of the radiant energy into other forms, mostly as heat which is absorbed by the water, and eventually some will be lost to the atmosphere during recycling. Scattering is the deviation of light from its path of propagation. Other factors include absorption and scattering of light by suspended particulate materials, both living and non-living, particularly plankton and non-living detritus resulting from the decomposition of organic materials.

Colour and Diurnal Changes in Transparency

As already discussed in Chapter Twelve, the short wavelengths of light are more rapidly absorbed than the long wavelengths. The

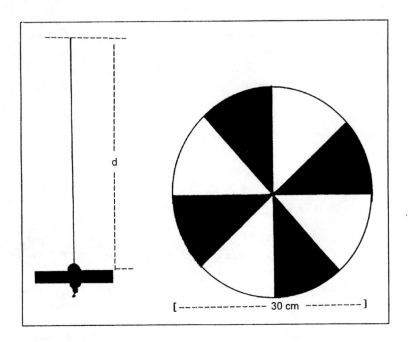

Appendix figure III.1   Secchi disc; it is usually about 30cm in diameter;
and d represents a measure of transparency.

colour of the oceanic waters, especially in the tropical regions, depends on the spectral distribution of relative loss (attenuence). Generally, the tropical waters are deep blue.

Measurement of Visibility or Transparency

A simple device to measure the visibility or transparency of the sea water is the Secchi disc. It is a flat disc about 30 centimetres in diameter with alternating quarters painted black and white to give good contrast. This disc (Appendix Fig. III.1) is fastened with a line, preferably at the centre. It can be lowered into the sea until it becomes hardly visible. The depth at which it nearly disappears from sight is noted as a measure of transparency of the sea water. This can be further converted into an extinction coefficient by making a simple calculation, using the following formula:

$$K = \frac{1.7}{d}$$

d = depth in metres

The depth decreases as the total attenuation coefficient of the sea water increases due to both absorption and scattering. Conversely, there is an inverse relationship between turbid waters and depth as the materials present in the water can greatly affect the light attenuated, and hence the depth increases with decreasing turbidity.

References

Jerlov, N. G., *Optical Oceanography*, Elsevier Publishing Co., Amesterdom, 1968, p.194

McFarland, W. N. and Munz, F. W., 'The photic environment of clear tropical seas during the day', Vision. Res., 15, pp.1063–1070

US Hydrographic Office., *Instruction Manual for Oceanographic Observations*, 2, Aufl., Washington, 1957

Appendix IV
THE pH

It is useful to express the acidity or basity of a solution by using an abbreviation known as pH. The body fluids of all living organisms consist of water. In marine environment there is a constant interaction between living organisms and the watery medium in which they live. As Claude Bernhard (1897) remarked 'fixite du milieu interieur'(ionic composition of internal fluid is fixed), the stability of the biological fluids is of paramount importance for the vital cell processes, since the biochemical reactions take place within the narrow range of pH.

The pH or hydrogen concentration, a term originally proposed by Sorensen (1909), is defined as the 'negetive logarithm to the base 10 of the hydrogen ion concentration':

$$pH = -\log(H^+)$$

In other words, pH represents the log of the reciprocal of the activity of hydrogen ions expressed in units of concentration. Because the hydrogen ion concentration can vary widely and because of the difficulty in estimation of it without some degree of error, it is usually convenient to express it in terms of pH.

When concentration of hydrogen ions and hydroxyl ions are equal, the reaction of a solution is said to be neutral; when the hydrogen ion concentration exceeds that of hydroxyl ions, it is acidic, and alkaline when hydrogen ion concentration is less than the value of hydroxyl ions. The hydrogen ion concentration of distilled water at 25°C is 10-7 Eq/L. The hydroxyl ion concentration is the same and hence distilled water is neutral. In the marine environment, the sea water is alkaline but the situation can easily be changed, depending on the ecological conditions of weather, drainage of terrestrial substances, heavy rains, circulation of water, the productivity of the euphotic layer, etc., and hence the pH fluctuations.

Measurement of pH

A variety of pH meters are available. Each type provides its own instruction manual. Read it carefully before measuring the pH level of the samples. Buffer solutions are usually required which may be purchased; often some portable meters come with these solutions. In order to standardise, place the electrode, after washing in distilled water, in buffer solution in a suitable container and allow the electrode to come to the temperature of the solution. Now balance the pH meter to the known pH of the buffer solution by adjusting the compensation dial and the standardisation knob on the pH meter. The pH of the sea water samples may now be measured. Before determining the pH the electrodes should be washed first with distilled water and then with the sea water samples. In some cases pH paper may be used. Small pocket pen-size pH meters are also now available.

The pH in the tropical surface layers varies from 7 to 8.6 and gradually drops in the deeper layers. The tropical estuarine waters show much low pH, varying from 6.8 to 7.2 almost from acidic to slightly alkaline throughout the year. The minute changes in the pH can bring about marked changes in the physical state of the marine organisms and may have the following effects:

(1) Like many higher animals, marine organisms have powerful mechanisms to regulate the internal pH level and a slight decrease of plasma pH tends to cause pH acidemia, and increases lead to alkalaemia.

(2) The pH should be optimal and an increase in pH enhances the precipitation of calcium carbonate.

(3) Autotrophic plants absorb large quantities of carbon dioxide during photosynthesis and release oxygen which would tend to increase the pH.

(4) Both animals and plants utilise the oxygen during respiration and reduce the value of pH.

(5) In some marine algae, the respiratory rate with increased rate of pH will have no noticeable change in the variation of oxygen content (Hoffman, 1929).

References

Grassof, K., 'Determination of the hydrogen ion activity', *Research Methods in Marine Biology*, (ed. Schilieper, C.), Sidgwick and Jakson, London, 1972, pp.12-16

Manuals of appropriate instrument. The manufacturing company usually provides.

Martin, 'Acids, bases and pH of natural waters', *Marine Chemistry*, New York, Marcel Dekker INC, 1968, pp.43-55

Rakestraw, N. W., 'The conception of alkalinity or excess base of sea water', *J. Marine Res.*, 8, 1949, pp.14-20

Appendix V
Unialgal Cultures in the Laboratory

Unicellular algae are often used as a satisfactory source of food for a wide variety of larvae reared in laboratory experiments. Although a single algal species can provide the necessary dietary requirement, a mixture of algal species proved to be healthier and to influence the vitality, particularly of oyster spats and many other bivalves and some fish larvae.

Algal Species

Dunaliella primolecta
Isochrysis galabana
Monochrysis lutheri
Phaeodactylum tricornutum
Prorocentrum micans
Sketleonema costatum
Tetraselmis suecica

Culture Media

Although a variety of culture media have been used by different laboratories, the modified Erdschreiber medium has been tested in our laboratory and found to be quite satisfactory and is given in Appendix Table V.1.

Table V.1 The Modified Erdscriber Medium for
Laboratory Culture of Unialgae

Components of the Erdscreiber medium	
1.*Soil extract	50.00ml
2. Sodium nitrate (NO_3)	0.20 g
3. Disodium mono hydrogen phosphate	
(NA_2HPO_4)	0.10 g
4. Sea water (pasteurised at &0°C)	950.00 ml
5. Distilled water	50.00 ml

   * Soil extract appears to have most trace elements, vitamins, especially B12, chelating and buffering agents as seen in the natural environment.

Preparation of Soil Extract

(1) Take 1 kg of finely sieved garden soil in a 5 litre round bottom flask.

(2) Add 2 litres of tap water.

(3) Autoclave for 1 hour at about 3 Kg (5lbs) pressure.

(4) Allow it to cool, and then decant off the supernatant cloudy fluid into a clean glass flask and plug the mouth with a cotton-wool bung.

(5) Resterilise the fluid extract by heating the contents to boiling point for a few minutes.

(6) Leave the soil extract to settle for about 6-8 weeks. It can also be centrifuged at 3,000 rpm. Usually a bright sherry colour liquid results.

Solutions of Culture Media

A. Soil extract (see above).

B. Filtered sea water, using glass wool.

C. Dilute this to 95% with distilled water.

D. Concentrated stock solution (see below) nutrient salts.

The stock solution is prepared to last for several culture generations as follows:

(1) Weigh precisely 20 g of Sodium Nitrate ($NaNO_3$).

(2) Weigh also 28 g of Disodium mono hydrogen phosphate (Na_2HPO_4).

(3) Transfer these to a volumetric flask, dissolve and dilute to 100 ml.

Note: Only 1 ml of this stock nutrient solution is to be added to a litre of sea water.

Mixing Sequence

Transfer each of the solutions of culture medium to a separate previously autoclaved flask. Using a sterile pipette, add an appropriate amount of nutrient salt solution to the soil extract and then add this mixture to the diluted sea water. Now allow the solution to come to room temperature before use.

Reference Culture

Reserve cultures are usually maintained in test tubes. These can be used to inoculate larger vessels. For routine feeding experiments, it is sufficient to use 1 litre capacity conical flasks. For regular subculturing, a constant source of light is necessary. Although daylight is desirable, when cultures are grown indoors, the use of warm white fluorescent tube light at a distance of about a metre will be satisfactory.

References

Baynes, S. M., L. Emerson and A. Scott, 'Production of algae for use in the rearing of larval fish', Fish. Res. Tecch. Rep, MAFF Direct, Fish Res., Lowestoft, 53, (3), 1979, pp.13–18

Gross F., 'Notes on the culture of some marine plankton organisms', J. Mar Biol. Ass UK, 21, 1937, pp.753–768

Heim, M. M., and Laing, I., 'Cost effective culture of marine unicellular algae' in (ed. Vogt, F.) Energy Conservation and Use of Renewable Energies in the Bio-industries, Oxford, Pergamon Press, 1981, pp.247–259

Singarajah, K. V., Moyse, J. and Knight-Jones, E. W., 'The effect of feeding upon the photactic behaviour of cirripede nauplii', J. Exp. Mar. Biol. Ecol., 1, 1967, pp.144–153

Ukeles, R., 'Continuous culture – a method for production of unicellular algal foods', in (ed. R. J. Stein), *Handbook of Phycological Methods: Culture Methods and Growth Measurements*, Cambridge University Press, 1973, pp.233–254

Appendix VI
Estimation of Plankton Density

As on the land, productivity of organic materials is continuous in the marine environment. It is often desirable to have some measure or estimate for the abundance or density of plankton population in order to relate to productivity and the transfer of energy of the given area in the sea. Depending on regions, seasons and fertility of the water, the species components of plankton often vary and their distribution is not uniform. One way of measuring the productivity is to know the biomass; i.e. the weight of organic materials in the living organisms and to convert into calorific value; the latter is not always simple. The current procedure will be restricted to our purpose of direct count of plankton, either phytoplankton, zooplankton, or a mixture of both.

Methods for Estimation

Many different methods are available; some are relatively simple, inexpensive and in common use, while others are sophisticated and more expensive. But, for our purposes, the direct count of plankton will be satisfactory. The planktonic materials in any collection consist of largely living organisms and often the dead and the decaying particulate matter. The latter is called detritus and not to be included in the plankton count.

Plankton Net

The simplest devices used traditionally for collection of marine plankton samples are the nets. Despite some disadvantages, they are widely used because of their relative ease to operate in the sea. Many types of plankton nets are now available commercially, but in smaller laboratories nets of desired size and shape can easily be designed at little cost.

A standard plankton net consists of a steel ring of about 30

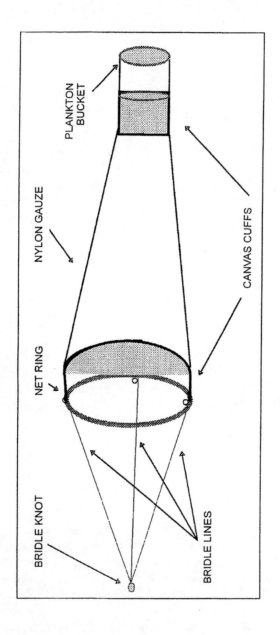

Appendix figure VI   A standard plankton net, particularly designed for tropical waters.

centimetres in diameter; the latter may vary in some situations. A cuff of flexible canvas about 20 centimetres wide is attached to the ring and thus the ring and the canvas covering the ring together form the mouth of the net. Three bridle lines, equally spaced, are attached to the ring of the mouth. The other end of the cuff is sewn to a cone shape nylon gauze of about 120 centimetres long with predetermined mesh size. The opposite narrow end of the gauze is again sewn to a tubular flexible canvas of about 12 centimetres in diameter and about 15 centimetres long, where an easily detachable tin or plastic plankton bucket of about one litre capacity can be secured by metal grip or thin rope (Appendix Fig. VI.1).

Plankton nets can be hauled horizontally near the surface or vertically or obliquely at desired depths with some weights attached to one of the bridle lines. However, the filtering efficiency of the gauze depends on the strength of the gauze itself, the mesh size of the gauze, the diameter of the mouth, the length of the net, and the towing speed during plankton haul, and the canvas segment which surrounds the mouth of the net. The canvas segment contributes to some extent to the filtering efficiency by diminishing the backward flow of the water near the mouth. Coarser nets with 62–72 meshes per inch and finer nets with 97–109 per inch will be adequate for the collection of both zooplankton and phytoplankton on surface water.

Collection of Plankton

In most tropical waters, a 10–15 minute haul will give sufficient plankton samples for a count. Nets can be towed near the surface or at desired deeper layers of water. Nets can be hauled from small boats with 1–2 knots towing speed. It is important to bear in mind, however, that for a fairly reliable count the speed of the boat and the duration of the collection must be carefully controlled. As soon as the haul is halted the net must be secured at its bridle line knot (see Appendix Fig. VI.1) and held vertically. The net can now be dipped 2–3 times in its vertical position in the sea water to rinse down and collect the plankton into the bucket. The samples may now be transferred carefully to clean bottles with lids and stickers with details concerning: date, locality, speed of the boat, duration of collection etc.

Appendix plate VI.1   The two plankton counting chambers.   (A)
Improved Nebauer Haemocytometer with the cover
slip. The rulings are not visible, except under the
microscope. (B) Sedgewick Rafter with thick cover
slip. The 1000 squares are clearly seen.

Washing the Nets

When the plankton collection is completed, the plankton bucket should be removed and the nets washed in large sink units in freshwater in the laboratory and dried and kept carefully until further use.

Preservation of Plankton

Samples of plankton collected routinely over a year or so can be preserved in transparent glass bottles with wide screw tops. However, before preserving a subsample of fresh collection is worth examination. This could be returned to the original sample. The concentrated samples are usually preserved in a 4% formalin, added with a neutral substance such as sodium borax powder. Each preserved sample should be properly labelled with date, place and duration of the haul, as already stated.

Counting Devices

The most common and useful devices for counting plankton, depending on the size of individual organisms, are the specially designed:

(1) Graduated glass slide or haemocytometer.

(2) The plastic ruled slide or Sedgewick rafter.

The former is more suitable to work with microplankton, while the latter is very useful for both micro as well as macroplankton. They are both easily obtainable from scientific suppliers in UK and USA directly.

Haemocytometer: the improved Neubauer haemocytometer is a thick glass counting chamber. It is provided with special cover slips. The haemocytometer has areas of two platforms each etched with small squares in the counting area. Each side of the small square has a length of 1/20 mm, i.e. an area of 0.0025 mm^2. An H-shaped trench, which separates the two platforms, serves to collect any excess overflow of the subsample (see Appendix Plate, VI.1A). When the cover slip is tightly pressed on the haemocytometer, a depth of 1/10 mm is formed between the platform and the cover slip. Therefore, the volume of a small square is 1/4000 mm^3. The number of organisms in each of the squares is counted, using a microscope with

20× magnification. When counting the cells, a convention is usually adopted; i.e. any cells which lie or touch the boundary line on the top or a left-hand side are *included* in the count, while cells on the other two opposing sides of the square are *ignored*. To get some reliable estimate, cells in at least 60–80 such squares should be counted.

Sedgewick rafter: the Sedgewick rafter is a plastic counting chamber designed to contain a volume of exactly 1 ml of subsample. It is provided with a thick plastic cover glass. The rectangular chamber is 50 millimetres long, 20 millimetres wide, and 1 millimetre deep internally. The chamber is divided into 1,000 small squares. When filling it with subsamples, place the rafter horizontally. Now gently slide the coverglass diagonally across the chamber, thus leaving a small space opening at opposite corners. Transfer an aliquot of 1 ml subsample, using a wide mouth pipette; with a pair of forceps, slide the cover slip to cover the whole chamber. Leave the rafter for 2–3 minutes for the organisms to settle before counting under a low power (20–40×) microscope. In order to get some results statistically, it is essential to count organisms in as many squares as possible. The Sedgewick rafter (Appendix Plate VI.1B) is relatively easier to count organisms than using the haemocytometer.

The Final Estimate

The final estimate of plankton population can be expressed as number of plankton per cubic metre. It should be remembered that however precise the method may be, the estimation will only be approximate – nonetheless it will serve a useful purpose.

Plankton Concentrate and Subsamples

As discussed briefly earlier, because of the enormous time involved in counting the entire sample of even a single collection, often carefully concentrated subsamples of plankton are counted. An aliquort of 1 ml subsample from the concentrate can be used for the Sedgewick rafter and even smaller fractions of subsamples need to be used for the haemocytometer.

Calculations

(1) Let us assume that the plankton collection was made at a given constant towing speed and over a certain duration.

(2) Let the distance through which the net was hauled be d, and

(3) the volume of water filtered can now be obtained.

(4) The total volume of the water column filtered by the net

$$= \pi r^2 \times d$$

r is the radius of the mouth of the plankton net.

(5) Multiplying the average count by the volume of the water filtered will give the total number of organisms per m^3.

References

Clarke, G. L., and Bumbus, D. F., *The Plankton Sampler – an Instrument for Quantative Plankton Investigations*, Spec. Publs. Am. Soc. Limnol. Oceanogr., 5, 1950, pp.1–8

McAlice, B. J., 'Phytoplankton Sampling with the Sedgwick-Rafter cell', *Limnol. Oceanogr.*,16, 1971, pp.19–28

Sournia A., (ed.), *Phytoplankton Manual*, Rome, UNESCO, 1978, p.337

UNESCO, 1968. *Zooplankton Sampling*, Paris, UNESCO, p.174

Appendix VII
Estimation Of Phosphates

Phosphate is one of the essential micro-constituents required by phytoplankton for growth within the euphotic zone. Chlorophyll, phosphate and radiant energy are the most important factors which determine the primary productivity of the sea. Phosphate varies in forms as inorganic, organic and particulate, but only the inorganic (reactive) dissolved form and the total phosphorus will be briefly discussed here. The dissolved content of inorganic phosphate varies in different regions of the oceans depending on the depth, effective water circulation and seasons. The concentration of phosphate in sea water may range from less than 0.01 µg atom P/l in surface waters to over 3 µg atom P/l in deep water (Armstrong, 1965). The overall mean value for the major oceans appears to be 2.3 µg atom P/l (Redfield, 1958). In tropical surface waters, the phosphatephosphorus as a measure of plant nutrient varies with monsoons within the range between 0.26 µg atom P/l and 1.05 µg atom P/l. In West Antarctic waters, an average of 2.55 µg atom $P-PO_4$/l at depths of between 10 metres and 25 metres has been reported (Bienati et al., 1971).

Since samples are often tested in laboratories or on board ship, a precise and quick method of estimating is of considerable importance. Although a few different methods are available, the calorimetric reaction is sufficient, but spectrophotometry method is more satisfactory and in wider use for dissolved phosphorus. Dissolved phosphorus content is defined rather arbitrarily as 'that fraction which will pass through a 0.5µ membrane filter' (Martin, 1968).

Reagents

(1) Sulphuric acid (H_2SO_4) (9M, 96%, sp gr1.84).

(2) Ammonium molybdate (NH_4) $6Mo7O_{24}.4H_2O$.

(3) Stannous chloride ($SnCl_2.4H_2O$) (12M, 37%, sp gr1.19).

(4) Standard phosphate solution – 'A' (see below).

(5) Solution 'B' (see below).

(6) Chloroform (as preservative).

(7) Perchloric acid ($HClO_4$) (60%).

(8) Sodium chloride (NaCl).

(9) Hydrochloric acid (HCl) (37%).

(10) Sodium hydroxide (NaOH).

Materials

(1) Glassware: all glassware, including 2 Erlenmeyer flask 250 ml capacity and/or Kjeldahl flask 75-ml capacity, must be thoroughly cleaned.

(2) Hotplate.

(3) Spectrophotometer.

(4) 1 cm spectrometer cell.

(5) 5 cm cell 2.

Procedure

1. Standard Solution 'A'

To prepare a standard solution 'A': dissolve 0.3400 g pure anhydrous potassium dihydrogen phosphate (KH_2PO_4) in distilled water or in phosphate-free artificial sea water and dilute to one litre. Add a few drops of chloroform as preservative. One ml of this solution contains 2.5 µg-atom PO – P (Barnes, 1959, Martin, 1968). From this standard solution suitable dilutions are made for the calibration curve, using 5 ml cell, at 705 mµ, (Martin, 1968) or at about 700 mµ (Harvey, 1963).

2. Solution 'B'

Now, dilute 10 ml of solution 'A' with 1 litre of distilled water or artificial sea water; and this is solution 'B'. One ml of solution 'B' contains 0.025 µg at PO_4–P (Martin, 1968).

3. Stock Solutions

(1) Carefully transfer 250 ml of concentrated sulphuric acid into 250 ml of distilled water and allow to cool.

(2) Ammonium molybdate-sulphuric acid mixture: dissolve 15 g of ammonium molybdate in 150 ml of distilled water. Filter and transfer the molybdate solution into 450 ml of sulphuric acid. Store in an amber colour bottle in the dark for future use.

(3) Stannous chloride solution: dissolve 4.3 g of stannous chloride in 10 ml of hydrochloric acid and dilute to 100 ml of distilled water, previously boiled out and cooled.

(4) Samples to be analysed must be filtered, using 0.5 membrane filter.

(5) Add 2 ml of ammonium molybdate-sulphuric acid mixture to a 100 ml sample in a 250-ml Erlenmeyer flask and mix thoroughly.

(6) Note time and temperature.

(7) Add, after allowing 3 minutes, 0.2 ml dilute stannous chloride solution to the mixture.

(8) Swirl the solution.

(9) Maximum colour develops after 7–12 minutes at a temperature higher than 20°C.

(10) Measure and note the maximum absorbency, using 1 cm-cell in a spectrophotometer.

If turbidity blank and reagent blank:

(11) Measure the absorbency at 705 μm with distilled water. Obtain turbidity blank and compare the absorbency with results with distilled water, using 100 ml of artificial sea water (see Appendix VIII) with known phosphate concentration. The turbidity blank and reagent blank must be deducted from the maximum absorbency.

Calibration Curve

Transfer appropriate ml of phosphate solution 'B' to 250-ml Erlenmeyer flask and dilute to 100 ml with artificial sea water. This solution now contains 0.25 μg at PO_4-P/litre (Martin, 1968). Using suitable aliquots, plot a graph of absorbency against phosphorus concentration. The concentration of the unknown sea water sample can be derived by interpolating the average of two readings into this graph.

Total Phosphorus

Total phosphorus is the difference between total inorganic phosphate and the phosphate-bound organic substances; the latter do not react with molybdate. Therefore, the organic materials need to be first oxidised with perchloric acid.

Add 3ml of perchloric acid to a 50 ml sample in a 100-ml Erlenmeyer flask. Carefully now transfer 3 ml of concentrated HCl to the flask. Allow the contents to cool. Add 30 ml of distilled water to the flask and dissolve the salt contents. Warm up the perchloric acid to near fuming temperature until evaporation of the contents is completed. Dissolve 40 g of NaOH in distilled water and make up to a litre. Neutralise the excess of acid in the flask by adding NaOH, drop by drop, until neutral to litmus. Add a few drops of HCl to the solution and make it a little acidic. Dilute the solution. Add 1 ml of ammonium molybdate-acid solution and dilute to 50 ml and mix thoroughly. After a pause of 3 minutes, add 0.15 ml of dilute stannous chloride solution and mix well again. After maximum colour development, determine the absorbency. Phosphate concentration is now obtained from the calibration curve.

References

Armstrong, F. A. J., 'The determination of phosphorus in sea water', Oceanogr. Mar. Biol. An. Rev. 3, 1965, pp.79-93

Barnes, H., 'Organic phosphorus in sea water and plankton', *Apparatus and Methods of Oceanography*, London, George Allen and Unwin Ltd., pp.157-162

Bienati, N. L., and Comes, R. A., 'Seasonal variation in the physico-chemical composition of sea water', in (ed. Cotlow, J. D.) *Fertility of the Sea*, New York, Gordon and Breach Science Publishers, 1971, pp.51-9

Harvey, H. W., *The Chemistry and Fertility of Sea Waters*, Cambridge, Cambridge University Press, 1963, p.240

Martin, D. E., *Total Phosphorus*, Marcel Dekker, INC., 1968, pp.121-125

Redfield, A. C., *American Scientist*, 46, 1958, pp.205-221

Appendix VIII
Useful Data

Linear Measurement

Length	Metric equivalent	Abbreviation
Kilometre	1,000 m	km
Statute mile	1.609 kilometres	km
Nautical mile	1.85 km (1.15 statute m.)	knot

Longitude and Latitude

1 degree of longitude = 69 statute miles or 111 km at the equator (0°).

1 degree of latitude = 68.704 statute miles or 110.569 km at the equator.

Diameter of the earth = 7,927 miles or 12,757 km (Equatorial).

Diameter of the earth = 7,900 miles or 12,714 km (along Polar axis), because the earth is slightly elliptical.

Weight

Ton (long)	2,240 lb
Ton (short)	2,000 lb
Metric tonne (Fr)	1,000 kg
Kilogram kg	2.2 pounds (lb)

Surface Area and Volume

Circumference of a circle	$2\pi r$
Area of a circle	πr^2
Surface area of a sphere	$4\pi r^2$
Volume of a sphere	$\frac{4}{3}\pi r^3$

Surface area of a cylinder $2\pi rh$
Volume of a cylinder $\pi r2h$

Chemical and Physical Constants

Avagadro's number	N_A =	6.022×10^{23}
Faraday constant	F =	96,476C/mol
Gas constant	R =	8.314J/K-mol
Plank's constant	h =	6.62 ergs/s
Speed of light (in vacuum) s	=	2.997×10^{10}

Chemical Solutions

1 mol = the mass in grams of substance equal to its molecular or atomic weight; i.e. It contains Avagadro's number (N) of molecules or atoms

Molar solution is the volume occupied by a mole of gas at standard temperature and pressure (25°C, 1 atm) 22.414 litre.

1 molar solution = 1 mol/1000 grams of solvent.

1 molar solution = 1 mol of solute in 1 litre of solution.

Temperature

Boiling point of Water

Water boils (under standard conditions) at:	100° [1]	=	373.15 Kelvin	BP
Freezing point	0°C	=	273.15 Kelvin	FP
	–17.8	=	0 Fahrenheit	
	–10	=	14 Fahrenheit	
	0	=	32 Fahrenheit	
	10	=	50 Fahrenheit	
	20	=	68 Fahrenheit	
	30	=	86 Fahrenheit	
	40	=	104 Fahrenheit	
	50	=	122 Fahrenheit	
	60	=	140 Fahrenheit	
	70	=	158 Fahrenheit	
	80	=	176 Fahrenheit	
	90	=	194 Fahrenheit	

---

[1]   In order to convert Celsius into Fahrenheit: multiply by 9, divide by 5, and add 32 [i.e. C=F×9/5 + 32]. To convert Fahrenheit into Celcius: substract 32, multiply by 5, and divide by 9. [F=(C-32)×5/9].

100 = 212 Fahrenheit

Saline Solutions

Substances	Weight in grams per litre of solution							
	NaCl	KC	CaCl₂	NaCO₃	NaHPO₄	NaH₂PO₄	MgCl₂	MgSO₄
Ringer's solution	0.9	0.042	0.024	–	0.02	+100 ml water		
Marine invertebrates	7.0	0.3	0.	1.5	–	–	–	0.3
Marine molluscs	23.38	–	5.55	–	–	–	7.62	–
Marine Crustacea	29.23	0.75	4.44	–	–	–	–	–
Elasmobranchs	16.38	0.89	1.11	0.38	–	0.06	–	–
Marine teleosts	13.50	0.60	0.25	–	–	–	0.53	–

Based on various sources

Artificial Sea Water

With salinity = 34.55 and chlorinity = 19.001

NaCl .... 23.470 g

MgCl₂ .... 4.981 g

Na₂SO₄ .... 3.917 g

CaCl₂ .... 1.102 g

KCl .... 0.664 g

NaHO₃ .... 0.192 g

KBr .... 0.096 g

H₃BO₃ .... 0.026 g

SrCl₂ .... 0.024 g

NaF .... 0.003 g

H₂O (add distilled water and make up to 1 litre)

(Based on Lyman and Fleming (1940))

Glossary

Absolute temperature

Temperature measured from absolute zero, and expressed in units of Kelvin (K) and the value of 1K is the same as 1°C; i.e. 0K is equal to –273.15°C.

Acuity

Resolving power, especially in vision.

Ahermatypic

Corals without symbiotic association with algae.

Atoll

A group of coral reefs and islands that surround a central lagoon.

Autotrophic

Mode of nutrition, ability to synthesise one's own food utilising inorganic materials, and can be considered as two types: holophyic nutrition and chemotrophic nutrition.

Baleen

Whale bone – horny plates of mysteceti.

Biomass

The gross weight of a given organism (or organic materials) of plankton in given volume or area of water in a given time.

Biosphere

That part of the earth where, due to constant flow of energy from the sun and interaction between abiotic and biotic factors, life can exist.

Biota

Plant and animal life of a given region.

Cenozoic

Era of geological history that extends from the Tertiary Period (65,000,000 million years ago) to the present.

Coralite

The skeleton of an individual coral polyp.

Cretaceous	Geological period between 65 and 135 million years.
Cyanobacteria	(=blue-green algae) – an ancient group of photosynthetic bacteria which appear to have played a decisive role in bringing about an increase in the amount of oxygen of the atmosphere and the subsequent evolution of eukaryotic life based on aerobic respiration; they are abundantly found as fossilised stromolites today.
Detritus	An accumulation of dead plants, animals and fine sediment.
Ecosystem	A system resulting from interaction between living organisms and their environment.
Epiphytic marine algae	Algae that grow attached to the mangrove plants, especially on the roots that are exposed to tidal fluctuations.
Engybenthic	Organisms living or crawling close to the bottom.
Flagellum (pi. flagella)	A fine thread-like organelle arising from the surface of the cell, capable of propelling the cell through fluid media.
Foetus	An unborn or unhatched offspring of vertebrate, especially of human embryo eight weeks after conception.
Fringing reef	The simplest form of coral reef in shallow waters of the Tropics.
Geomagnetism	Study of earth's magnetic properties.
Habitats	Geographical locality where plants and animals occupy or live.
Herbivore	Plant or algae eating organisms.

Homeostasis	The maintenance of relatively stable physiological conditions of the internal environment of an organism.
Hydrostatic pressure	Fluid pressure exerted over an area, especially in sea water.
Interstitial	The space between adjacent cells or tissues.
Isogamous	Both sexual reproductive gametes are alike; of similar shape and size.
Jurassic	A geological period relating to about 160,000,000 years ago of the Mesozoic era.
Kelvin	Unit of absolute temperature (see above).
Larvae	The newly hatched out actively feeding stage of invertebrates and some fish.
Lithosphere	The outer solid part of the earth considered to be about 50 miles in thickness and composed of rocks essentially like that of the surface of the earth.
Littoral	Relating to shoreward region of the sea.
Mesoglea	An undifferentiated or non-cellular layer between the ectoderm and the gastroderm in lower group of animals, including coelenterates.
Metamophorsis	Cycle of transformation of life, e.g. from larval stage through juvenile to adult.
Nernst Equation	An equation for calculating the ionic potential difference across a membrane, usually in excitable cells such as nerves and muscles.
Osmolarity	The effective osmotic pressure exertable.
Osmole	Universal unit of osmotic pressure.

Osmoregulation   Maintenance of constant salt and water concentration of internal body fluid of an organism.

Oviparous   Release of unfertilised eggs by the female; fertilization and development take place outside the body.

Pelagic   Relating to open waters of the sea.

Permeability   The passage of diffusable substances across a membrane.

Philopatry   Species of animals which return to the same place or locality where they were born.

Phylum   The major taxonomic division of animals and plants related to have the same plans, phylogeny and origin.

Planula larvae   Free swimming larvae of coelenterates, including those of corals.

Pleistocoene   A geological epoch, about 1.8 million years ago, of the Quartenary Period of the Cenozoic Era.

Production   The rate of carbon fixation, usually expressed as $gCm^{-2}y^{-1}$.

Quaternary   Geological period comprising both Pleistocene and recent time.

Ringer solution   Physiological saline solution used in animal experiments.

Saprophytic nutrition   Organisms feeding on solutions of dead and decaying tissues.

Scleractinia   Hard corals.

Sessile   Organisms that are fixed to substratum and not free to move about.

Standing crops	The quantitative expression of population of planktonic organisms in numbers or weight for a known volume of sea water in a given time.
Stenohaline	Ability to tolerate only a minute range of variations in salinity.
Stromatolites	Hemispherical mounds of blue-green filamentous algae of the pre-Cambrian Era.
Silicious	Glassy, composed of silicon compounds.
Symbiosis	An association between two dissimilar organisms for their mutual benefits.
Substrate	The surface layer of the seabed consisting sand, mud, silt, cobbles, clay, rocks and corals where animal and seaweeds are attached.
Swim bladder	Gas-filled bladder possessed by many teleosts for floatation mechanism.
Tectonics	The geological feature of continental plates which is concerned with structure, specially with folding and faulting.
Tethys Sea	An ancient tropical ocean that existed before the continents came to the present positions.
Tight junction	The site of fusion between two membranes of adjoining cells to prevent the extracellular fluid leaking out.
Triassac	Geological period which lasted from 225 until 180 million years ago.
Ultraviolet (UV) light	Just beyond the violet of the visible spectrum with light wavelengths between 5×10^{16} and 10×10^{16} cm.

Urea	An organic molecule $CO(NH_2)2$ formed in the liver and a primary nitrogenous excretory product in the urine, especially of mammals.
Vacuole	A small fluid-filled cavity in the protoplasm of the cell.
Visible light	Visible part of the spectrum with light of wavelengths lying between 400 nm and 750 nm.
Viviparous	Organisms producing live young ones.
Wavelength	The distance between two successive crests (Lambda =).
Xerophytes	Plants adapted to dry conditions.
Xyleum	The vascular tissue in plants which conducts water and dissolved nutrients upwards from the tip of the roots to the leaves, as opposed to phloem, the latter conducts the food elaborated in the leaves to the different parts of the plant.
Yellow substance	Collective name for a humus-like yellow complex mixture of compounds in the sea with certain components displaying a distinct absorption peak between 250–265 nm.
Zooxanthellae	Symbiotic algae which associate with corals.
Zooplankton	Animal plankton.
Zooxanthellae	Yellow-brown unicellular algae living symbiotically with different animals, especially corals.
Zygote	The fertilised cell resulting from fusion of the male and female gametes.

Index

A

B

D

Z